James Van Allen

James Van Allen

The First Eight Billion Miles

Abigail Foerstner

UNIVERSITY OF IOWA PRESS Iowa City

University of Iowa Press, Iowa City 52242

www.uiowapress.org
Printed in the United States of America
Design by Richard Hendel

The University of Iowa Press is a member of Green Press Initiative and is committed to preserving natural resources.

Printed on acid-free paper

Library of Congress Cataloging-in-Publication Data
Foerstner, Abigail, 1949–
James Van Allen: The first eight billion miles / by Abigail Foerstner.
p. cm.
Includes bibliographical references and index.
ISBN-13: 978-0-87745-999-6 (cloth)
ISBN-10: 0-87745-999-1 (cloth)
1. Van Allen, James A. (James Alfred), 1914–2006. 2. Astrophysicists—Iowa—Biography. I. Title.
QB36.V357F64 2007 2007008870
523.01092—dc22
[B]

07 08 09 10 11 C 5 4 3 2 1

Dedicated to

ELSIE FOERSTNER

my aunt and inspiration

Contents

Preface and Acknowledgments

The corner of Jefferson and Dubuque Streets in Iowa City surely marks one of the world's epicenters for physics and metaphysics. Within a block, St. Mary's Catholic Church, the First United Methodist Church, the Congregational United Church of Christ, and Gloria Dei Lutheran Church give voice to metaphysics. At the intersection itself, Van Allen Hall anchors the University of Iowa's Department of Physics and Astronomy.

The cement walls of the seven-story physics building convey the quality of a citadel. Step through the west entrance and the legacy of the building becomes clear. Display cases in the hallway document astrophysicist James Van Allen's space missions and seminal models of cosmic ray detectors later refined for America's first satellites—detectors that discovered the Van Allen radiation belts. To the uninitiated, the maze of circuitry stacked in the tubular detectors suggests an overzealous design for a flashlight. But scientists who had to pick experiments to launch in a satellite knew Van Allen's designs would work. He had a reputation as a practical physicist who applied the lessons of small-town Iowa ingenuity to outer space.

Finding Van Allen's headquarters from the hallway was easy. Scientists, students, friends, and reporters just took the elevator to the top floor and followed the slight drift of pipe smoke through the open door across the hall. Only weeks before his death, on August 9, 2006, Van Allen worked in the office beyond this door every day, participating in the fields of space science he helped establish and fulfilling constant requests for interviews, articles, and forewords to books. He loaded hand-programmed cards into a Hewlett-Packard 1972 calculator/plotter to appraise findings from his own space missions in the light of ever closer recent encounters other missions have made with the planets. He remained an influential voice in the continuing crusade for more robotic space probes to reach deeper into space.

At the front of the office, Van Allen kept his diaries, journals, bound copies of papers he and his former students have written over sixty-five years, and the latest photographs from space. But the columns of floor-to-ceiling bookcases that led to the back of the room cut a path through space history. Here, yellowing paper tapes documented data from instruments loaded in balloon-launched rockets from the earliest days of space exploration. Here, bound volumes of charts and graphs plotted a new map of the solar system, planet by planet, and field logs dated back to the days of shooting reassembled German

V-2 rockets in New Mexico starting in 1946. Remnants of rockets, models of spacecraft, and backup detectors found their way to the corners and shelves of the room.

Van Allen's office occupied Room 701. A sign on the door gave visitors the exact location of the office: Longitude 91° 31' 55.1" W; Latitude 41° 39' 43.6" N. I first went up to Room 701 to meet Dr. Van Allen in November 1998 with my aunt and University of Iowa alum, Elsie Foerstner. His door was open, as usual, and we walked in. She reminded him about their very casual acquaintance from years past and introduced me. Van Allen asked her about herself and she asked about his children. I couldn't help inquiring about the Hewlett-Packard plotter as it charted rolling hills of cosmic ray counts on graph paper. The plotter, filling one end of his desk, graphed the latest messages from Van Allen's cosmic ray detectors aboard *Pioneer 10*, launched in 1972.

"Where is *Pioneer 10*?" I asked.

"About 73 astronomical units from earth," he said.

"And that's how far in miles?"

"Say, 6.8 billion miles. But we don't use miles. Around here, miles are Detroit units."

I had clearly stepped into the Big Picture. Room 701 became a good place to reflect on a time when most people believed in the clear progress of the human race and of science. Van Allen's life and work compellingly reflect those ideals. The world may feel less optimistic now but space still beckons us to look up. In space at least, the dreams still await us—we have barely reached the front gate of our own solar system.

It has been my privilege to follow Van Allen's journey and write this biography. My heartfelt thanks to all who participated in interviews and supported the research effort. My thanks first and foremost to Dr. James Van Allen for his time, his teaching, and his trust in giving me carte blanche to research all his papers, diaries, and other documents in Room 701. He was a great teacher and he taught me sixty years of space science and technology to supplement the one year of college physics I studied as a journalism major. He explained the intricacies of radiation belts, magnetospheres, and the solar wind—cornerstones for new branches of physics he helped establish. We met at 1:30 P.M. on Monday, Tuesday, and Wednesday on the third week of every month for interview sessions for more than two years. There were numerous trips to his office after that to fill in gaps and check facts. Van Allen turned hard-nosed science into a riveting space adventure.

My thanks to Abigail Halsey Van Allen for the many invitations to the Van Allen home where she filled in the personal side of the story from the vantage point of sixty-two years of marriage. Thanks to their children Cynthia Schaffner, Sarah Trimble, and Peter Van Allen for lengthy interviews about their parents and family life and to nephew David Van Allen for providing three generations of family journals, letters, and photographs.

Dozens of family members, friends, and professional colleagues of James Van Allen participated in interviews. Many of those friends and colleagues are Van Allen's former graduate students. Among them, Les Meredith, George Ludwig, Carl McIlwain, Stamatios (Tom) Krimigis, Don Gurnett, Larry Cahill, Michelle Thomsen, and Bruce Randall offered extraordinary help with this biography. Special thanks to Les Meredith, former associate director of NASA's Goddard Space Flight Center, for sharing his detailed journals and photographs of the first two "rockoon" expeditions that established the merits of economical balloon-launched rockets. Many thanks to George Ludwig, retired assistant to the chief scientist at NASA headquarters, for sharing the memoir he is writing of the early space program at Iowa. His time line for the years of 1953–1959 often gave day-by-day accounts, an invaluable resource for which I am deeply grateful.

Thanks to astrophysicist Carl McIlwain, professor emeritus at the University of California at San Diego, for photographs of the rockoon missions as well as several interviews. Thanks to Tom Krimigis, director emeritus of space science at the Applied Physics Laboratory (APL) of Johns Hopkins University for a set of APL's photos of Van Allen during World War II when he trained gunners to use precision antiaircraft shells that he and APL helped develop. Thanks to Bruce Randall, Van Allen's research associate on *Pioneers 10* and *11*, for answering a thousand questions, for tours through hidden nooks and crannies of the old and new physics buildings, for old department directories, and for samples of vintage electronics, computer chips, and other devices that helped with technical aspects of the story. Thanks to NASA space physicist and University of Iowa professor Don Gurnett for contributions to several chapters of this book and to his assistant, Kathy Kurth, for providing archival photographs.

At the Smithsonian Institution's National Air and Space Museum, I am deeply grateful to David De Vorkin for his advice on the draft manuscript and for the invaluable oral histories he completed with Van Allen and other space pioneers.

Thanks to Alan Rogers of the Haystack Observatory at the Massachusetts Institute of Technology (MIT) for providing photographs of the ham radio operation run by his father, John Rogers, a station that Van Allen helped turn into the Rhodesian Space Center. Thanks to retired Jet Propulsion Laboratory (JPL) director William Pickering and Ernst Stuhlinger, chief scientist for the

Saturn V rocket that took Americans to the moon, for copies of speeches and articles that recapture "the good old days" of space exploration. Dr. Stuhlinger also provided photographs.

Institutional support for this biography at the University of Iowa has been an overwhelming tribute to Van Allen and a tremendous help to his biographer. David Dierks of the University of Iowa Foundation championed the project from the start. Former university president Sandy Boyd, among Van Allen's closest friends, offered insights into Van Allen's character as an administrator, teacher, early computer advocate, and companion. At the University of Iowa Department of Physics and Astronomy, many thanks to Wayne Polyzou, former department head, and Tom Boggess, current department head, for their support during all the times I worked in the building. Thanks to Chris Stevens, department assistant, for providing contact information and countless services during my stays. Thanks to Agnes Costello McLaughlin, retired department secretary, for interviews and access to her voluminous collection of news clippings and photographs. Thanks also to the nighttime maintenance staff—it was nice to have people check up on me at eleven o'clock at night when the physics building often seemed ominously empty.

Research beyond Room 701 led to hundreds of boxes of records in the Van Allen Archive in Special Collections at the University of Iowa Libraries and to archives at the Mount Pleasant Historical Society and the James Van Allen House Heritage Center in Mount Pleasant. It led to archives at the Jet Propulsion Laboratory, the Kennedy Space Center, the Marshall Space Flight Center, the U.S. Space Center in Huntsville, Alabama, the National Air and Space Museum, the Florida Institute of Technology in Orlando, the Lyndon Baines Johnson Library, and the Dwight Eisenhower Library. My thanks to Kathy Hodson, Sid Huttner, Amy Cooper Cary, David McCartney, Denise Anderson, and the entire staff in Special Collections at the University of Iowa. Warm thanks to Doris Onorato at the Van Allen House Heritage Center in Mount Pleasant. Special thanks to John Bluth at the JPL archives, Gerald Bennett at the APL records center, and Rosemary Kean at the Evans Library, Florida Institute of Technology.

My thanks to my friend and author Cissi Falligant, who offered insights and advice on the initial draft of the book. Casandra Profita, Sarah Stone, Ted Rosean, and Bridget Suse, student editorial assistants, helped transcribe the tapes of all the interviews. Editorial assistant Anne Lee indexed the book.

Other than Van Allen himself, my family and my editor, Holly Carver, contributed to making this book a reality. Carver, director of the press, offered encouragement, friendship, direction, and writing suggestions over the past six years. A million thanks to her for saying, after she published my last book, "Well, what do you want to write now?" I knew immediately—this biography. Alex Meyer, husband to my cousin JoAnn Foerstner Meyer, offered his services as ambassador when I shared the book idea. One of the many boards on which Van Allen served included a scholarship board established by the late George Foerstner, founder of Amana Refrigeration, and directed by Alex after his father-in-law's death. He set up the seminal meeting in July 2000 where Van Allen and I first discussed the biography.

My husband, Art Caudy, scanned all the photographs for the book and took over the task of running our family during countless research trips and marathon writing sessions. My son William Caudy earned undergraduate degrees in physics, math, and classical languages at the University of Iowa while I was researching this book and was a willing technical consultant. My son John Caudy received outstanding awards at two Illinois state science fairs with a junior high school project comparing the flight patterns of model rockets. The challenges he faced in gathering data with model rockets heightened my deep respect for space exploration and the improvisational spirit of the space pioneers such as Van Allen. My daughter Elizabeth Caudy provided books about art, politics, and even spiritual searching that put scientific discovery in perspective. And warmest thanks to Elsie Foerstner, who supported this book in a thousand ways after that day she suggested we drop in on Van Allen in Room 701.

Introduction

Pioneer 10 and *Voyager 1* raced toward opposite ends of the solar system, each on a path to find its edge. The vastness of space separated James Van Allen's cosmic ray detectors on *Pioneer* and Don Gurnett's radio receiver on *Voyager* while the offices of the two physicists stood only a few doors apart at the University of Iowa. *Pioneer 10* fell silent after sending home a final faint signal on January 22, 2003, from nearly 8 billion miles away. The last streams of data from his detectors told Van Allen that the probes were getting close to the intermediate boundary of the solar system, a zone called the *termination shock* where the solar wind and galactic cosmic rays crash and begin to mix. Physicist Norm Ness found this boundary with an instrument on board *Voyager* in 2005, a momentous discovery in space science.

Van Allen pioneered the field of space science, discovered the earth's radiation belts and helped draw a new map of the solar system on his journey across 8 billion miles. His instruments on board more than two hundred rockets, satellites, and space probes transmitted data over six decades. *Pioneer 10* alone, launched for a twenty-one-month mission in 1972, sent Van Allen more than thirty years of readings that helped us recognize that the boundary of the solar system extended billions of miles past Pluto.

Van Allen's career crystallizes the entire history of space exploration. He and a group of freewheeling colleagues launched the era of space exploration with captured German V-2 rockets reassembled and launched at the White Sands Proving Ground [now the White Sands Missile Range] in New Mexico after World War II. Partnering with the U.S. Army, the scientists replaced the warheads of the "vengeance machines"—that had terrorized London and Antwerp—with scientific instruments little more than a year later. Van Allen helped found and later chaired an informal rocket panel that met monthly to divvy up the V-2s—and later rocket missions—and to provide a forum for the new field of space science. For twelve years, the panel filled a key role that would be taken over by NASA in 1958.

With his detectors on the V-2 rockets, Van Allen pierced "the cosmic ray ceiling of the atmosphere," as he called it. The rocket had given him the tool to make his first discovery in space. He had found the cosmic ray plateau, the stream of incoming cosmic rays before they began to disintegrate in the earth's protective atmosphere and splinter into showers of secondary particles. Van Allen now wanted to map the changing intensities of cosmic rays across the globe, a change controlled by the earth's magnetic field. It was the perfect project when he returned to his alma mater, the State University of Iowa (now the University of Iowa), in 1951, as head of the physics department and developed a rocket program on a shoestring budget.

During the summer of 1952, Van Allen and a cadre of graduate students hitched surplus rockets to navy balloons and launched them from the U.S. Coast Guard Cutter *Eastwind* as it headed toward the Arctic. Each "rockoon" mission cost less than $2,000 and carried a single Geiger counter to count cosmic rays. The launches went poorly at first until Van Allen heated juice cans in the gondolas of the balloons to keep the rocket igniters warm enough to fire. Finally, with rockets reaching altitudes of more than 60 miles, Van Allen and the students painstakingly recorded cosmic ray levels across journeys to the Arctic and the Antarctic. Each launch filled in a dot on the map. Subsequent trips over the next five summers added more dots and made groundbreaking discoveries about the earth's magnetic field and the aurorae. Yet a single launch involved government grants, a ship, crew members, Van Allen, graduate students, other Iowa staff, and a floating laboratory of gear.

Then on January 31, 1958, Van Allen's cosmic ray detector went into space on board the first American satellite—*Explorer I*. A satellite could map cosmic ray intensities around the globe—data from more points in a single orbit than all the rockoon missions had covered in six years.

In addition, Van Allen's instruments on *Explorers I* and *III* discovered the earth's radiation belts, areas in the earth's magnetic field that trap intense concentrations of electrons and protons. This major scientific discovery with America's very first satellites had tremendous political impact as hearings got underway in 1958 to create a U.S. space agency. The debate raged during Senator Lyndon Johnson's Senate hearings on the benefits of a civilian versus a military space agency. Scientists pointed to the instrument on *Explorer I* as a prime example for the potential of a civilian space agency program. It added another plus to the prevailing sentiment in Congress that resulted in the civilian-controlled National Aeronautics and Space Administration—NASA.

From a scientific standpoint, the discovery spawned a whole new field of physics—magnetospheric physics. The new field helped remap the solar system in coming years, identifying the shape and dramatic influences of the magnetic fields—or lack of them—of Earth and the other planets.

An ongoing debate in history is to what degree movements shape individuals or individuals shape movements. Are individuals the deciding factor in events or do events create roles that any number of people could fill? The discovery of the radiation belts offers a compelling case study for how individuals and events interplay in the drama of human affairs. Van Allen's discovery of the belts relied on hard scientific know-how as well as quirks of individuality and fate.

America's first satellite, *Explorer I,* was an orphaned space mission called upon to rescue American self-esteem in the aftermath of *Sputnik*. And Van Allen had the only major scientific experiment available to load on board because of a gamble he made. In the first lap of the space race, he recognized that America had bet its money on the wrong rocket to compete against *Sputnik*. Van Allen prepared his instrument so it could fly on either the official rocket and satellite or on a clandestine satellite project developed on the sidelines by rocket pioneer Wernher von Braun in Huntsville, Alabama and the Jet Propulsion Laboratory (JPL) in Pasadena, California. Von Braun's rocket, JPL's satellite, and Van Allen's cosmic ray detectors saved the day for American prestige in the face of *Sputnik*, but Van Allen's major scientific discovery added significantly to that prestige. Without Van Allen, *Explorer I* would have gone into space virtually empty-handed. Discovery isn't just in the data, however. Van Allen and his students had to recognize in unsettling gaps of their *Explorers I* and *III* data that the gaps themselves held the secret of the radiation belts. "They told us our instruments had stopped working. We knew better and realized we had encountered a whole new phenomena in space," Van Allen said.

In the years that followed, Van Allen juggled space missions with administrative duties as head of a university physics department, with a legendary introductory astronomy course he taught, and with membership on dozens of national committees and panels. He mentored thirty-five students through their PhDs with research often derived from full partnership in his space missions. The graduates shouldered leadership roles in space programs here and abroad. However hectic the schedule, he came home for dinner at 6 P.M. sharp whenever he was in town. James and Abbie Van Allen raised five children—their proudest achievement of all.

Fundamentally, Van Allen's life encapsulates the drive of the human mind and heart to explore. Curiosity about the stars and the heavens inspired religion, art, and basic astronomy tens of thousands of years before human beings settled in the first villages—or sent the first satellites into space.

Abbreviations and Acronyms

ABMA	Army Ballistic Missile Agency
APL	Applied Physics Laboratory (Johns Hopkins University)
AGU	American Geophysical Union
ARPA	Advanced Research Projects Agency (Eisenhower administration)
BuAER	U.S. Navy Bureau of Aeronautics
BuORD	U.S. Navy Bureau of Ordnance
CSAGI	Comité Spécial de l'Année Géophysique Internationale
DTM	Department of Terrestrial Magnetism (Carnegie Institution)
GALCIT	Guggenheim Aeronautical Laboratory, California Institute of Technology
ICBM	Intercontinental Ballistic Missile
IGY	The 1957–1958 International Geophysical Year
IRBM	Intermediate-Range Ballistic Missile
JPL	Jet Propulsion Laboratory
MIT	Massachusetts Institute of Technology
MOUSE	Minimum Orbital Unmanned Satellite of the Earth
NACA	National Advisory Committee for Aeronautics
NASA	National Aeronautics and Space Administration
NRL	Naval Research Laboratory
NSB	National Science Board
NSF	National Science Foundation
OGO	Orbiting Geophysical Observatory
ONR	Office of Naval Research
PSAC	President's Science Advisory Committee (Eisenhower administration)
SSB	Space Science Board

SUI	State University of Iowa, now the University of Iowa
TPESP	Technical Panel on the Earth Satellite Program
UARRP	Upper Atmosphere Rocket Research Panel (initially called the V-2 Panel)
V-2	Vergeltungswaffe 2, or Vengeance Weapon 2
WGII	Working Group on Internal Instrumentation (for the TPESP)

James Van Allen

1 Frontier Roots

Pillar Point, New York—February 24, 1848. Seventeen-year-old George Van Allen relaxed after a day of backbreaking labor in the cold on his family farm. But instead of picking up a book, he began a lifetime habit of writing a journal. The pages to come introduced his parents, his ten brothers and sisters, and their struggle to eke a living from the rocky fields at the shores of Pillar Point, New York. But most of the writing wasn't about his work. Like the first passage written on February 24, the journals sustained an inner life of the mind and the dream of leaving the farm to get a good education.

"If you would have your machine start easy, you must use it often and keep it greased. If you would—would—would write readily, [then] write often and think to some point—get some subject in the morning—and think of it all day. Get the subject clear to your own mind, thoroughly understanding it yourself . . . and by evening you can write it off as fast as you can make letters," he began. "I have made up my mind that it is very beneficial to a person whose time is mostly spent in outdoor labor to write a little evenings. If he does it every evening, so much the better. It keeps his thoughts from rusting and there is something saved. It gives strength to his mind, inculcates regularity and there is something gained on the whole. For if a person intends to write evenings, he will think at some point during the day and not let his thoughts run to seed, as reverie is sometimes called."

• • •

George and his brothers and sisters attended country winter schools that generally opened after the harvest and closed in time for spring planting. The grueling demands of work on family farms necessitated the help of all the children. Between the close of school and the time it reopened, George ploughed the rocky North Shore acres, salted the cattle and sheep, laid in a fence, chopped timber, cut rails, and brought the rails to Brownville to sell. He harvested the barley and potatoes and helped butcher the hogs. He hauled timber to the shipyards and moonlighted to earn some extra money, working at the shipyards and for the blacksmith in town.

He rarely complained about his daytime tasks in his journal except to write that the hog butchering "has been very unpleasant." He took the work for granted and it never let up, not even for illness. "Had the toothache. Went to Dr. Hunter's office. Had two extractions," he reported on July 17, 1848. But that afternoon, he was back home pitching hay. Sometimes he combined work and play, taking "a load of girls and butter tubs to the harbor in a sailboat." He loved to sail in Sherwin's Bay for a rare evening outing, he noted in one passage. His social life centered around services at the Methodist Episcopal Church that his father had helped build and the increasingly boisterous temperance meetings that combined grassroots politics with a good dose of religious revival.

Most pleasant of all, he read his new books when his family could spare money for him to buy one. The books fed an idealistic and poetic vision of life that he poured into his journal. And the journal expressed his curiosity about nature and his love of study. George Clinton Van Allen bequeathed the gift of journal-keeping and the love of learning to his son and to a grandson he never met–James Alfred Van Allen. Equally eager for adventure and success, George Van Allen abandoned generations of family heritage in the East and headed for a new start in Iowa.

His forefathers came to America from Holland in colonial times and settled in upstate New York. The Van Allens joined the fight for American independence and intermarried with the Ackermans and the Childs, two other patriot families from the same area of the state.

It's unknown when Robert Ackerman of Saratoga County joined the freedom fighters—he had just turned thirteen in 1776. But young boys and old men grabbed their rifles to reinforce the regiments and aided trusted officers such as Captain Increase Child when the British invaded their county in the fall of 1777. The Northern Continental Army surrounded the British at Freeman's Farm and General John Burgoyne surrendered his troops. While the British still held New York City, the stunning victory at Saratoga marked the turning point in the Revolutionary War and convinced the French to join the American cause.

Ackerman served in the thirteenth regiment out of Albany after the war but eventually returned home and married Child's daughter Roxalana. The young couple moved to a farm near Pillar Point on the rocky New York shoreline of Lake Ontario where they raised thirteen children. A frequent visitor at the busy household was Cornelius Van Allen. Cornelius moved to Pillar Point with his mother, Catharine, and stepfather, Adam See. He could barely remember his natural father, Cornelius Van Allen, Sr., who had also served in the Revolutionary War.

Young Van Allen farmed with his stepfather but also acquired informal training to draw up legal papers and settle estates. The legal services brought him enough income to marry his sweetheart Lory Ann Ackerman in 1829. George Clinton Van Allen, the first of their eleven children, was born the following year on July 6, 1830. The young couple stayed at Pillar Point and bought their own farm on the north shore of Black River Bay in 1831. But their growing family and hard times soon ate into their prosperity.

George longed to free his life from the land. He could see only two escapes—an education to prepare him for a profession or a move to the western frontier. He grabbed for both. His break came when he attended yet another country school session in 1850 at the age of twenty. His age and academic advancement made him a better candidate to teach school than attend it. He was offered a teaching position for the winter school at Watertown, New York.

The job went well. With new confidence in his ability to earn a living, he began taking classes in the spring of 1851 at Falley Seminary in Fulton, New York, a Methodist school that his father had early on helped to support. The seminary offered the equivalent of a high school education and the training needed to secure choice teaching jobs. Van Allen was nearly twenty-one years old when he started classes, uncertain how long he could stay given his family's unrelenting need for help on the farm and the difficulty in raising the $6 for a semester's tuition.

"One term was all that I dared hope for and there was a prospect that even this short period would be shortened, for at the middle of the term our folks were somewhat desirous that I would return and resume my labor on the farm, which I (had) more than half a mind to do," he wrote in a new journal. The headmaster convinced him to stay, offering to help him privately with the tuition. In return, he asked only that the young scholar take classes in Latin.

George readily agreed. He applied an ironclad self-discipline and a love of knowledge for its own sake to his studies, chiding himself for "getting lazy" if he slept in till 5:30 A.M. Often, he rose in the dark at four. "'Tis pleasant to watch the stars and revel in the beauties of nature while others sleep," he wrote. "The same routine of study today as yesterday. I sometimes wonder that I am not weary of study, it is so constant. Yet when I remember that new truths continually make the heart glad, I cease to wonder. There is a beauty in truth."

He helped out on the farm over the summer but returned to Falley in fall and managed to convince his family to send his brother Martin along with him for the new semester. The Van Allens faced the same dilemma as other farm families: they couldn't get along without the help of their children working the spread yet knew the children needed some provision for future employment. The typical farm just wasn't big enough to divide up and support them all as adults. On the other hand, the population in upstate New York swelled in the 1840s and the opportunity for teachers seemed limitless. Teaching and the pulpit were the primary careers attracting students to Falley, where George and Martin shared a room in a boardinghouse with several other students.

George studied Latin, algebra, geometry, and ancient geography that fall. Martin studied geometry, algebra, grammar, and vocal music. Their sister Sarah joined them for the spring term and other brothers and sisters revolved in and out of the scene over the three years that George studied at the seminary.

Social life at the seminary centered on religion. Firebrand ministers held revival meetings in the clearings near the towns, and Van Allen reported in a letter home how he camped out for a two-night stay with Brother Bird, one of his teachers, to hear Reverend C. P. Bragdon preach. "The Lord's presence was manifested in the conversion of a good number of souls," he wrote in a letter home. Two burning movements—temperance and the abolition of slavery—fueled the religious revival of the 1850s, and the revival gave both movements the fervor of crusades in God's righteous name.

George Van Allen's personal religious beliefs were more reflective. "At 6 A.M., I went to the prayer meeting—had a good time. There is indeed a joy in religion, a power in prayer. At 9, went to my class in the Sabbath School. It was rather small yet quite interesting. How pleasant to be engaged in the work of structuring in things which concern their souls," he wrote in his journal of one typical Sunday. He composed his own prayers in the journal as well. "Oh God, parent of all good, graciously permit thy spirit to direct while we offer to thee our mite of praise and thanks. For all thy mercies past and crave thy blessings in times to come from our infancy to the present moment. Thy guardian care has been over us!"

But Falley Seminary also sharpened George's ambition and his taste for nice things. He asked his mother in one letter to make him fine linen shirts instead of cotton ones and to sew him a sweeping cloak instead of a coat. The handmade clothes helped him keep up appearances on a tight budget. George and Martin were constantly on the move and on the lookout for a cheaper living situation. "I have $2.55 and Mart has 38 cents of our money left," George wrote home on September 2, 1851. Classes hadn't even started yet and they were nearly broke. Money loomed as a major preoccupation. "It cost me about $10 to come [here], $5 for a bed, $4 for pants, $6 for books besides some other necessities. Everything I have bought at the very lowest price so that it is remarked by those who ask (as they all do) what a thing costs," he wrote home in an undated letter. "Now if you can send me some $8–$10, it will last till I have to pay my board, though they will not let a fellow get far behind." George asked for the money apologetically, with a constant sense of embarrassment, obligation, and expectation to make repayment. His out-going and light-hearted brother, Martin, simplified such matters by cheerfully writing his parents to "please send as much as you can spare." He noted he would leave it to George to fill in details. "I have let him have control of everything and followed his directions in all things. Ain't I dutiful?"

As difficult as it was for George to ask for money, he had no choice but to do so in letter after letter. There was no established allowance from home and the brothers were constantly getting behind in bills. George took periodic jobs as a teacher in the winter schools to defray costs. Winter school meant rising at 5 A.M. and walking to the school from whatever farmhouse offered the teacher shelter for a week. On a snowy day, the teacher rose even earlier to dig out a path to the road and then hurried to start a fire in the cast-iron potbellied stove in the corner of the one-room schoolhouse. One teacher often had thirty to sixty pupils, at different grade levels. Yet the winter school offered $20 and a way to earn an entire year's tuition of $12 at Falley Seminary.

George Van Allen left home before dawn on December 26, 1851, for the two-mile walk up the bay from home to his winter school for the season. He started up the fire so the schoolhouse would be warm when the children arrived and then had plenty of time to himself for prayer and lesson plans. He lamented that his students seemed "intent on everything more than education" at first. Their attention improved despite numerous absences among the thirty children due to severe storms. The storms continued into March and, despite the cold, George reveled in the beauties of nature that he observed on his walks to and from school. He read poetry and studied the Bible and the life of Martin Luther during his free hours. But he suffered a severe cold during the winter and found it hard to recover. He believed his illness cost him his health, the excuse he offered for declining to return home to the farm during the following summers after he returned to the seminary. "I am afraid if I go home I shall work too hard and make myself sick," he wrote to his father. "I am at present little better than an invalid." George bluntly refused help in another undated letter. "You will shorten your life if you have the care you did last summer and I see no way of making it lighter for I can be of no use to you. For reasons you understand, the family cannot all live at home. I think you had better put things in snug shape, do a small business and not try to carry on your farm or, if you do, leave other men's farms to other men. I'll try as indeed I have all along to pay my way."

Sarah, too, feared the trap of returning to the farm and Martin, after a short return, moved to Syracuse in 1853 to "take a job in the Syracuse No. 4 Globe Building selling dry goods." George took yet another winter school in 1853 in Fulton County, attracted by the unusually handsome sum of $30—as compared to the usual $20—for the term.

"Though my school is not yet as full as I expect it will be, I have 69 names on my roll: Two in algebra, five in grammar, two good classes in arithmetic and a variety of other branches. I am getting the confidence of my scholars, and the most of them, especially the older ones, are becoming interested in their studies," George wrote home. "Without that spark for learning, teaching is a perfect bore," he lamented.

The death of a student due to symptoms that suggest acute encephalitis cast a pallor over George Van Allen and all his students that winter, however. "One of my little girls, Julia Rice, aged 11 years, was detained at home by a slight headache. Then I heard she was sick, then that she grew worse. I called to see her but she did not know me, nor anyone else. She was very drowsy so that she slept most of the time. The doctor called it congestive fever, but her symptoms changed and he thought it was erysipelas. But she grew still worse. I called again. She was very low and had no sense of her pain. Her tongue was swelled in her mouth, her throat had become sore from chafing to get her medicine down. There she lay in intense suffering till, as I went to school one morning, a little girl said, 'Mr. Van Allen, Julia is dead!' And the big tears stole down her cheeks for she, like the rest, had lost one of their favorites. I could not refrain from tears as one after another gave vent to grief on being told Julia was dead. That was a mournful group that sat before me that morning," he wrote. "At night I called on the afflicted family. Her mother was inconsolable and incessantly repeated some hope or wish for her child, now blasted forever."

Even though Julia Rice and her family had no strong religious ties, George concluded that the little girl's daily life and the love everyone felt for her reflected her grace and ensured her a place in heaven. His liberal views on human goodness stood in sharp contrast to the near obsession with sin and evil that dominated so much religious thought of his day. Despair over his faulty endeavors to follow God's path found its way into many a letter home, yet George's journals celebrated the spirituality of natural forces and the mysteries of God evidenced in the discoveries of science.

George returned to Falley for the spring term in 1854 and completed his studies in June. That fall, he enrolled at Wesleyan University in Middletown, Connecticut. A biographical review written many years later suggests he left school due to illness but his letters home emphasize continued financial duress. Costs at college quickly outdistanced his resources. With transportation, tuition, room and board, books, and other costs, he estimated a tab of $100 a year—far beyond what he could earn. He taught another winter school and considered a full year of teaching at a private academy to save money to return to college in fall 1855.

But now another option beckoned. Martin had settled in Dubuque, Iowa, a frontier town on the Mississippi River. He urged his brothers to join him. So George and Cornelius went west, traveling by ship across Lake Ontario, through the Erie Canal and across Lake Erie to Detroit. There, they hopped a train that traveled through Chicago en route to Dubuque.

George delighted in the trip, relishing adventures that reached beyond his books. Indeed, few books of the era could have prepared him for Dubuque. Cattle and oxen roamed freely down the muddy main streets of the town where homesteaders flocked to outfit covered wagons as they headed west to buy up land still available for as little of $1.25 an acre. Speculators bought land too, driving up prices by the day in Dubuque and supporting the town's rollicking riverfront red-light district by night. The brash boomtown exhilarated Martin, who made handsome profits from land speculation. Impressed, George urged his parents to come west, promising they could double their money in no time. With little savings to buy land himself, George took a job as a bookkeeper for the *Express and Herald* newspaper, later the *Dubuque Herald*. An "ignoramus beast," as he described his boss in a letter home, swiftly ended his media career and he decided to pursue a career in mapping and surveying instead.

The building boom in Dubuque promised surveyors the handsome salary of as much as $5 a day at a time when many workers made $1. George learned the trade at the surveying firm of Webb and Higly and, in May 1856, his skill won him independent job offers from the railroads as competing lines raced west, laying track with immigrant labor almost as quickly as the surveyors could map a route. "Mr. Smith, the resident engineer on the Dubuque & Pacific Railroad called to offer a job with the railroad," George wrote home that May. But Martin got wind of a better-paying job in the land department of the Illinois Central Railroad and the income gave George the means to secure loans and begin speculating on land purchases himself. Soon, the railroad promoted George and sent him to Effingham, Illinois, to sell land for the company. Cornelius joined him and settled in the town permanently as a real estate agent.

Now twenty-seven years old, George's success seemed assured and he returned home that summer to marry former Falley Seminary classmate Jane (Jennie) Mariah Wright on August 6, 1857. Then disaster struck. The land bubble burst in the panic of 1857 and land prices plummeted in a swift and fatal spiral just two months after the Van Allens married. George Clinton Van Allen lost $6,000 in the crash, the equivalent of almost five years' pay from his job as a surveyor. The disaster "became an impetus for renewed labor," George philosophically recalled years later, though the loss left him in debt for the mortgage on the land he had bought at inflated prices. He remained upbeat and generously came to the aid of needy neighbors and his own brothers and sisters in years to come. But financial duress, so evident in his student letters, now haunted his entire adult life.

George returned to surveying in the East. Often separated from his bride, he stayed in the hotels along the railroad tracks occupied by an itinerant culture of salesmen, gamblers, and speculators. He spent his evenings reading law books borrowed from Judge F. W. Hubbard in Watertown. Judge Hubbard hired him as a law clerk in 1859 and he attended classes at the Albany Law School as well. George and Jennie lived in the nearby town of Pamelia for the next two years until the Supreme Court of New York certified him as a "counselor of law" in 1861. He returned west and, in 1862, took another surveying job, this time with the Burlington and Missouri River Railroad. The job brought him to Mount Pleasant, Iowa, a growing town crowned by the Henry County Courthouse and the steeple of Old Main at Iowa Wesleyan College. Farms sloped across the hilly terrain where old Indian trails had converged into the roads that now made Mount Pleasant a hub of county activities.

Mount Pleasant originally lay west of the Black Hawk Purchase of 1832 that ceded Illinois and eastern Iowa lands to the United States government. Early pioneers such as Presley Saunders quickly stepped across the boundary. In 1834, Saunders built a log cabin on the scenic high grounds rising from the Skunk River Valley. As more homesteaders settled in, the U.S. government extended its claim and sold the land to the settlers. Saunders bought 400 acres surrounding his cabin but, instead of farming it, he platted out the town he called Mount Pleasant. The town incorporated in 1842 but stayed small with a population of only 758 in 1850. Then the railroad cut tracks through the town in 1855 and the population burgeoned.

Mount Pleasant still resembled the Wild West in some ways when George Van Allen settled there. City fathers had to allocate funds to wet down the roads in summer or the dust clouded every storefront, blinded horses, and smudged the hand-washed laundry hanging to dry behind nearby homes. Business leaders also built a wooden fence around Central Park in the town square to protect the herd of sheep corralled there to keep the prairie grass "mowed."

But Mount Pleasant had public schools, newspapers, a library, an opera house, and a string of churches that followed the Methodists to town. The Methodists chartered the Mount Pleasant Literary Institute in 1842 and renamed it Iowa Wesleyan University (later, Iowa Wesleyan College) in 1855, one of a string of Wesleyan colleges across the country. Old Main, with its neoclassical columns and curving stairways, gave the town a towering monument to higher learning that still stands today.

The town had a social conscience too. People took turns nursing in the military hospital serving Civil War wounded. And the "hatchet-ladies" of the temperance movement hauled out and smashed several kegs of spirits from stores selling liquor in violation of city ordinances. George Van Allen, an avid temperance supporter, must have been as outraged as the ladies when the town trustees issued liquor permits to several businesses, making saloons legal in Mount Pleasant. George backed the abolition movement of the fledgling Republican Party and became active in county politics.

Most important for George Van Allen, Mount Pleasant occupied the county seat, the perfect place for him to establish an abstracts and law practice specializing in land titles. George purchased a storefront for an office near the county courthouse, located in the middle of Central Park. He also bought land on a hillside along Washington Street just a few blocks west of his office. Here he built a two-story frame home in the fashionable Italianate style and planted a grove of maple, oak, hickory, elm, pine, and spruce trees to shade it. He built a barn and planted a garden and an orchard of cherry, peach, apple, and pear trees on the adjacent acre of property.

George and Jennie had been separated off and on during the panic years of the 1850s and the Civil War era. Finally, established in his new career at age thirty-six, he went back east in spring 1867 to help Jennie with the move to Mount Pleasant. "She had some 600 pounds of stuff boxed and we had two trunks and a valise besides," he wrote his mother from Mount Pleasant in April. Jennie brought her mother's large mirror as well. Her father David Parks Wright was expected to visit that May and George urged his parents to visit as promised as well.

He confided to his parents in July 1869 that Jennie "expects trouble in October early," his first mention of the birth of their only child Alfred Morris Van Allen. "Do not mention it for she does not know that I have written you of it and would complain if she knew it. For my own part, I would be glad to have you here if you can happen along. She is squeamish about it. I suppose women generally are." Having a first baby at age thirty-five in an era when many women died during childbirth or from complications that followed explains Jennie Van Allen's state of mind. But Alfred was born without mishap on October 3, 1869.

His father proudly penned reports of the baby's progress. "Freddie knows his name quite well and knows all the family," he wrote to his mother in February 1870. "For all his trouble with salt rhemn [eye discharge] he is a strong healthy boy. Everyone speaks about how strong he is and how bright he looks."

The little boy also suffered from asthma and hay fever, though the worst of the symptoms subsided by mid-September when school started. Mount Pleasant's elementary school had a class for each grade by now and the town had a high school where Alfred enrolled in 1883. The first network of telephone lines connected thirty-six homes and businesses that year. The town had two banks to provide capital for a burgeoning business and farm community. Young people enjoyed dances, hayrides, and the new sport of bicycling.

But a depression triggered another financial panic in 1883, ending a relatively carefree time for the community and the Van Allen family. Land purchases dropped off and foreclosures claimed dozens of farms. The population of Mount Pleasant reached 4,410 in 1880 and dropped to 3,997 by 1890 as people headed farther west once again to start over with new land and better hopes.

George, already caught in the nationwide depression, suffered a further blow that year when a fire destroyed his office at a loss of $2,000. Worse yet, all his title records went up in smoke. He spent the next several years recovering information, rewriting his books, and reestablishing his law practice in a new office across the street from the spacious new county courthouse, built in 1872.

Ever optimistic, he figured out a way to capitalize on the fire. "One of the novel improvements of our city recently is the new fireproof abstracts office of Geo. C. Van Allen," reported the *Mt. Pleasant Daily News.* The arched brick ceiling of the office would protect his records even if the building roof burned through and collapsed, the paper reported. Both the office and the law practice he founded in it remained in the family for three generations. But the construction cost him nearly all of the $8,000 in savings accumulated during the good years. He had $388 left.

Despite the adversity, George soon developed and standardized title search procedures adopted throughout Iowa. Ever compassionate, he continued to provide frequent pro bono services to area farmers even when he himself was in distressed circumstances, adding to his financial woes. The lovely hilltop house on Washington Street, his profession and law office, and his position in the community provided a prosperous veneer. But he was always pressed for money. From his own perspective, a perspective that dramatically influenced his son Alfred, he settled into a life of genteel poverty.

George's financial circumstances never robbed him of his poetic descriptions of life, his love of learning, or the spiritual searching evident in his letters and journals. Alfred shared his father's love of learning and his habit of journal-keeping but approached life with hardheaded business acumen. The contrasts evident in comparing the journals of father and son mirror deep-seated differences in their characters. George wrote his journal sporadically, whenever ideas or events inspired him to pick up his pen and he pursued a topic in lyrical detail. The pragmatic Alfred wrote every day with relentless self-discipline, summarizing life in one or two lines of cryptic facts.

No matter how tight the budget, George Van Allen, like his father before him, sacrificed to give Alfred an education. Alfred enrolled at Iowa Wesleyan University in 1887, opting for the "classical" curriculum in classical languages, history, natural history, ethics, and literature rather than the "normal" curriculum that trained students to be teachers. He was an editor and writer for the *Philo Star of Hope*, a journal that tackled social and political issues. Alfred was a fierce writer, whether attacking the barbarity of the communist regime in Russia or heralding the mysteries of the universe as the inspiration for human learning. He loved debates and oratory as well and won a oratory competition his junior year of college. Such skills might have predicted his fifty-one-year career in law and county politics.

Alfred planned to become a lawyer early on and quit Iowa Wesleyan after his junior year in 1890 to work as a law clerk in his father's office. The following winter, tragedy struck the family again. Jennie Wright Van Allen died at age fifty-six on January 27, 1891. The bereaved father and son continued to live and work together until Alfred started law school in September 1892 at the State University of Iowa [now the University of Iowa] 50 miles north of Mount Pleasant. Three years at Iowa Wesleyan and two years of clerking gave him a lead over most students in both academic and professional preparation. Prior college study was recommended but not required for the two-year law program as well as for the medical school, since the university had no official graduate program as yet. It had no library, museum, auditorium, or dormitories either. Alfred settled into a rooming house at 529 South Clinton Street, about three blocks from campus, paying room and board of $2.50 per week.

Alfred's letters reveal that he understood the value of networking early on. His father, though ambitious for a profession, had pursued learning for learning's sake. Alfred made connections: He sacrificed some study "to pursue social contacts with men from all over the state," he wrote home, "contacts which will result in financial and political preferment." Social fraternities were setting up chapters on campus and Alfred quickly joined Sigma Nu, writing home about the elaborate welcoming party hosted by the fraternity for new students. He had the honor of answering to the inflammatory "toast" offered by the "anti frats," he wrote his father.

Alfred hopped the train to Chicago that summer to visit his Uncle Martin, who sold real estate there. It was a well-earned break between semesters but his father warned him to steer clear of any of his uncle's business deals. "Some fishermen hang out the hook with intentions to catch something and some [do so] unconsciously. But if you get the hook in your mouth once, it will be hard to get away," George had advised Alfred when his Uncle Martin made a visit to Iowa City. Perhaps in deference to his father's warning, Alfred stayed in a rooming house at 235 South State Street, rather than with his uncle on the North Side.

But the focus of the trip that summer was the World's Columbian Exposition in full bloom in Chicago's Hyde Park neighborhood, surrounding what is now the Museum of Science and Industry. Alfred took in the technological wonders and an international bazaar of cultural exhibits filling rudimentary frame buildings camouflaged by alabaster classical facades decorated to resemble marble. The brilliant "White City" encased a network of lagoons traversed by picturesque bridges and Venetian boats. The Midway offered souvenirs, imports, exotic performances, and even tawdry fan dances. A giant Ferris wheel gave tourists a bird's-eye view of the whole dramatic scene.

Alfred returned from the fair for his second year of law school and wrote his father entertaining accounts of events on campus. When the medical school acquired the body of the famous French 6′8″ giantess, "Anne," who died in Des Moines in January 1894, he wrote home. Law students quickly filled the front rows of the lecture hall where a professor prepared to discuss the unusual features of the body. "Angry med students booed as they took seats in back of us," Alfred reported in a letter. "The prof seemed to share the medics' dislike for law. When he came in after the body, his first words were, 'Where there is a carcass, there is a law student.'"

Alfred graduated in the midst of the 1890s depression and dutifully returned to Mount Pleasant to help with his father's law practice. "I don't see the inducement to come here—I think you might do better in some other town but I am getting old and want you here with me," his father wrote. He redecorated the office with fancy wallpaper and a new stove in celebration of his son's return. Alfred moved back home with his father and new stepmother. George had married Anna Watters, a widow and a neighbor who had grown up in Mount Pleasant. He was nearing seventy now and Alfred's return and partnership allowed him to slow down, though he worked until the day he died on September 4, 1902.

Far from resenting the fate of joining a small-town law firm and living with his father and stepmother, Alfred Van Allen immediately threw his energy into local politics. He was elected to the GOP central committee in Henry County the same year he graduated. Alfred had a knack for local politics in conservative Henry County—he had a good family name, a law degree, and a no-nonsense oratory style. And he had a platform—calling for cost-cutting government reforms. Government reform became the big issue as the 1890s depression worsened and foreclosures on farms multiplied. Farmers blamed their problems on government debt and patronage spending. They vented their anger against do-nothing politicians and the "ring" rule that guaranteed two terms to county politicians regardless of performance. The outcry came to a head at the 1897 GOP county Republican convention that met inside the Mount Pleasant courthouse with 134 delegates.

Temperatures reached an unseasonable 85 degrees that September weekend in southeastern Iowa and emotions steamed far higher in the courthouse chambers. Some delegates wore linen suits and some wore overalls, but they all geared for a fight over the nomination of Wash Mullen of Scott Township for a second term as state representative. "Mullen was nominated to order to pay a few old political debts," the *Mt. Pleasant Daily News* reported the following week.

Reformer William Jones of Baltimore Township had been passed over and Baltimore Township delegates backed a political coup. They got one with the help of the newly elected county GOP Chairman Alfred Van Allen. The normal course of a county convention was clear cut. The chairman recognized one of the party faithful to renominate first-term officials for a second term by acclamation. That was the "ring rule," a long-standing political courtesy paid to the ring of county political bosses who Alfred and other young delegates now defied. Instead of recognizing a member of the ring as the Mullen nomination came to the floor, Alfred recognized the delegate from Baltimore Township. "Mr. Archibald, a bright young man arose, and in a clear determined voice moved that nominations for representatives be made by an informal ballot," the paper reported. A clash of voices seconded the motion that passed by a comfortable majority. The balloting stripped Mullen of the nomination.

The *Mt. Pleasant Daily News* took full credit for the victory, citing a two-year editorial battle against the ring. But they lauded Alfred Van Allen's convention as a "model convention" that "rewarded faithful and efficient officers."

The ring rule "was declared off and the delegates made a new rule to the effect that one term was enough for poor officials."

Alfred led the county GOP as party chairman off and on for the next thirty-three years. He served intermittently on the Mount Pleasant school board and on the city council. In 1904, voters elected him to his first term as a Mount Pleasant city alderman and he turned his attention to the city's power outages. The city had bought out the privately owned electric plant in town because outages were so common and rates so changeable. Alfred and a committee of two other councilmen studied the crisis and recommended that the city purchase a $12,000 boiler and steam engine generator. Such a power plant required a steady and reliable water supply, however, and the one well serving Mount Pleasant provided barely enough water to put out fires. Serving on both the water and light committees, Alfred pushed plans to quadruple the number of wells and connect them to a new city water works to supply the power plant.

As he helped rebuild the infrastructure of the town, he also rebuilt the Van Allen law firm. The financial instability Alfred had witnessed at home in his youth haunted him and he practiced strict frugality and made investments in land while land was cheap. Desperate farmers were willing to sell their land and stay on as tenants. The rents supplemented Alfred's income. He was determined to build a secure financial future before he considered marriage. Finally, as he neared forty, he began to court Alma Olney, a young woman enrolled at Iowa Wesleyan Academy, a high school associated with the university that offered the benefit of college level instructors and college credits.

Her father James was a kindly, warmhearted Dudley, Iowa, farmer whose family had settled in Iowa as pioneers. Her mother Martha (Mattie) Eyre died when she was young and her father remarried. Alma grew up learning to speak softly and work hard, trying to appease her quick-tempered and critical stepmother. But Alma was an avid student, eager to escape home by becoming a teacher. While the towns already had public school systems, county districts still relied on the old winter school system and Alma took a job in one during the winter of 1906. She lived with Mrs. Thomas Cook, who took her to school in a sleigh for the start of classes on New Year's Day. The school she described in a journal she wrote that winter matched almost exactly George Van Allen's experiences fifty years earlier. She rose before dawn to walk through the snowdrifts to the school most mornings so that she could light the stove and shovel the pathway to the door. She almost quit in the middle of the term due to a severe cold but made it through to February 24, the close of the term. Another teacher took over the classes for the spring term that ended in late April.

A corps of maiden lady teachers taught at established elementary schools in the towns and rural districts, but securing one of these full-time jobs required additional education. On January 2, 1907, Alma set out to prepare for such a career, taking the train to Iowa Wesleyan Academy. Within a week, a host of activities with new girlfriends fill the pages of her diary. They studied, went shopping in town, and joined the glee club. She and a friend dressed up as ghosts as a prank in the dorm and, she notes, she got "a double dose of scolding" for complaining about the food. Religious services claimed lots of time. She went to the Pentecostal church on Saturday evening January 5 and Sunday school the next morning. That Sunday evening, she went with a friend living in town to visit the Van Allens. "Had a good time," she noted in her diary. Apparently, Alfred did too. "Mr. AMVA called me up and invited us down to spend the evening," she wrote in her diary on January 19.

She was twenty-three years old. The couple courted as she finished her education and then taught in rural schools near Dudley for three years before marrying on June 14, 1911. Alfred's stepmother, Anna Watters Van Allen, had remarried a taxidermist and moved to Des Moines so the newlyweds had the family home all to themselves as they started a family.

2 Heartland Boyhood

Mount Pleasant, Iowa—November 11, 1918. Four-year-old James Van Allen and his family pressed into the crowd that packed the town square in Mount Pleasant. The great war in Europe was over, his parents told the child and his brother George. The Germans had surrendered under terms of the Armistice signed that day. In the chilly twilight, orators hailed the righteous victory. Preachers prayed for the Mount Pleasant Boys of 1917 who lay dead in the trenches of the Argonne Forest in France. And the high school band played patriotic songs. It almost seemed like the Fourth of July when flames leapt against the night sky and riveted James's attention. The flames outlined the figure of a man with a pointed helmet. The straw body burned quickly and Alfred Van Allen explained that the figure represented Kaiser Wilhelm II, the German leader who had led the world into war.

• • •

The garish mock execution traced an indelible memory in the little boy's mind, one of a series of earliest memories from that fall and winter. More than eighty years later, James Van Allen vividly recalled his walk to Saunders School on his first day of kindergarten that fall. He proudly wore a military campaign hat like the one worn by General John Pershing, who led American troops to victory in Europe. And James recalled thinking he was dying as he drifted in and out of consciousness with a raging fever from the flu epidemic that spread across the globe that winter, killing millions.

The little boy healed, but the world around him did not. The war ignited revolution in Russia and planted the seeds for the Nazi Party in Germany. The staggering political and social changes would catapult research into the realm of "big science," with national laboratories and a national space program in which Van Allen would play so momentous a role. Closer to home, rapid-fire inventions remolded everyday life in Mount Pleasant. The community-run power plant that Alfred Van Allen and his fellow aldermen spearheaded brought electricity to the town and surrounding farms. Telephone lines, cars, and an occasional glimpse of an airplane overhead made distant places seem closer and brought more people to the county seat.

• • •

Alfred Van Allen entrenched his political power base in Mount Pleasant shortly after his marriage. He watched Mount Pleasant's new courthouse take shape in 1913 right across from his office, aware that the political scandal surrounding the whole project meant a certain win for the county Republicans in the next election. The Democrats had won enough seats in the previous election to control the county board of supervisors and gave a politically connected brick plant the contract to make 700,000 bricks for the courthouse construction. But none of the bricks were used on the building. Van Allen led the GOP back to a majority on the county board in 1914, the year his second son, James, was born.

Alma Olney Van Allen gave birth to four sons in the first ten years of her marriage. George Olney was born on May 28, 1912, James Alfred on September 7, 1914, Maurice Wright on April 3, 1918, and William Albert on September 6, 1921. She delivered them at home in the cherrywood bedstead her husband ordered custom-made from the trees in their own orchard. The rugged, hard-working Alma recovered quickly from each birth and relied on advice for new mothers in Procter and Gamble's 1912 edition of *How to Bring Up a Baby.*

Handyman Slim Norton and a succession of country girls hired on to assist Alma with chores as she raised her four boys. Her household ran on a regimented schedule. The family usually rose at 7 A.M. and Alma made a simple breakfast of oatmeal or Cream of Wheat and fruit. Alfred left for the short walk to his office every morning at 8:30, about the same time the boys left for school, and everyone returned home for lunch at noon. The family had dinner at 6 P.M., eating chicken that Alma butchered herself or catfish after Alfred's fishing expeditions to the Skunk River.

On summer nights, the family took walks to the railroad depot where the eastbound *Number 9* arrived in Mount Pleasant each evening. Alfred got acquainted with the engineer, who invited James to climb into the cab of the locomotive and pretend to be an engineer. Though diesel trains such as the *Denver Zephyr* began to roll through town, James loved the power and majesty of the *Number 9* and the other huge locomotives that roared into town, belching smoke.

The city's power plant was the next stop on the walk. Coal shoveled into heaving blast furnaces powered the mammoth generator at the plant. James and George drew dramatic electric sparks from the long leather belt that powered the generator, further sparking their curiosity about all things mechanical. Back home after the walk, Alma mended slacks, shirts, and underwear. George wore new clothes ordered from the Sears catalog and Alma kept them neatly sewed, ready to be handed down to brother after brother. She darned wool socks at night, too, using a darning egg that she replaced only once when James made her a new one for his first shop project in high school.

While Alma sewed, Alfred read to the boys from the *Book of Knowledge*. George and James read *Popular Mechanics* and *Popular Science* from cover to cover when the latest issue arrived each month. But comic books were forbidden and Alfred frowned on storybooks and novels. He frowned on athletics and many social activities as well, so the boys focused their free time on studying and chores. Alfred Van Allen's stern credo was clear: "Don't waste your time, don't waste food, don't waste your money, and don't waste your breath arguing." James started kindergarten just after his fourth birthday because his father considered it a waste of time for the little boy to spend another idle year at home.

The brothers studied hard and James learned early that persistence gave him an academic edge over other smart students who simply weren't willing to work as hard. By second grade, he realized he had a knack for arithmetic, multiplying quickly and carrying over numbers in his head as the teacher posed oral arithmetic problems such as "What is thirty-seven times eight?"

When not studying, the boys helped with chores—lots of chores. As the oldest two, James and George each owned their own flock of chickens, though the privilege meant they had to clean out the coop. They winced from the smell of ammonia as they cleared soiled straw and replaced it for the families of Rhode Island Reds and Plymouth Rocks. And they collected eggs for breakfast. The cost of food rose dramatically in the early twentieth century. Eggs cost 5¢ a dozen in 1900 and 13¢ during World War I. Butter prices rose from 10¢ a pound to 22¢ a pound. Alma made her own butter and homemade bread, and the Van Allens had a plentiful garden harvest as well.

All spring and summer, family attention focused on the vegetable garden with neatly weeded rows of corn, beets, string beans, snap peas, radishes, asparagus, and tomatoes. The boys husked the corn, shelled peas, and cut beans on the porch in the hot afternoons of late summer. Alma boiled pots full of the vegetables for canning and boiled more pots of water to sterilize the mason jars that she filled and sealed. The kitchen steamed on canning days and the house barely cooled overnight before Alma started the process again early the next morning. She lined all the jars up on shelves that Jennie Wright Van Allen had built in the basement. She prepared bottles of homemade maple syrup as well—syrup gleaned from the spring sapping of the eight sugar maple trees on the property. Alfred Van Allen bored a hole in the trunk of each tree, inserted a spile into the hole as the sap began to run in spring, and hung a bucket from each spile. Sap dripped into the buckets over the course of a day and Alma emptied them into a larger barrel in the barn. She slowly boiled down some 50 gallons of sap until about a gallon of maple syrup was left to be canned.

Black walnuts filled baskets in the basements. The boys picked gunnysacks full of hickory nuts and black walnuts in the woods. The black walnut shells turned from green to black, with a gooey inside layer that stained clothes and hands as the brothers shelled them. Most of the boys in town had brown stains up to their elbows from shelling the black walnuts in late fall.

George and James loaded their coaster with the surplus vegetables, eggs, and nuts and sold them in the neighborhood. Alfred recorded their earnings one summer in the back pages of his journal—$9.80. The two boys also had paper routes. They spent their money on electronic kits and parts they saw in *Popular Mechanics* to make whirligigs in the barn.

"We built a crystal set (an early radio receiver) from raw components, though the crystals themselves came from a local radio store. One of the thrills of my life was picking up KDKA, a famous pioneer station in Pittsburgh, Pennsylvania." They also made a spark coil used in the ignition system of early cars, James recalled. "We would draw off sparks several inches to a foot long and they'd make your hair stand on end. There was a ball on the end to wind up the coil and you could release several thousand volts, yet there was practically no current so it really wasn't dangerous. But my mother didn't know that. She was horrified to come into the barn and see my hand on this coil and sparks shooting off from my hair."

James described his mother as a "one-woman army." She served on the library board. She gave spring and summer parties for the library staff and for the girls of Phi Mu sorority at Iowa Wesleyan. But mostly, she worked seven days a week on an endless round of cooking, baking, cleaning, and making sure the hired girls refilled the water tank from the rain barrel. The stove heated the water in an attached tank that provided hot water for baths and cleaning. But huge tubs of extra water had to be heated on the coal-burning stove for laundry day.

Housewives throughout Mount Pleasant reserved Mondays for laundry day, the hardest day of Alma's week. She scrubbed load after load of clothing with a washboard and pungent lye soap. She wrung out the soapy water with a hand-cranked wringer, rinsed the items in clean water, and wrung each piece out again. Everything had to be starched. Then everything had to be hung on lines. Then everything had to be ironed with pairs of 5-pound irons heating on the stove so one could be used while the other reheated. The labor remained unchanged until Alma got an unbelievable luxury in the late 1920s—an electric washing machine with an electric-powered wringer attached on top. It was the first electrical appliance most families bought. Despite the long hours of her days, Alma needed the nights to nurse family illnesses.

Asthma plagued Alfred and the boys. Alma kept tins of Power's Asthma Relief on hand to help ease the symptoms. The tin contained dried herbs that resembled tobacco. She put a small pinch in the lid of the tin and lit it with a match. The smoke usually checked the dizzying coughing spells. But, toward the end of August, the blast furnace of hot afternoons dried and shredded the ragweed pods until a cloud of pollen sifted everywhere. Ragweed intensified the boys' hay fever and allergies. Alfred Van Allen bought a Dodge sedan in the early 1920s, packed up his family every August and weathered out the ragweed season in the woods of northern Wisconsin or the resort town of Brainard, Minnesota. There, the family boated and fished, catching sunfish that Alma cleaned and fried.

Winter brought colds and flu and the boys often fell ill in quick succession. Alma set up a hospital ward for her patients in the living room where she could keep a better eye on them and where the stove provided heat. The unheated upstairs bedrooms were icy cold. But all Alma could really do as the fevers shot higher was to wash the boys down with cool water. She hovered over the children with water and chicken broth and dosed them on castor oil—the purgative cure-all of the day. She darkened the living room so they would simply sleep as much as possible during the day and slept near them through the long, worrisome nights. Dr. Smith frequently came in the morning, making the rounds throughout the town. Pneumonia and strep throat remained serious risks as the boys recovered in an era before antibiotics. The risks weighed heavily on Alfred's mind. He recorded every illness and every raging fever in his daily journal, filling a single ledger with cryptic entries that covered the years 1920–1931.

"James is delusional—fever 105½," he wrote on one occasion in 1922. "Fever's better today—down to 104," he reported a day later, a familiar litany throughout his journal.

With nothing but his own imagination to occupy the hours he lay ill, James built a house. He mentally etched a place with a shabby exterior hidden by overgrown bushes, a place people would pass without taking notice, a place with shades over the window so he could work in private. Inside, the rooms contained lathes, saws, hammers, planes, and all the metal and wood parts needed to build furniture and electronic gadgets. The workshop had phonographs, crystal sets, and lots of space where he could take devices apart to see how they ran or to build inventions of his own. The room had workbenches and a big desk with a bright light to illuminate his studies. With each illness, he added new details to his secret haven. And there were lots of illnesses.

Alma nursed her sons through measles, smallpox, and chicken pox and watched for signs of the most feared illness of all—polio. Polio crippled and paralyzed the legs and arms of children Alma knew—and killed when it paralyzed the lungs. George's bout with polio in high school started out resembling a simple case of the flu. But the stiffness of his neck and back persisted—symptoms of polio. After a long bed rest and Alma's rigorous applications of hot compresses, he recovered, left with a mild limp. Fortunately, no one else in the family caught the disease.

Aside from the childhood illnesses, financial stability remained an overriding concern of Alfred Van Allen's life. The polite veneer that hid the poverty of his youth cut a deep impression, and he ran a tight ship where money was concerned. He weathered the slow periods of a small-town law firm with a sideline business in land investments and tenant farming. Farmland had skyrocketed in price, selling for $50 to $350 an acre depending on location and improvements, and tenant farmers rented property from $6–$10 an acre. He indulged in some luxuries, building a spacious addition on the house shortly after he married and adding a big new kitchen. But, in general, James Van Allen recalled, "We lived like we were poor and that's why we weren't."

Considering Alfred's successes, his frugality seems extreme. But ministers, politicians, and newspaper editors of his era all expressed admiration for the "pioneer spirit" of hard work, self-discipline, and thrift. Alfred valued these traits and isolated his children from challenges to them. The garden, the orchard, and chickens that provided so much of the family food reflected labor-intensive thrift. Like most children of the era, the boys went barefoot all summer and, in winter, climbed the stairs to the frigid, unheated second floor. Each boy carried a flat soapstone bed warmer, about the size of a book, warmed from heating on the stove. The boys rubbed the heated stones across the cold sheets and then placed them down by their feet, "a great comfort" on frigid winter nights, James Van Allen said.

The boys shared a single upstairs bedroom across from their parents' bedroom. As children, they played together in the house, the barn, and the shaded yard that rambled across a city block with the orchard and garden occupying the adjacent block. They took their coaster wagons down the hills of Mount Pleasant in summer and sleds down the hills in winter. They fished with their father on the Skunk River near the Oakland Mills.

As the county seat, Mount Pleasant attracted the county fair and a host of other entertainments that the family enjoyed each year. The Van Allens also attended the annual Chautauquas, traveling variety shows that combined religious revival, opera, theater, and lectures on political or social controversies. Evangelist Billy Sunday gathered huge crowds with his fire-and-brimstone sermons ringing through the tents of this road show that originated in Chautauqua, New York. The Ringling Brothers Barnum and Bailey Circus rolled in every summer and the whole community turned out as the circus parade of wild beasts, clowns, and performers paraded through the streets to the big top at the fairgrounds southeast of town. Every Sunday, the Van Allen boys went with their parents to the Presbyterian church and attended Sunday school. On Sunday afternoons, they sometimes drove to Dudley for dinner with Grandpa Olney and his third wife, whom everyone called Aunt Neva.

The family stayed home for Christmas and Alfred reflected on the Christmas of 1925 with the longest passage in his eleven-year diary. He spoke proudly of his older sons' progress in school and then described the holiday festivities. "George furnished two small cedar trees which his mother decorated together with some room decorations, all in the parlor. We held our celebration Christmas Eve and Ma was the distributor of the gifts. George received his long coveted erector set and a necktie; James a set of tools—two planes, a coping saw, chisels, bits, etc., etc.; Maurice his much coveted scooter and a lot of small toys; William a lot of toys including a train, tinker toy, a 'sing and spin' as he called (the) revolving music box and top. Anna Steppan, [Alfred's remarried stepmother] and Aunt Neva Olney and the Candy Kitchen made us fine presents of candy. We had a big chicken dinner with oyster dressing and everybody was happy."

Alfred's mostly terse descriptions of 1920s life in his journal focus on a timeless world of family, business, and farming familiar to all his neighbors. But, unlike most parents of his times, he expected all four of his sons to go to college and set their educational course early in life. James stayed at the Saunders School from kindergarten through sixth grade, and then followed George to the junior high. James loved school and spent two or three hours a night in problem solving, writing, and studying the next day's assignment. His seventh grade assignment notebook shows the emphasis his history class placed on the European explorers. Columbus, Balboa, Magellan, Cortez, Pizzaro, and

Leif Ericsson, were the epic heroes of many lessons, and James memorized long lists of explorers and their accomplishments.

He liked writing essays, too, yet an almost mathematical approach to narration shines through his junior high essays such as one written about the history of Halloween. "The background, or lining of the quilt, was furnished by the Romans and the Druids while the bright-colored patches on top were added from time to time by the Irish, Scotch, English, Welsh, Bretons, French and Teutons," James wrote, tracing the holiday rituals to this patchwork of cultural contributions. "Exactly in the center [of the quilt] is worked a huge yellow pumpkin with glittering eyes, fantastic nose, grinning mouth—the Jack o' Lantern—America's contribution to the Halloween quilt." All of James's teachers through junior high had been women—dedicated professional teachers who weren't allowed to work after they married. They reinforced the rigorous approach to study that James learned at home. They grounded him in reading, spelling, history, and mathematics to prepare him for high school.

James started Mount Pleasant High School in 1927, a tumultuous year for the town. The *Mt. Pleasant Daily News* reported successful raids on bootleg distilleries as the Roaring Twenties ushered in the age of gin, jazz, and bobbed hair. Though the paper refrained from giving locations or the names of those arrested, reporters had a field day when Sheriff Hannah walked past a Mount Pleasant garage and noticed the mechanic repairing a car with the springs sagging under the weight of the load in the trunk. The sheriff popped the trunk, found 126 gallons of spirits, and promptly arrested the owners when they returned. Sheriff Shepherd took over Hannah's job a few years later and chased bootleggers all the way to the Illinois state line where they escaped his jurisdiction. Federal agents solved the problem by swearing him in as a federal marshall so he could continue his dragnets into both Missouri and Illinois.

The colorful 1920s vignette of chasing bootleggers down county roads reflected that cars had revolutionized personal travel. And while speed limits had to be enacted to slow cars down, muddy roads stalled them completely. Mud collected under the mud guards of the car until the wheels couldn't turn. A spade or heavy stick had to be carried in the car to free the wheels, which soon clogged again. Alfred Van Allen campaigned for paved roads early on but the North Henry County Anti-Hard Roads Association worked just as hard to preserve the dirt roads. They predicted it would ruin more than half the farmers of the county if they were forced to pave 840 miles of county roads at a cost of $16–$17 million.

"If you bond the county, please go out in the future with me 20 years and see your children and grandchildren toiling in the fields to pay for the roads that are worn out," one flyer warned ominously in 1916. Car owners were still too few in number to push their cause. But the Model T, the best mud car on the road, kept dropping in price due to Henry Ford's production innovations. At a cost of $260 in 1925, the average family could afford one. Half the cars in Henry County were Model Ts in the 1920s and car owners hit the testing ramp for their new vehicles at the Oakland Mills rise across the Skunk River. Cars raced each other up the hill and one Mount Pleasant driver in his Winston Six even raced the train heading to Chicago. The car had three flats along the way, the *Mt. Pleasant Daily News* reported, but the automobile won.

With 3,000 registered vehicles in the county, mud made good politics by the mid- 1920s. People drove to Mount Pleasant early on Saturdays so they could shop and secure a good parking spot at the town square for evening concerts played from the bandstands. As the county began to gravel and grade the local roads, farmers traveled to market.

But Alfred and other local officials lobbied for federal help to pave the long stretches of dirt highways connecting towns and counties. In 1925, Alfred's friend and law school classmate, GOP Congressman William F. Kopp, secured support from the Calvin Coolidge administration for a pork barrel project—the paving of Harding Highway, the major east-west thoroughfare through Mount Pleasant. But the county still needed a bond issue to float its share of the costs, and Alfred's leadership role in the bond referendum anchored his bid for mayor in 1927. "Elected mayor of Mt. Pleasant: I received 666 (votes) [compared to 378 for the closest rival]," he noted in his journal on March 26, 1927. The "hard roads movement" named him chairman in June and the countywide referendum passed three-to-one on July 19.

James and his brothers learned to drive at ten or twelve years old but he and George were even more interested in the mechanics of a car than in driving one. They pooled together $50 from working odd jobs at 25¢ an hour and bought an old Model T with an old-fashioned running board and open top. They completely stripped it apart to see how it operated. The clutch, pistons, and three-pedal transmission lay systemically arranged across the backyard lawn. "We wanted to see how the clutch worked and how the transmission worked. There was a very tricky transmission on the Model T," James Van Allen recalled. "We got it back together and we drove it but it was given to having flat tires, as I recall, so we sold it." The boys recovered their $50 on the sale.

Maurice began to join George and James in their electrical projects by then, but Bill shied away. Even as a kid, Bill was more interested than his brothers in the family subscription to the *Chicago Tribune* and the political issues of the day. James also crafted simple, wooden household items in the basement with the set of tools that he received Christmas 1925 when he was eleven. High school gave him access to a shop with electrical-powered drills and saws, and he took the class all three years. He made a magazine stand and a fashionable footstool known as a taboret. With increasing skill, he built elaborate pieces, such as a bedstead with lathe-turned spindles and bedposts, which was still in the guestroom of his Iowa City home when he died. He made a desk and chair that his grandson Andrew Cairns still uses.

The shop class also covered mechanical drawing, a skill James would use throughout his career. The drawings involved skill, patience, and ink pens that had to be filled with an eyedropper. The drawings were stretched across a brass frame. A piece of photosensitive paper was laid on top, exposed to sunlight, and then washed to make a blueprint.

High school opened vast new academic horizons. James discovered an interest in Latin and enjoyed translating classical works into English. His early talent for math easily met the new challenges of ninth grade algebra and geometry. In geometry, he learned to visualize the world in three dimensions, a perspective that he later applied to celestial bodies and space itself.

The Van Allen brothers avoided theater, athletics, and the general social life of high school just as they had in grade school. "We were still isolated. We'd get to school, do the chores when we got home, and study," James Van Allen recalled. But his routine changed dramatically his senior year when he discovered physics. The class included lots of lab assignments and he got caught up in the excitement of experiments designed to teach basic principles about mass, velocity, and energy. The science teachers permitted the quiet and serious James to work in the lab after school. "Just close the door when you leave," they told him.

James designed his own experiments to test the principles of physics with equipment in the lab. He explored different lens configurations to make a telescope modeled on the one his father had ordered from the Sears catalog. He often hurried home at 6 P.M., raced through chores, dinner, and homework and then returned to the lab to continue doing experiments until about ten o'clock at night. By graduation, James's fascination with experiments and making components for those experiments had established a passionate empirical approach to science he would pursue all his life. He liked problems that could be defined, measured, and retested with equipment he could build.

"James graduated tonight with first honors. A big crowd," Alfred wrote in his journal on May 21, 1931. Alma, dressed in her best black linen dress for the occasion, and Alfred and the boys joined her in the school auditorium where sixteen-year-old James, the class valedictorian, recited his essay *Pax Romana,*

Pax Americana, which compared America's greatness to the greatness of ancient Rome before the taste for luxury and entertainment weakened the moral strength of the empire. "The poor are becoming poorer while the great and rich are becoming greater and richer. The robot and other cunning machines are taking the place of the common man as did the foreign slaves in ancient Rome," he said. "The common honesty and respect for the law, the industry and frugality that made our forefathers great are being replaced by the wantonness and idleness of unearned wealth. Are we following Rome? If not, what is our future? What is to follow the Pax Americana?"

"I was well coaxed in this point of view by my father and by Colonel McCormick, the publisher of the *Chicago Tribune,*" James laughed some seventy years later as he pulled the essay from a drawer of George Clinton Van Allen's desk—a desk where he worked every evening in the study of his Iowa City home.

As a graduation present, James and George went to Washington, D.C., to stay with Representative William F. Kopp. They toured the sites of the capital, and the congressman's driver chauffeured them to Mount Vernon, the home of George Washington. The congressman arranged for them to join a presidential receiving line where President Herbert Hoover shook hands with select constituents of select politicians. "It was the midst of the depression and he was pummeled on all sides for being the cause of it. He was standing up to shake our hands and there was no strength in his hand and no expression in his face. His face was totally drained of color. He looked about as close to being a corpse as anyone I've ever seen," James Van Allen recalled.

Most high school graduates from the class of 1931 joined their families in running farms or businesses in town. But higher education remained a "fundamental expectation" of the Van Allen household, even in the depths of the depression. James had hoped for an appointment to the U.S. Naval Academy at Annapolis, and Congressman Kopp sponsored him. He easily passed the academic examination but failed the physical due to his asthma. Instead, James followed George to Iowa Wesleyan, less than a mile from home.

He walked through downtown Mount Pleasant to the campus. Two neoclassical gems—Pioneer Hall and Old Main—preside over the campus, landmarks of the early years when Methodist Episcopal ministers founded the Mount Pleasant Literary Institute in 1842.

The institute welcomed its first students to a curriculum of composition, natural science, moral science (ethics), classical languages, and literature. James Harlan, later a U.S. senator and secretary of the interior, took over as president of the institute in 1853 and quickly expanded the curriculum. The school soon offered mathematics, theology, piano, drawing, French, German, and "political economy." Harlan changed the name of the school to Iowa Wesleyan University to emphasize its connection with the Methodist Episcopal church. And he built Old Main with its impressive auditorium that attracted national lecturers. Frederick Douglass came to Mount Pleasant as did suffragists Elizabeth Cady Stanton and Susan B. Anthony and temperance crusader Sojourner Truth. Babb Mansfield, a Wesleyan graduate and the first woman lawyer in the country to be admitted to the bar, brought the Women's Enfranchisement Convention to town in 1870 and invited Anthony, Stanton, Truth, and many other prominent women to speak.

With all the activity, tiny Mount Pleasant ranked among the most popular lecture stops in the country during the 1870s and 1880s. The cultural life prompted Universalist minister W. R. Cole to conclude in a speech that Mount Pleasant had done more than any town in Iowa to develop "the wealth of the soul." He toasted the town and the college as the "Athens of the Midwest." Iowa Wesleyan's new chapel, built in 1893 with a 1,200-seat auditorium surrounded by stained-glass windows, reflected the prestige of the college and the town as a spiritual and intellectual center. Women of the town founded the secret sisterhood of P.E.O. on the second floor of Old Main in 1869. The organization added to the prominence of the community as local chapters sprung up across the nation and thousands of women joined to socialize and Promote Educational Opportunity for women, the official name for the group.

Alma joined the organization while she was still in school at Iowa Wesleyan, where George began college in 1930. James started with an eager agenda to sign up for physics and astronomy classes his first semester. By the 1930s, Iowa Wesleyan was developing a reputation for training good scientists and physics teachers in Iowa. Professor Thomas Poulter deserved much of the credit.

James Van Allen met Poulter just a few days after his seventeenth birthday when he walked into the introductory physics class at Iowa Wesleyan. Like a coach recognizing the star athlete in the freshman lineup, Poulter took James under his wing. He taught him to think of science and scientific research as a way of life. "Your father was the chief inspiration for my professional career, dating from 1931 when I became his student and laboratory assistant," James wrote to Poulter's son Thomas in 1999.

Poulter grew up in Mount Pleasant where his father was a tool and dye machinist. He enrolled at the Iowa Wesleyan Academy in 1914, the high school associated with what was by then Iowa Wesleyan College, the final name change for the campus. Minnie B. King, who headed the academy, allowed students to take as many classes as possible where they could get credits both toward a high school diploma and a college degree. Sophomore year, Poulter petitioned to take college physics and soon became the lab assistant for the college program, the same job Van Allen now held.

Poulter joined the navy in 1918 during World War I and returned to college in 1922, earning a degree in physical chemistry from the University of Chicago. He returned to Mount Pleasant and, during the next five years, he married, had two children, and took over as head of the physics department in quick succession. He added new classes and a rigorous "blackboard session" devoted entirely to solving problems.

Poulter also began testing the physical and electrical properties of solids—crystals, in particular—at high temperatures and pressures. This was critical testing in the developing field of solid state physics where innovations later paved the way to electrical components such as transistors and semiconductors. But by 1930, Poulter had a third child and found his job and his college in precarious circumstances. "There was much beating the bushes for students as the depression had hit in 1929 and many of them were without funds. Salaries took a drop and there were many people out of work," Poulter wrote in his unpublished memoir, sending a copy to Van Allen. "PhD graduates were desperate for jobs of any kind and some were doing janitor work. It was a tough period." Still, at a time when the entire student body at Iowa Wesleyan numbered about four hundred, Poulter attracted sixty-five students to his introductory physics class alone, partially because of his dramatic lab demonstrations. Students and townspeople alike considered him an affable local genius who could build or invent just about anything. He built a circular section of the floor in the class mounted on ball bearings to dramatically demonstrate angular momentum. His spectacular displays demonstrated principles such as the amount of energy stored in the magnetic field of the transformer core. He took a rectangular core 20 x 24 inches divided into 4-inch cross sections. He wound ten coils with four hundred turns of 1/8-inch wire and created a switchboard to connect the coils in any combination of series. If he closed the circuit with the coils that were connected to a series of storage batteries and then pulled the switch he could draw off an arc of electrical current.

The college couldn't pay for this kind of dedication and ingenuity but, in 1932, the board rewarded Poulter by placing him at the helm of the divisions of physical sciences, mathematics, and astronomy. Poulter's salary remained $100 per month, however, and he used some of it to buy and fashion lab equipment so that each student could be working on an experiment at the same time.

He paid Van Allen's lab assistant salary of 35¢ an hour out of his own pocket as well. If they didn't have a part in the lab, Van Allen learned to make it in the physics department shop where Poulter taught him to cut and weld metal sheet and bar. Best of all, Poulter and other professors at the college enlisted Van Allen to do field research. He loved it from the start and his very first research assignments involved the two specialties that would drive his career—space and magnetism.

First, Van Allen helped test a "reticle" device Poulter had invented and built to observe meteors. The device amounted to a series of concentric circles mounted along a 6-foot long conical device made of welding rods with an eye-ring at the vertex. The conical fields corresponded to meteor altitude ranges and the device could help track a meteor's trail. Van Allen located one of the devices in his backyard in Mount Pleasant, and fellow student Raymond Crilley made a second set of measurements with a reticle he set up in Iowa City. This allowed them to plot the path from two different perspectives. Van Allen and Crilley watched the skies simultaneously in the early morning hours of August 7 and 12, 1932, and then went into action measuring trails of the Perseid meteor shower. They delivered their observations to Professor of Astronomy C. C. Wylie at the University of Iowa so that he could determine the rough heights of the luminous trails and their appearance and disappearance. Poulter credited Van Allen and Crilley in his subsequent publication, Van Allen's first professional scientific credential.

Next, Poulter entrusted Van Allen with the task of taking magnetic field readings across Henry County with a field magnetometer from the Department of Terrestrial Magnetism (DTM) of the Carnegie Institution in Washington, D.C.

"It had beautiful construction with brass fittings and ivory knobs and a small telescope for measuring the angle of the sun [at the time of taking a measurement]. I set up a survey station on campus and learned how to use it," Van Allen said. The magnetometer could continuously measure both the direction and intensity of the magnetic field at any given point and it was the most elegant instrument Van Allen had ever seen. The instrument essentially followed the design developed by Carl Friedrich Gauss one hundred years earlier. It relied on the rate of oscillation of two small bar magnets in a horizontal plane. The oscillation rate corresponded to the intensity of the magnetic fields and the angle of deflection between the magnets showed the direction of the magnetic field.

The log of the trial readings recorded on campus on October 10, 1932, show Van Allen's persistence and precision in his approach to measurements. He took sixty separate readings at one-minute intervals from 3:43 P.M. through 4:42 P.M. Individual measurements varied by as much as 5 percent but averages of five-minute clusters of readings varied by less than 1 percent.

Van Allen applied the procedure he worked out for the trial readings to the ones he took in the fields and towns across the county. He copied his results and sent them to DTM for incorporation in a worldwide survey of magnetic field measurements. To his chagrin, DTM sent his work back, explaining that only original field log entries could be used for the survey. Van Allen sent them the original entries and learned a lesson he never forgot.

Van Allen's measurements proved that the magnetometer was properly calibrated for a critical mission—Poulter's expedition to the Antarctic with Admiral Richard Byrd in 1933. Van Allen helped Poulter construct a tiltometer to measure the angle of a glacier for the same mission. They pieced together the device in an open lab of the physics building, adding to the high adventure of Poulter's trip with the famous explorer. An avid follower of Byrd's first Antarctic expedition in 1929, Poulter sent the retired admiral a letter with a description of his reticle and a suggestion that Byrd include meteor observations on his next trip. Byrd not only accepted the suggestion but enlisted Poulter to make those observations as chief scientist for the second Antarctic expedition departing in fall 1933. He named Poulter second-in-command as well. In his later book *Discovery,* Byrd described the reticle as "one of the brightest memories of the expedition."

Poulter invited his young protégé to join him on the legendary Byrd expedition but Alfred and Alma wouldn't give their consent. They had other plans for their precocious son. George intended to start law school at the University of Iowa and Alfred suggested that James apprentice with the town dentist and then go on to dental school at the university. For the first time in his life, James challenged his father. With Poulter as his mentor, he had made up his mind to become a physicist. Alfred, uncertain what his son could do to earn a living in such a field, reluctantly agreed. James had little time to ponder the possibilities as he was attending classes full time and running Iowa Wesleyan's physics lab and workshop in Poulter's absence.

Another Iowan, Cedar Rapids inventor Arthur Collins, developed a short-wave radio that enabled Van Allen to follow the Byrd expedition from afar. Chicago radio commentator Howard Von Zell got one of the short wave sets from Collins Electronics and used it once a week to talk to Little America, the base station near the South Pole for Byrd's expedition. He broadcast the conversation during his regular radio show and Van Allen faithfully tuned in, elated when the voice of Poulter came across the radio. Other stations across the country carried broadcasts transmitted from Little America with dramatic reports of dangers and new discoveries.

But the men at the base withheld the harrowing story of Byrd's near death at a secondary base he established closer to the pole. Byrd broke the silence in his book *Discovery,* telling how Poulter led the rescue of the great explorer from an isolated field station where carbon monoxide was slowly poisoning him.

The potential for disaster haunted the expedition from the start. Byrd originally targeted his departure for 1932 during worldwide collaboration in research for the Second International Polar Year. Despite his hero's acclaim, Byrd found limited interest in funding for his expedition in the middle of the depression. The Armour Institute in Chicago and DTM became the major sponsors, but working-class Americans sent Byrd a total of $150,000 in cash and government agencies and universities turned over $100,000 in instruments. Still, Byrd found himself overextended. His expedition costs involved pilots, geologists, meteorologists, biologists, radio operators, navigators, an aerial photographer, a surveyor, a physician, carpenters, an artist, a newspaper reporter, and a crew from Paramount News. He needed truckloads of sensitive instrumentation and provisions, ships, crews, an airplane, tractor trucks, and dog teams.

Byrd picked up one of his two ice ships at an auction for $1,050 from the City of Oakland, California, and renamed it the *Bear of Oakland.* But damage from a tornado dry-docked it for over a month in North Carolina. Byrd finally set out from Boston on October 11, 1933, to make the 13,323-mile trip to the Ross Ice Barrier, where the iceberg-studded Ross Sea froze into the impassable ice mass that extended over the Antarctic continent. Poulter and two colleagues took cosmic ray measurements from the deck of the ship all the way south through the Panama Canal and past Easter Island and New Zealand. The other scientists on board pursued research in meteorology, magnetism, cosmic ray studies, astronomy, studies of the aurora, oceanography, seismology, plankton studies, bacteriology, and medical studies of the men themselves. They docked at the Bay of Whales at the shelf of the ice barrier on January 17, 1934, the closest a ship could get to the South Pole. Little America lay just ahead.

Byrd could only pray that the base of cabins and shacks, known as Little America, had held intact since the hasty evacuation of the site in 1930 as storms threatened to cut off escape. He spotted the 70-foot radio towers first but everyone had to help dig their way through the ice and snow to reach the roof of the administration building with the shelter of bunks, the science lab, and the mess hall. The men found parkas, dirty underwear, boxing gloves, and even dinner frozen on the stove from the previous frantic departure. They heated up the dinner, and the frost on all the ceiling and tunnels melted somewhat over the next few weeks, though never completely since indoor temperatures often dropped to 25 degrees at night. The phones between the buildings and the lights worked, though. Scientists thawed out the lab, set up their experiments, and got to work. They embarked on eighteen months of life as "sardines in a goldfish bowl," scientist C. J. V. Murphy described the scene, adding, "Little America lived, ate, slept, and talked in noisy congestion." Poulter introduced meteor observations into the Antarctic for the first time. He had a revolving platform suspended in one of the cabins and fitted with outriggers for four folding chairs. Four observers sat back to back with their legs dangling and their heads elevated into a transparent dome built into the roof. There was just room enough for four heads, each observing a different quadrant of the skies through Poulter's reticles. With the team settled in at Little America, Byrd flew to the Bolling Advance Weather Station 123 miles south of the base station on March 28, 1934. The base consisted of one tiny subterranean shack where Byrd stayed for seven months. The sun set for the last time on April 19 and then he lived in the perpetual darkness of the Antarctic winter. Little America and Byrd communicated three times a week by radio. But by June, Byrd's cheerful communications became slow, spotty, and then ceased for days at a time. Concerned, Poulter asked for Byrd's permission to make a trip to the advance base and extend his meteor research. Byrd granted permission for a small team to visit on the condition that the men travel only if they could follow the trail of flags anchored in the permafrost. The expedition set out in tractors on July 20. The tractor trucks circled for more than 100 miles trying to locate the flags and then turned back to Little America as temperatures plunged.

After the first failed journey, Byrd's upbeat messages became more uneven, and the men at Little America agonized about what could be wrong. Byrd himself didn't know that he was dying of carbon monoxide poisoning from the gas stove and gas generator at the advance base. In July, when the generator failed, the level of fumes began to drop and he began to improve ever so slightly as he switched to a hand-cranked generator. It took all his energy to crank up enough power to operate the all-important radio. He barely ate, fought off waves of dizziness, and collapsed from weakness after one transmission. Yet he continued to crank out comforting lies, fearful that any hint of his suffering would trigger a reckless rescue mission. His proud pretense finally crumbled in a ragged message that read "Please don't ask me crank anymore. I'm okay." Poulter decided to make a run for the advance base against all odds. He and a few others agreed to ignore Byrd's direct order not to proceed unless they could follow the flags. They set out on August 4.

The journey had taken Byrd only an hour by airplane but flight was out of the question during the stormy August winters. Poulter and his team slowly clawed their way south in the tractor trucks for the next six days. The generator on one tractor failed and the men jury-rigged the well-worn spare with tiny blocks of wood to keep it running. For the last leg of the journey, the team lost radio contact with Little America and lost visibility in the storms. An eerie green cast at the southern horizon offered the first hint of the coming spring and of hope. They crept along for two hours at 5 knots (5.75 miles per hour) to traverse the last 12 miles to the advance station. Then, suddenly, a shaft of flashing light broke through the darkness, dim at first but ever brighter as they followed it—Byrd's beacon light. He knew they were coming. Byrd himself, dressed in furs, came out to greet them, every movement a victory of will.

"Come on down, fellows," he said warmly, and promised hot soup inside.

Despite his comforting welcome, Byrd's emaciated appearance told the story of his prolonged suffering. The rescue team repaired the stove and generator and departed, all except Poulter. Byrd was too weak for the overland trip and Poulter stayed with him as he slowly recuperated. Antarctic winter ended in October, making it safe for the biplane to return to the base to fly both men back to Little America.

The expedition started the three-month trip home in February 1935. Scientifically, the trip had been a tremendous success. The firsts of the mission included the meteor and cosmic ray readings from the Antarctic and seismic readings that established where the land mass of the continent ended and the mountain range of ice began.

Poulter arrived back on campus in May. He had been away for all of Van Allen's junior and senior years, leaving his young protégé to pretty much run the physics lab and workshop. Yet Poulter's absence taught Van Allen lessons that he would instinctively apply to his own students—lessons that catapulted many of his own protégés into leadership positions in space exploration. Instead of pushing his students to their limits, he would give them room to push themselves to their limits under his guidance and give the self-starters a partnership role in his work. He designated important research responsibilities to students while he left for extended periods on scientific expeditions, just as Poulter had done.

Van Allen got his first taste of life at the University of Iowa while Poulter was gone as well, studying differential calculus there during the summer of 1934, just as his brother George returned home from the campus after completing his first year of law school. George went to work with Alfred for the summer. James checked into a dorm in the Quadrangle on the west bank of the Iowa River where he had a room of his own, a luxury devoted to the expectation of nonstop study. He caught on quickly to the differential equations that enabled him to describe the world of classical physics in mathematical terms.

The class met weekday mornings in MacLean Hall, home of the physics and math departments as well as to astronomy, electrical engineering, and fine arts. He stayed at MacLean to dig into his homework after class and then returned to the dorm for dinner. Unaccustomed to a social life and fascinated by the math, he spent evenings in his room completing every problem in his textbook. He earned an A in the calculus class and came home to prep the physics lab for the fall semester at Iowa Wesleyan as he started his senior year 1934–1935.

Word was soon out after Poulter's return that the great Admiral Byrd, now a national hero, had accepted the invitation to Iowa Wesleyan for graduation in 1935. The whole town turned out for the parade in Byrd's honor before the commencement. Van Allen graduated summa cum laude and he saw the look of pride and satisfaction as Poulter rose to cross the stage and recognize him with a handshake. But Van Allen simply kept in step and walked back to his seat, too shy to wait even momentarily on the stage for any special acclaim.

Van Allen was twenty years old and had been accepted as a PhD candidate in physics at the University of Iowa. He had never been out on a date and never gone to a football game. But he could deftly compose experiments to test how electricity, mechanics, and optics—the cornerstones of classical physics—shape the physical world. And eight years of shop made him a master at building parts, wiring components, and retooling industrial castoffs into cutting-edge tools for research. He took all these skills to Iowa City and stepped into a pivotal period of international research for nuclear physics.

3 The Making of a Scientist

Iowa City, Iowa—September 1935. On a hot, Sunday afternoon in September 1935, Alfred Van Allen parked the Dodge at the side of the house and James loaded two suitcases into the trunk for a more permanent stay at what was then called the State University of Iowa. One suitcase held clothes and the other held books and, of course, a slide rule. The Dodge pulled away from the town where James had spent most of his life and headed north 50 miles to Iowa City.

• • •

That fall, Van Allen rented a room for $14 a month on the second floor of a rooming house at 223 Linn Street, a short walk from MacLean Hall, on the southwest corner of the hilltop hub of the campus known as the Pentacrest. Van Allen took all his classes during four years of graduate school in that same building. The five-story building offered state-of-the-art laboratories and a shop where master instrument maker Joseph Sentinella could create almost any gadget the physicists needed. "He was the graduate student's best friend," as Van Allen described him.

Van Allen barely saw his dorm room except to sleep. He spent twelve hours a day in classes, doing research and learning the shop, sometimes catching a quick meal at the soda fountain of nearby Pearson's Drug Store. He felt both inclined and obligated to ignore the rest of campus life. In the depths of the depression, only four high school classmates went on to college, let alone graduate school. Most took jobs in family businesses or on the farms and many had started families of their own. Alfred paid all his sons' college expenses and, naturally frugal, James had little time or desire to purchase much of anything else. He didn't even incur a monthly laundry expense. He simply boxed his dirty clothes in a cardboard suitcase and sent it home to his mother who washed everything and shipped it back. But the mundane world slipped away in MacLean Hall.

Physics was turning the seemingly precise world of space and time into a grand illusion. Within days of Germany's defeat on the battlefields of World War I, German physicist Max Planck received the Nobel Prize for his quantum theory of radiation. Energy and light didn't flow in a steady stream, as scientists, poets, and philosophers had assumed for centuries. Instead, it burst forth in packets, or quanta, at distinct frequencies regulated by a universal constant—Planck's constant. And energy was interchangeable with matter—interchangeable at the rate of $E=mc^2$ as Albert Einstein proved in 1905 in his Special Theory of Relativity. Releasing all that energy from a small amount of matter opened the door to staggering powers of destruction and optimistic hopes of unlimited cheap energy. Few people understood the promise or the peril.

James Van Allen started graduate school with a solid grounding in classical physics but knew almost nothing about the breakthroughs in quantum mechanics and nuclear physics that revolutionized the work of many physicists in MacLean Hall.

MacLean Hall carried the name of former university president George Edward MacLean, a man who didn't need the start of a scientific revolution to turn the campus upside down. University of Iowa historian Stow Persons compared life at the college before MacLean took the reins in 1899 to a "a sedate game of croquet just before the eruption of a volcano." But the quiet, courtly MacLean adapted the changing East Coast ideas about rigorous academics to the University of Iowa. He created the blueprint for the professional graduate schools out of undergraduate departments specializing in law, medicine, homeopathy, dentistry, and pharmacy.

MacLean revamped the catchall collegiate program into modern undergraduate departments and added new disciplines in economics, statistics, Scandinavian languages, physical education, speech, Greek art, and archeology. He also went on a building spree, erecting the four buildings surrounding Iowa's Old Capital building on the Pentacrest. The gray limestone and neoclassical formality of all five buildings created a coherent hub from which the university mushroomed in all directions. Each building had singular architectural flourishes such as the exterior cornice on the 1910 physics building that bears MacLean's name, a building carved with the names of great physicists and astronomers—Galileo, Newton, Ohm, Cavendish, Faraday, Fresnel, and a host of others. These men laid the groundwork for an orderly universe where physical laws applied without exceptions. The laws described the operations of electricity, magnetism, mechanics, acoustics, and optics, a world of classical physics that had been taught at the university since the first science classes began there in 1856 as part of the Department of Natural Philosophy.

Physics department head George W. Stewart directed every detail of the construction of the $225,000 MacLean Hall, including placement of an elevator shaft. The unfinished elevator did the heavy lifting for his laboratory budget: every time he needed more apparatus, he went to the university with a request for the elevator that was never installed. The elevator shaft offered the perfect location instead for a Foucault's pendulum, a heavy metal ball suspended from a wire that swung in a constant, rhythmic pendulum motion. The plane of the sweep changed continuously as the earth rotated on its axis, a sort of sundial for the earth's rotation. The pendulum, devised by Jean Bernard Foucault in 1851, demonstrated the uniform rotation of the earth and was a permanent fixture in the physics department for decades. The pendulum didn't deter Stewart from using the elevator shaft as a public address system, calling out a name through it if he wanted to locate someone fast. Smaller, unseen shafts and tunnels that honeycombed MacLean Hall hid 20 miles of electrical wiring for all the laboratories and workshops.

The physics department underwent a dramatic transition in the early 1920s when the university lost a series of professors tempted by higher paying jobs, department chairmanships, or pure research elsewhere. Stewart, who had mentored many of these professors into distinguished careers, shopped some of the best universities in the country for new professors to catapult the department into new arenas of physics. He hired Edward Tyndall, a native of South Africa with a PhD from Cornell University who ranked as an international authority in the new field of solid state physics. Solid state physics emerged as a separate science in 1912 when German physicist Max von Laue laid the groundwork for modern electronics through his discovery that crystals could diffract X-rays, implying that crystals can carry and control an electrical charge. Transistors and semiconductors evolved from von Laue's discovery for which he won the Nobel Prize in 1914. Stewart, himself a researcher in X-ray diffraction, found a kindred spirit in Tyndall.

Alexander Ellett came on board from Johns Hopkins University, bringing with him research in the trailblazing new area of atomic physics. John Eldridge, a gregarious optimist lured from the University of Wisconsin, made science accessible and entertaining in his 1934 book, *The Physical Basis of Things,* with chapter headings such as "The Dance of the Molecules" and "Atoms à la Bohr." Claude Lapp, from the University of Illinois, brought to Iowa his interest in physics education and the development of learning materials and student testing.

The burly, autocratic Stewart was the terror of the graduate students. But he brought a luster to the physics department with his research in acoustics and X-rays, his appointment to the National Academy of Sciences, and his innovative academic programs. He teamed up with the four younger professors to teach popular summer courses on the latest innovations in physics. The 1926 course on X-rays and the constitution of matter attracted teachers from colleges in eight different states. Stewart also brought in eminent physicists such as Konrad Lorenz, Erwin Schrodinger, Arthur Compton, and Paul Dirac to lecture and give colloquiums. Van Allen arrived on campus just as the roster of visiting guest speakers merged into a popular summer colloquium. Like many graduate students, he stayed on campus during summers, and the world of physics came to him in the colloquium that presented sweeping, multidisciplinary themes such as "Physics and Society."

Physics had been a matter of learning and applying rigid Newtonian laws for Van Allen up until graduate school. Now he realized it was a volatile subject, full of controversy, with bedrock theories challenged. He studied in the basement of MacLean Hall where he had a desk and worked on math problems and physics homework until midnight or 1 A.M. The smell of liver and onions lingered there, souvenirs of Professor Art Rouse's Saturday night suppers cooked over the Bunsen burners in his lab. Rouse invited graduate students to share the feast, a welcome treat for a hungry group of bachelors. "Almost none of the students were married then. No one had a car. We didn't feel handicapped. That's just the way things were," Van Allen said.

While James worked on his PhD, George graduated from law school and moved back to Mount Pleasant. Now Maurice joined him, beginning studies at the university medical school. The two brothers often got together for dinner. James frequently took a break for lunch and a game of ping-pong with Bill Furnish, who was earning a PhD in geology and took over as head of Iowa's geology department during Van Allen's tenure heading the Department of Physics and Astronomy. Both of them had meal contracts at the Gamma Alpha fraternity house.

Early on, James Van Allen recognized his own direction as a scientist. His first class in quantum physics convinced him that he didn't care for theoretical topics or quantum mechanics. But experimental physics meant a rigorous initiation in Sentinella's shop class. "He was a terrific guy, small—about 5-feet-5—and slender but a real dynamo of activity and affability and extraordinary skill." Van Allen said. He had little formal education but trained with a British company that made weapons and mechanical computer equipment for the British Navy before he moved to Iowa.

"He took me on as a kind of apprentice," Van Allen told David De Vorkin of the National Air and Space Museum. "I knew very well how to do wood working—I was really pretty good at that. But things like silver soldering, bronzing, using the metal turning lathe and milling machines, power saws, all these things—he instructed me in all of those. And I also got instruction in glass blowing, making vacuum systems, and this sort of thing. Graduate students these days just write a purchase order."

Four to five students a year earned a PhD in physics at the University of Iowa and Van Allen was one of approximately twenty students in the department's graduate program. Students pretty much had their pick of advisors and Van Allen paired up with Tyndall by mutual agreement. Tyndall taught him the art of growing zinc crystals and Van Allen researched the deformation of crystals under high tension, a problem in solid state physics. Van Allen built a tabletop device to pull the zinc crystals straight downward and used an arrangement of mirrors to measure the change of lengths and the point at which the crystals lost elasticity and began to deform. In most previous experiments, the crystals bent before they reached full extension and Van Allen's thesis problem was to develop a device to overcome this defect. His inventive approach earned him his master's degree in 1936.

As a graduate student, Van Allen also taught physics to undergrads. "There are 21 students in the physics class that I teach—a pretty good-sized class but I have only one or two first class students," he wrote home. He sent some newspaper clippings home, one with a contribution "for Dad's collection of freak names—I think this one runs a close second to Harriet Ketchum Loving." He unfortunately didn't mention the name in his undated letter.

Meanwhile, Ellett recruited graduate students to help him build the hottest new tool in nuclear physics—a Cockroft-Walton particle accelerator, an "atom smasher." The device gave scientists a way to generate and study nuclear particle interactions and laid the foundations for experimental nuclear physics. It also gave scientists a window into the cosmos, simulating the kinds of particle reactions that occur in nature's own laboratory. English physicist Sir John Douglas Cockroft and Irish physicist Ernest T. S. Walton invented the accelerator and achieved the world's first artificially induced nuclear reaction with it in 1932. Scientists around the world quickly copied their design for the machine. It stripped off protons from hydrogen atoms, accelerated them along a linear path, and hurled them into a target to allow scientists to study the subatomic debris and the building blocks of matter. Modern-day particle accelerators occupy huge spaces, typified by the 4-mile circumference of the accelerator ring at the Fermi National Accelerator Laboratory in Batavia,

Illinois, or the 16.5 mile ring of the collider at CERN in Geneva, Switzerland. By contrast, the Cockroft-Walton fit in a single room. Ellett commandeered space in the attic of MacLean Hall where the entire accelerator was built from scratch.

Building it became Van Allen's passion. "I wasn't very social and I didn't belong to a fraternity as my father did," Van Allen recalled. Instead, he joined the small cadre of students, guided by Ellett, who pieced together the accelerator out of improvised parts, industrial castoffs, and generous applications of glyptal, a goop made by General Electric. The accelerator required a vacuum to operate and the glyptal, combined with a vacuum gauge, allowed the team to detect the sources of frequent leaks of air into the chamber.

Van Allen and the other students worked hours each day for nearly two years to build the accelerator. They hauled the parts up six flights of steps to the attic. They wired a circuit of capacitors and rectifiers (vacuum tubes that converted ac current into dc current) to create the high voltage needed to accelerate streams of protons focused into a sharp beam. They glued aluminum foil to plates of window glass from the local hardware store to make the capacitors. They recycled glass cylinders junked from old gasoline pumps to create the accelerator tube and the rectifiers.

The machine filled the room with a noisy whine, eerie coronal discharges, and bolts of sparks—when it ran. Keeping it running "was a real nightmare in experimental physics, let me tell you," Van Allen said. The leaks had to be routinely sealed off and the air pumped out to restore the vacuum. High-voltage sparks ripped through the machinery, causing constant damage. Burned-out filaments in the vacuum tubes, like burnt-out filaments in a lightbulb, shut down the machine until the tubes were replaced. Wafers of New-Skin, a liquid Band-Aid ordinarily painted on wounds, were made in beakers and then used to seal off the target area from the vacuum in the accelerator. No one could do anything about the summer heat and humidity that short-circuited the current and closed down operations, however.

"I finally got everything to work at the same time in the fall of 1938," noted Van Allen. "Another graduate student, Stanley Atchison, helped me take data, and I made a continuous run of forty-eight hours of readings. I caught naps on a cot right next to the accelerator, being unwilling to turn it off because of the well-founded expectation that many weeks might be required to restore full operation."

The run provided the basis for Van Allen's PhD thesis on the nuclear reactions of deuterons, a "heavy" hydrogen nucleus with a proton-neutron pair instead of just a proton. Van Allen stripped the deuterons off of deuterium gas and accelerated them to a high energy before bombarding them into a target of more deuterons. The infinitesimally thin gas target chamber, custom-built by Sentinella, gave the experiment an innovative twist. It allowed Van Allen to more accurately measure the effect of escalating energies on particle interactions because the particles didn't lose energy passing through a thick target. His nuclear physics experiment demonstrated the effect of a wide range of bombarding energies on the well-known reaction of:

$$H2 + H2 \rightarrow H3 + H1.$$

The number of nuclear reactions increased dramatically at higher energies, converting deuterium (H^2) into tritium (H^3), an unstable hydrogen isotope with a proton and two neutrons, and hydrogen (H^1). Testing nuclear reactions played a key role for physicists hoping to harness nuclear power on Earth. The cross-section of energies measured in Van Allen's experiment represented a significant advance over what had been achieved before and he confirmed his results in a second trial a few weeks after the marathon run.

Van Allen defended his results before Ellett and his thesis committee in the spring of 1939 and headed to an American Physical Society meeting in Washington, D.C., for the first major presentation of his research as a physicist.

Scientists in the audience expressed keen interest in both Van Allen's technique and his results, and he answered numerous questions capably. But one scientist saved his comments for a private conversation with Van Allen after the presentation. The scientist who buttonholed him needed no introduction. He was nuclear theorist Hans Bethe of Cornell University. Bethe was only thirty-three years old at the time but his work describing how stars produce energy had already made him one of the elder statesmen of physics. He "found that the trend of my curve of cross-section vs. bombarding energy was impossible to believe at the lower energies," Van Allen wrote in a 1989 autobiographical sketch. "The criticism was unsettling to put it mildly. Ellett and I went over the entire matter critically and eventually realized that my method of measuring the beam current through the reaction chamber was faulty."

They determined that Van Allen had failed to take into account the partial neutralization of the beam as it reacted with the target gas. The effect was most pronounced at low energies, so that the current measured was too low. Van Allen feared for his PhD. But Ellett, a tough task-master who kept a cool distance from his students, backed him and shrugged off the error as an opportunity for another graduate student to refine the results. Atkinson, who had helped Van Allen collect his data, did the follow-up experiment with a modified version of the apparatus to correct the results.

At that same American Physical Society meeting in Washington, D.C., that summer, Ellett ran into his colleague Merle Tuve in the lobby of the Wardman Park Hotel. Tuve had recently launched a nuclear physics program at the Department of Terrestrial Magnetism in Washington, D.C., where he was now the director. Tuve and Ellett knew each other through Johns Hopkins and had a shared interest in developing accelerators at their respective institutions. Now Tuve was scouting for post docs for a new venture. Ellett introduced Van Allen as the young man who helped build the Cockroft-Walton accelerator at Iowa and recommended him as a good researcher.

Van Allen had the perfect credentials to impress Tuve, who had just installed a newer, more powerful accelerator at DTM. Tuve liked the young physicist immediately and told him he'd let him know about a job within a few weeks. There was no résumé and no formal interview. "It was done on a handshake," Van Allen said.

He returned to Iowa, graduated with his PhD, and received the formal invitation to join DTM. "My appointment now awaits only the O.K. of the president of the institution [Vannevar Bush]," he wrote home. "I will hear definitely within 10 days to two weeks. The salary will be between $1,600–1,800 for the first year." Van Allen already had a job offer from his old mentor Tom Poulter who invited him to join the applied physics team at the Armour Research Institute in Chicago. Poulter had left Iowa Wesleyan to direct the team that did contract research for a variety of small companies. Another offer came from electronics maker Raytheon, which sent a recruiter to the physics department. The job involved a project to improve the coating on fluorescent lights. He declined both offers. He knew he wanted to stay in research and accepted the offer from DTM. He packed all his possessions in the same two suitcases he had brought to Iowa City and spent two months in Mount Pleasant before hopping one train to Chicago and a second to Washington, D.C. He arrived at DTM in late summer, just shy of his twenty-fifth birthday and just before Europe went to war again.

Almost immediately, DTM became one of the central brain trusts for developing weapons for World War II, a broad leap beyond the department's original mission envisioned by industrialist Andrew Carnegie. Carnegie sold his steel empire to financier J. P. Morgan for $480 million in 1901 and retired to pursue philanthropy full-time. He opened 2,500 libraries across the world, built Carnegie Hall in New York, and established the Carnegie Institution in Washington, D.C., with research arms that included the Department of Terrestrial Magnetism, founded in 1904. DTM opened shop in the Ontario Apartments, 2853 Ontario Road NW, and built the *Carnegie,* a yacht customized with non-magnetic properties that took DTM scientists on a four-month, 8,000-mile geomagnetic tour of the North Atlantic and the Caribbean.

DTM's programs mushroomed and the Carnegie Institution moved the facility to a sprawling, forested campus five miles from downtown Washington, D.C. Surveys of the geomagnetism, cosmic rays, and the earth's atmosphere branched off in new directions. Then, in 1925, Tuve and researcher Gregory Breit developed a technique that bounced radio pulses off the ionosphere, the electrically charged layer of the earth's atmosphere. Their technique created echo patterns that laid the foundations for global radio transmissions and for the invention of radar. But attention had shifted to the threat of a nuclear bomb when Van Allen started there.

The research at DTM brought Van Allen into the forefront of nuclear physics at a time when physicists worked feverishly to identify and control nuclear forces necessary to develop the atomic bomb and sources of nuclear energy. War in Europe gave sudden urgency to unlocking that might. Bethe, Enrico Fermi, Edward Teller, and other physicists who had fled the Nazis in Europe lent expertise that soon gave America an edge in developing the bomb. Fermi had actually split the uranium atom in experiments in 1934, though he hadn't realized his success. Back in Germany, scientists Otto Hahn and Fritz Strassman achieved a fission reaction using uranium in 1938 by extending Fermi's research. That event alone prompted Einstein to urge President Franklin Roosevelt to develop a fission bomb before the Nazis did. He sent the letter just as Van Allen arrived at DTM. Roosevelt heeded his advice, and government funding of the research began to flow to DTM and other laboratories.

"The fission of uranium was the hottest topic in nuclear physics when I arrived," Van Allen said. "Dick Roberts, a physicist on staff, had just discovered the delayed emission of neutrons from uranium fission products, an essential feature for the control of a nuclear chain reaction for power generation." That discovery at DTM resulted in Fermi's gamble in December 1942 to artificially trigger—and hopefully stop—a controlled chain reaction in a bunker laboratory burrowed beneath the athletic field at the University of Chicago.

Van Allen's graduate research with deuteron collisions sparked theoretician Gregory Breit's interest and he wanted Van Allen to study the collisions over a larger energy range—ranges that had to be precisely nailed down to help understand nuclear reactions. Van Allen extended his PhD research using DTM's new Van de Graaff accelerator, a machine that increased the acceleration of the particles and, thus, the "bang" of the collisions with deuterons.

Van Allen's research had increasing significance in nuclear physics. While hydrogen has a single proton for a nucleus, the proton-neutron pair in the deuterons form the "heavy hydrogen" nucleus of deuterium. And when deuterium replaces hydrogen in water, it creates heavy water that can be used to control a nuclear chain reaction. Van Allen suggested new ways of bombarding deuterium to study nuclear reactions and he joined a team of equally enterprising young researchers that included Norman Heydenberg, another former student of Ellett's now in charge of the Van de Graaff machine. Norman Ramsey, a graduate student at Columbia University, shuttled in and out of New York City, bringing the DTM group insights from his research, under physicist Isidor Rabi, concerning the mechanics and magnetism of atoms.

Van Allen, Ramsey, and another DTM staffer, Dean Cowie, discussed the findings of the day over the nightly dinner specials nearby at the Hot Shop. They also pooled their money and bought a sixteen-cylinder Cadillac for $150—a gas guzzler that purred along on rides to Mount Vernon and into D.C. "It was a monstrous coupe that drove like a truck and got about 5 miles per gallon, so we sold that and got a Buick. Another old clunker," Van Allen said. He received $150 a month in pay and rented a one-room apartment with a bathroom down the hall in a private home. He walked across the street each day to the sprawling DTM campus.

The informal atmosphere at DTM gave Van Allen time for a social life for the first time, though much of it centered around the rapid-fire progress in research. DTM and George Washington University held frequent joint seminars to keep everyone abreast of the latest advances. Van Allen attended every seminar where brainstorming frequently spilled over into the coffee breaks and luncheons planned as part of the events.

Early on, Van Allen introduced himself to DTM's director John A. Fleming. Van Allen had submitted his geomagnetic readings from Henry County to Fleming years before and they talked about DTM's time-honored field surveys. Fleming introduced the shy, young PhD to geophysicist Scott Forbush, and Forbush, shy himself, took a liking to him. Quiet and admiring, Van Allen stopped by to ask about the latest research and Forbush rewarded him with detailed explanations, swinging down the thick magnifying lenses clipped to his glasses and pulling together graphs to make a point as he talked—and always with a scruffy, hand-rolled cigarette dangling from his mouth. The studious scientist enjoyed one major diversion—weekly poker games. "I arranged to play poker with his people and got well acquainted with him and my interest sort of shifted over to geomagnetism even though I was technically part of the nuclear physics high energy lab," Van Allen said. The congenial social evenings helped Van Allen cross the divide between two research worlds at DTM.

While the "young Turks" at DTM unraveled the secrets of the atom, the "old guard" continued the arduous and less glamorous task of carrying on the laboratory's mission in geophysics, Van Allen said. The traditionalists occupied a separate sphere at DTM as they studied atmospheric electricity, earth currents, aurorae, geomagnetism, cosmic rays, and the ionosphere. Van Allen felt drawn to the group and especially to the soft-spoken Forbush, son of an Ohio farmer and native son of a small town much like Mount Pleasant. At thirty-five, Forbush had a shock of unruly, prematurely graying hair that matched his "passionate and solitary (but not lonely) devotion to his work," Van Allen wrote in his introduction to his book gathering together Forbush's lectures. Forbush sat at his desk hour after hour, pulling down the magnifying lenses over his glasses to better decipher long lists of numbers of geomagnetic and cosmic ray measurements from across the globe.

Forbush had established a worldwide grid of instruments used to collect cosmic ray readings through an international network of colleagues. Hundreds of readings from around the world poured across his desk each month. He copied them by hand, compiling thousands of entries organized by date, location, and latitude. Armed with these readings, Forbush quietly and relentlessly graphed his findings into a worldwide mapping of shifting levels of cosmic rays. Without ever leaving his desk at DTM, he established the fundamental relationship between the eleven-year cycle of solar flares and sudden, temporary declines (known as Forbush decreases) in cosmic ray levels. Cosmic ray levels decreased because the flares kicked up a blasting gale of solar plasma that blew outward from the sun and pushed back incoming streams of galactic cosmic rays. Van Allen graphed the most recent Forbush decreases picked up by *Pioneer 10* in data transmitted from 7.5 billion miles from Earth in May 2001. "I followed in his footsteps with all these graphs," Van Allen laughed.

The *Pioneer 10* space probe to the outer planets carried Van Allen's Geiger tubes among its experiments—instruments similar to but more sophisticated than the Geiger tubes and ionization chambers that Forbush and other geophysicists had used to measure high-energy traces of cosmic rays in the earth's atmosphere. Accelerators in the 1930s couldn't come close to matching the energy levels of cosmic rays. Nature had the sun, the stars, and the explosions of supernovas as its accelerators. Forbush realized that cosmic rays held fundamental clues to the nature of matter and the evolution of the universe.

Forbush sparked Van Allen's interest in cosmic rays. His columns upon columns of numbers that pieced together the forces of the solar system inspired Van Allen. Tuve had little patience for it all. "He regarded the old timers at DTM as hopelessly out of touch with modern science and he considered Forbush an outstanding example of a person who was filling massive notebooks with numbers that no one would ever find of interest," Van Allen wrote. Tuve viewed nuclear studies and accelerators as the wave of the future and took DTM in that direction when he later succeeded Fleming as director.

Van Allen recognized a compelling field for research in those notebooks. In one of the pivotal moves of his life, "crossing the culture gap at DTM" focused Van Allen on his future career. The new path would take him into space where he tracked cosmic rays and magnetic fields across the solar system. But at the time, cosmic rays were still largely a mystery. Some scientists viewed them as remnants of the creation and others thought they might hold the secret to an energy source to rival the atom.

• • •

Cosmic rays are high-energy nuclear particles streaking across the galaxy after a dying star explodes in one final cataclysmic blast of a supernova, the final stage in the life cycle of a massive star. Young stars, like the sun, generate energy by fusing hydrogen and deuterium nuclei into helium in a thermonuclear furnace that releases the energy equivalent of 9,000 tons of coal from a pound of nuclear fuel. The outward thrust of the energy radiates across the solar system, counterbalancing the intense gravitational pull of a star's mass. But as stars far more massive than the sun age across billions of years, they deplete the nuclear fuel of the lighter elements and continue releasing energy through the fusion of ever heavier elements until the fuel becomes iron.

Even a thermonuclear furnace can't fuse iron, and the star begins to lose the battle with gravity. The core of the star collapses and, as gravitational collapse continues, pressure mounts until the star explodes. The blast provides the building blocks for new stars and shock waves that accelerate charged particles into cosmic rays. The cosmic rays that shoot toward Earth disintegrate into showers of secondary particles in the earth's protective atmosphere. When Van Allen came to DTM, scientists couldn't reach high enough to find the stream of primary cosmic rays that had remained a matter of heated debate and conjecture for decades.

In 1912, Victor Hess, a young Viennese physicist, made a startling discovery about radiation in the earth's atmosphere. He measured the radiation from the gondola of a balloon and demonstrated that the intensity increased with altitude. He concluded that the earth was bombarded by ionizing radiation from space, with only the earth's atmosphere shielding the flow. Scientists lined up to discredit him. One of Hess's most vocal critics was University of Chicago physicist Robert Millikan who had discovered the exact charge of the electron.

Millikan became the guiding light for the popular press and readers who simply couldn't accept that extraterrestrial radiation was piercing the atmosphere. Millikan launched a series of measurements of ionizing radiation that penetrated the waters of mountain lakes. The levels of radiation were higher in lakes situated at higher altitudes, confirming Hess's findings and convincing Millikan that the variable stream of radiation Hess measured must indeed come from outer space.

Millikan coined a phrase for the energy stream—*cosmic rays.* Science fiction writers couldn't have invented anything better. Cosmic ray guns and lethal cosmic ray clouds became stock in trade for space adventures. In scientific circles, physicists began taking their instruments up in balloons to extend the altitude of the cosmic ray readings. Millikan sent his counters up in unmanned balloon flights called *balloon sondes* and made a series of new measurements from 16,000 feet in the 1920s after taking the post as director of the Norman Bridge Laboratory of Physics at the California Institute of Technology in Pasadena.

Millikan found that he could account for the tremendous cosmic ray energies only "if he assumed the total annihilation of matter into energy," noted William Pickering, former director of Jet Propulsion Laboratory (JPL) at Caltech. As a graduate student he helped Millikan measure cosmic rays and later collaborated with Van Allen as the head of JPL. Based on his calculations, Millikan theorized that cosmic rays were ultra high energy photons—particles of light. He concluded that the ionizing radiation he measured was a secondary shower of particles produced when the photons collided with oxygen and hydrogen in the earth's atmosphere. Back in Chicago, physicist Arthur Compton claimed the particles found with his own cosmic ray measurements *were* primary particles, made up of electrons and protons. His counters couldn't differentiate between these particles and others to prove the point, but many scientists accepted his view. Soon, the composition of cosmic rays locked the two Nobel laureates in a heated and escalating debate.

Compton saw a chance to settle the matter once and for all at the 1933 Century of Progress Exposition in Chicago that he was helping to plan. The Exposition's opulent pavilions set out to restore luster to the wonders of science and technology, a luster badly tarnished by the 1930s depression. The depression enveloped the United States and Europe despite the innovations of the telephone, electricity, and automobiles. Now Exposition planners needed a showstopper and hammered out a deal with French balloonist Auguste Piccard to orchestrate a repeat of his 1932 balloon flights that had reached a record-breaking altitude of more than 35,000 feet. Piccard touched the stratosphere for the first time on two trips, traveling in sealed, spherical aluminum gondolas made by a company that built beer brewing vats. He carried cosmic ray detectors with him on the flights, fascinated by the idea that cosmic rays could potentially be harnessed for a limitless energy supply. The rigors of balloon travel in a sealed gondola, problems with a gas valve needed to control altitude, and the extremes of hot and cold he endured left Piccard with little time or energy to record cosmic ray readings. But the public didn't need science to glorify the intrepid balloon adventurer.

For the flight planned in Chicago, Compton invited Millikan to join him in sending instruments to study cosmic rays up to new heights as part of the event. Willard Dow of Dow Chemical Company in Midland, Michigan, offered to craft a gondola for the flight out of a newly invented magnesium alloy that was lighter and stronger than aluminum. The *Chicago Daily News* backed the flight financially and turned the enterprise into a spectacular scientific competition between two Nobel laureates. Reports that the Soviet Union planned to beat Piccard's record escalated interest in the flight launched from the center of Soldier Field on August 5, 1933, at 3 A.M.

Thousands of spectators bought tickets and began gathering in the stands on the afternoon of August 4 with their children, picnic lunches, and a general carnival spirit. They watched for hours as the balloon slowly filled with hydrogen gas and rose against the sky. They cheered when navy captain Thomas Settle climbed into the porthole of the gondola that Piccard had designed, that Dow had built, and that Compton and Millikan loaded with their detectors. Danger added to the excitement. Settle had to clear the crowd and clear the towers and cables of the Skyride installed for the fair. One commentator compared the takeoff to a steamship navigating a rocky reef.

Near 3 A.M., the balloon rose swift and sure above the Chicago skyline, clearing the obstacles as the crowd roared. But just as swiftly, Settle lost control of the balloon when a valve malfunctioned and he aborted the flight seconds later, landing near fourteenth and Canal Streets, a short walk from the stadium. Compton and Millikan recovered their instruments intact and the flight got an unceremonious second chance to launch from an army base near Akron, Ohio, on November 20, where it rose to a record-breaking 61,237 feet. Yet even at that height, Millikan realized he still hadn't reached high enough to identify primary cosmic rays.

Millikan provided scientific instrumentation for two more piloted flights in 1934 and 1935, the famous *Explorer I* and *Explorer II* expeditions, names applied generations later to America's first satellite flights. *Explorer I* fell a few hundred feet short of Settle's November 20 record but *Explorer II* rose above 72,000 feet. Another record, another set of measurements at new altitudes, and another disappointment. Millikan still found the telltale counts of secondary particles.

Now he focused his attention on robotic balloon sondes and a key cosmic ray mystery. Compton's collaborative worldwide balloon survey made at sixty locations, Millikan's worldwide airplane survey, and the work of other scientists suggested that cosmic rays dramatically increased in intensity on a steady curve as latitudes approached the North and South poles. The earth's magnetic field couldn't affect a neutral particle so cosmic rays had to be charged particles such as protons and electrons, Millikan agreed. He slowly filled in points on his famous cosmic ray intensity curves showing the relation to latitude. He took measurements by balloon sondes as far north as Saskatoon, Canada, and as far south as Madras, India, where Pickering and Victor Neher helped collect data at the end of the 1930s. "Vic Neher and I had developed a balloon system to carry cosmic ray measuring instruments almost to the top of the atmosphere [nearly 100,000 feet], and send their measurements back to earth by radio," Pickering said. Yet, they still could not determine the composition of the primary cosmic rays.

That's where the debate on cosmic rays stood as Van Allen's interest took shape. Then, on December 7, 1941, the Japanese attacked Pearl Harbor. DTM galvanized for the war effort and Van Allen worked on secret weapons projects. World War II gave Van Allen an expertise with missiles that allowed him to make a new generation of research instruments to load in a new generation research vehicle—the rocket.

4 Physicists to the War Effort

Battle of Saipan, the Philippine Sea — June 19, 1944. Navy Lieutenant James Van Allen watched the first wave of sixty-nine Japanese fighters and bombers swarm in over the Philippine Sea. The Japanese had the advantage of lighter planes with greater range. The American Hellcats had maneuverability and ace pilots who shot down most of the incoming enemy planes. But several penetrated the cover and headed in a kamikaze dive toward the USS *Washington* where Van Allen stood on the bridge as the gunners fired with a new secret weapon Van Allen had helped invent. The secret was in the detonation fuze that used radio signals to sense the presence of a plane and destroy it within a 70-foot range.

Van Allen had spent nearly two years hopscotching across the Pacific to convince gunnery officers to use the new proximity fuze. The capture of the Japanese stronghold of Saipan convinced even the skeptics. The fuze repelled attacks on an armada of U.S. ships mobilizing the invasion from the Philippine Sea in the summer of 1944. That day on the bridge, the bombers reached such close range before they exploded that Van Allen could see the terrified face of one of the pilots in the first wave. Another wave of Japanese fighters swarmed in and another. American fighter pilots and gunners on the ships armed with the fuze combined forces to destroy almost all of them.

"We went down after the battle and had roast beef and strawberries for dinner on a table set with a tablecloth. It was a strange feeling. We were at risk of being sunk or shot that day and a few hours later we were eating in the wardroom. I've never forgotten the contrast," Van Allen said.

• • •

Saipan was the beginning of the end for the Japanese. Shortly after the Japanese surrendered on September 2, 1945, the navy released the story of how the fuze saved thousands of lives in the Pacific and helped end the war. "The war's second most startling invention is a fuse that knows when a shell is close enough to a target to explode," reported the Associated Press, using the army's spelling of the term. "The 'radio proximity fuse' works so well that it once shot down 35 enemy planes in 30 minutes and antiaircraft gunners got 68 out of 72 buzz bombs bound for London."

"The discovery and development of the radio proximity fuse and its success in war, rated second to the atomic bomb, were made known today by the Navy Department," reported the *New York Times.* "The fuse, no larger than a pint milk bottle, packs its power in a glass tube."

The new weapon captured headlines in Europe as well, where Winston Churchill wrote that the American-made proximity fuzes "proved potent against the small unmanned aircraft [V-1 buzz bombs] with which we were assailed in 1944."

"The funny fuze won the Battle of the Bulge for us," said General George Patton, commander of the Third Army. He summed up the impact best: "I think that when all armies get this shell, we will have to devise some new method of warfare." For all the publicity after the war, scientists developed the fuze during the war under the same cloak of secrecy that veiled the atomic bomb. They raced against time as enemy onslaughts killed their sons, brothers, and friends in battle.

America entered World War II with a dangerously outdated arsenal of weapons. The country quickly rearmed itself with state-of-the-art guns, tanks, and battleships, but it was the mobilization of hundreds of scientists in top secret labs that allowed the United States to achieve victory with three new weapons—the atomic bomb, radar, and the radio proximity fuze that Van Allen helped invent. The proximity fuze gave America the first "smart" precision weapon in the history of warfare and gave gunners six hits for every one that could be expected with conventional antiaircraft shells and artillery guns. The fuze sniffed out targets in the proximity of 70 feet and destroyed them. Though the Germans had a head start on developing the weapon, they never perfected it and focused on rockets instead. The fuze destroyed the Nazi V-1 buzz bombs that laid siege to a panic-stricken London in the last stages of the war. And the fuze "broke the back of the Japanese navy," according to navy historian Dean Allard, quoted in the PBS documentary *The Deadly Fuze*. But the weapon was so secret that few Americans had heard of it at the close of the war.

The proximity fuze came to the rescue of thousands of American servicemen and women because Tuve didn't wait for a declaration of war to mobilize DTM scientists. He started work on it in 1939 as Germany invaded Poland and

France, and England declared war. The United States revoked laws that limited arms sales to foreign powers and tooled up war production to aid Britain in 1939. Though opposition to fighting remained strong in the United States, Carnegie Institution president Vannevar Bush, Tuve, and their associates pushed for a high-tech rearmament of the United States. They recognized the potential for scientists to partner with the government to parlay the revolutionary advances in experimental physics into military advances. President Roosevelt created the Office of Scientific Research and Development with Bush at the helm. Tuve joined the key branch of the office, the National Defense Research Committee, and immediately met with navy brass to identify weapons needs.

As the German Luftwaffe blitzed London in nightly bombing raids, antiaircraft guns proved to be pitifully ineffective against the new generation of high-speed aircraft that could attack and escape into high altitudes. The weapons did little more than raise a good show of fireworks to perk up the morale of the British. Battleship and artillery guns relied on timed fuzes that simply exploded a few seconds after firing, seconds that allowed pilots to deftly maneuver out of range. The British had taken the seminal steps to develop a proximity fuze that could sense a target and trigger a deadly blast, but development proceeded slowly in the war-torn country. Tuve, a master at applying electronics to practical needs, and other researchers at the Department of Terrestrial Magnetism recognized the fuze as an immediate project they could undertake.

"Tuve seized on this as a priority to which he could devote his own staff and pull in others he needed," Van Allen recalled. Bush backed Tuve to form Section T ("T" for Tuve) at DTM. Tuve pulled in physicist Charles Lauritsen and other staff and recruited William Fowler from Caltech, Wilbur Goss from the University of New Mexico, and ham radio operators. Tuve also fielded out some of the research to trusted colleagues such as Ellett at the University of Iowa.

Van Allen's position as a Carnegie research fellow isolated him from the war effort as he split his time between nuclear physics and geophysics during his first eighteen months at DTM. But he noticed that more and more of his colleagues were drifting off to another end of the building to work under classified security conditions and the work sounded exciting and essential. In spring 1941, as the nightly blitzkrieg of London sent thousands of demoralized Brits to burrow in the subway tunnels, Van Allen volunteered to join Section T. Tuve immediately agreed.

Section T was developing the new fuze to fit in a nose cone that would screw onto existing 5-inch shells, replacing the existing cones containing timed fuzes. As Van Allen entered the scene, size and durability remained key hurdles. A timed fuze required only a small clock to trigger ignition but the Section T fuzes required radio triggers and, in a day before transistors, radios relied on vacuum tubes the size of lightbulbs. Then rumors of a tiny, rugged glass tube caught the attention of Section T. They heard that Percy Spencer, a Raytheon scientist in Newton, Massachusetts, delighted his kids with a remote-controlled model airplane guided by a radio transmitter using miniature vacuum tubes. Raytheon had developed the tiny vacuum tube for hearing aids.

Tuve contacted Raytheon and Van Allen set to work adapting the miniature tubes to a light-sensitive photoelectric proximity fuze that would be triggered by changes in light levels. He reasoned that the shadow cast by a plane or the light reflected from it could trigger a photoelectric cell to fire the fuze. He devised a circuit sensitive to a wide range of light levels, so sensitive that the fuze fired even when he passed his hand over the cell in a semidarkened room to demonstrate it for Lauritsen and Fowler. "The exuberant response not only made my day, it propelled the photoelectric fuze into the realm of serious consideration," Van Allen recalled. Still, the photoelectric fuze could be set off by many spurious signals and didn't work at night when enemy pilots often preferred to fly. As Japanese conquests in Asia and the blitzkrieg of London continued, a radio proximity fuze promised a weapon that could fire day or night with greater reliability. Tuve's staff pursued a fuze with a battery-powered radio transmitter and receiver controlled by a circuit of four vacuum tubes, each measuring only 5/8" long. The fuze transmitted a radio signal that echoed off an aircraft and the receiver in the fuze picked up the signal. If the target was close enough, the signal grew strong enough to trigger the fuze and fire the projectile. The fuze—about the size of a can of pop—could explode a projectile packing enough fire power to destroy any plane within a proximity of 70 feet.

The fuze was now small enough for the guns but not rugged enough for battle conditions. It had five hundred parts—three hundred of them so sensitive that when one failed, the fuze failed. "If individual parts worked 99 percent of the time, you would only get the fuze operating about 4 percent of the time," said Ralph Baldwin, a veteran of the Section T team and historian of the proximity fuze. The navy wanted at least 50 percent reliability before deploying the fuzes into battle but they fell below that goal in test after test. The team dissected them and repeatedly found that the vacuum tubes failed in any conditions that simulated shooting it out of a barrel of a gun at 20,000 g-force acceleration and spin rates of 300 revolutions per second. Oddly, it wasn't the glass casing of the tube that shattered but the filament inside.

Van Allen and his colleagues worked around the clock to cushion the vacuum tube or make the filament stronger. For once, money was no object. "I don't want you to waste your time saving money," Tuve told his team.

By now, the proximity fuze took over every available nook at DTM and the work spilled into a new building earmarked for a new atomic particle accelerator laboratory. DTM director John Fleming balked at building any more space but going it alone with a maverick operation wasn't an option for Tuve: the navy liked to work with established companies or institutions. Tuve solved the dilemma by moving the whole program to an abandoned Chevrolet garage in Silver Spring, Maryland, where he founded the Applied Physics Laboratory (APL) of Johns Hopkins University, his alma mater. "We were 'plank owners' of APL, as the term is used in the navy for a member of a crew who places a new ship in commission," Van Allen said.

Van Allen transferred there in April 1942 and began taking box loads of experimental results to the navy test sites Stump Neck and Dam Neck along the Potomac River in Maryland. He loaded them into small cylinders and fired them vertically out of a "four-pounder" gun from beneath sandbag-covered bunkers. He craned his neck to listen as the cylinders whistled downward from 10,000 feet to estimate where each one fell on the 20-acre field at Stump Neck. Then he went around the grounds with a posthole digger to salvage the fuzes and take them back to the lab for "postmortems." The navy's security-conscious Bureau of Ordnance (BuORD) feared hijackers might accost him on these runs as he ferried a top secret weapon between the test site and APL. Tuve dreamed up a simple solution. He asked Montgomery County to swear in Van Allen as a deputy sheriff so he could legally tote side arms to ward off spies or thieves.

"I'd never fired a pistol before in my life," Van Allen said. "A crusty marine drill sergeant provided training." Trips to the test sites with batch after batch of the tubes continued. At Raytheon, engineer Ross Wood made adjustments to the tube design after each trial. He and Van Allen experimented with different materials and thicker components in a trial-and-error process to develop a hair-fine filament that wouldn't shatter. But no matter what they did, it kept breaking. After one failed mission, Van Allen decided it wasn't the filament but the metal support anchoring the filament that buckled under the force of acceleration. "In a moment of inspiration, I sketched a minute coil spring to hold one end of the filament," said Van Allen. The spring provided a suspension system for the proximity fuze, like shock absorbers in a car.

Wood gave specifications to a corps of women who built the tubes with the skill and dexterity applied to fine embroidery. Van Allen took a batch of the new tubes to the Potomac. He shot them and then retrieved each one with the posthole digger. Van Allen stemmed his optimism as he drove back to APL. There had been so many disappointments. He hauled the box of tubes into the lab and dissected one. Even with the naked eye, he could see it remained intact. He quickly stripped away the casing of another tube and another. Nearly every one remained intact.

Van Allen's scheme for the all-important vacuum tube had worked while Section T refined dozens of other engineering details. The new spring improved the tube's performance markedly in subsequent tests. But duds and premature firings still plagued the test shots targeting aircraft suspended between two huge towers at a testing ground in New Mexico and drones flown at the Aberdeen Proving Ground on Chesapeake Bay in Maryland. With Van Allen's spring in place, Section T exceeded the 50 percent reliability goal in 1942, and Captain S. R. Shumaker, research and development director for BuORD, ordered Tuve to get the proximity fuze into production immediately. Hundreds of women at Raytheon and in factories across America hand-assembled eighty million tubes for the war effort. The spring design later earned Van Allen his first patent.

• • •

In autumn 1942, the first proximity fuzes rolled off production lines, ready for shipment to the South Pacific. The Section T team supervised loading them into projectiles for more test shots in New Mexico. Even with the duds and premature bursts, the proximity fuze worked five to six times better than the fuzes in the projectiles that the U.S. Pacific Fleet had to fire at the Japanese. "Ten guns became the equivalent of sixty guns," said Baldwin. The new weapon promised a strategic edge to halt Japanese advancements in the South Pacific and then recover territory with far fewer ships.

After Pearl Harbor, the Japanese conquered British-held Hong Kong and rolled through Southeast Asia staking claim to Indochina, Thailand, Burma, Sumatra, Borneo, and the rugged mountain jungles of northern New Guinea. They captured the U.S. islands of Wake and Guam, and General Douglas MacArthur threw the full weight of the Pacific Fleet and his ground forces to halt the Japanese advances at the Philippines in April 1942. But a foothold in northern New Guinea gave the Japanese a base to move south and invade Australia. When Roosevelt ordered MacArthur to abandon the Philippines and rush to the defense of Australia, MacArthur made his famous promise: "I shall return."

Battles at the Coral Sea and New Guinea in May halted the Japanese advance on Australia. That turned the Japanese offensive to the west once again and their battleships set sail for an assault on Midway, the northern doorway to the Hawaiian Islands. The Americans had to stop them there. The U.S. forces sank four Japanese aircraft carriers and downed two hundred of the 272 Japanese planes involved in the Midway invasion, the first Allied victory in the Pacific. Dog fights between the Japanese and American pilots downed most of the enemy planes. The Japanese lost their best aircraft carrier and fighter pilots. The Americans lost 150 pilots out of 348 in the forays. Midway was the turning point in the war in the Pacific, but the victory exposed a major tactical flaw—the 5-inch, .38-caliber antiaircraft guns on the ships simply couldn't provide effective cover when Japanese planes made it through the gauntlet of fighters for their attack runs. The Americans lost the aircraft carrier *Yorktown* and lost hundreds more lives as the Japanese bombed the five thousand ground troops on Midway.

The story replayed itself when U.S. destroyers and carriers returned to the Coral Sea in August to capture Guadalcanal in the Solomon Islands before the Japanese could finish constructing an airbase there. In a crucial October ground attack on Santa Cruz Island, part of the Solomon chain, American fighter planes downed one-third of the 212 Japanese fighters and bombers unleashed against them. But again, antiaircraft guns on the ships proved ineffective as the Japanese fighters darted in for their attack run. Back in Washington, Commander William S. "Deke" Parsons reviewed the reports knowing that the proximity fuze could provide six times better cover for the planes, ships, and thousands of ground troops. As the bloody marine assaults on Guadalcanal erupted in November, the Naval Bureau of Ordnance decided to ship the proximity fuzes to the Pacific, and Parsons, the BuORD liaison officer for the fuze project, asked for three volunteers for duty to introduce the new weapon. Van Allen and APL physicists Neil Dilley and Robert Peterson agreed to take on the job and were commissioned in the U.S. Navy Reserve as lieutenants junior grade. Parsons advised the three men that the commission was essential to their safety.

"The discussion was that if we were civilians and caught by the Japanese, we would be immediately executed as spies, whereas if we were naval officers, we would be subject to the Geneva Convention for the treatment of troops and just be starved to death," Van Allen said. The commissions probably provided little safety, in fact, the Japanese summarily executed three pilots who landed on Japanese positions in the aftermath of Lieutenant Colonel James Doolittle's famous attack on Tokyo in April 1942.

But Van Allen, Dilley, and Peterson were eager to join the navy if that would put the fuze to work. "I was steeped in the virtues of the fuze and I was confident in the prospect that we could save American lives in the Pacific where the fleet was under murderous attack," Van Allen said. Parsons informed the trio that they would have less than a week between receipt of their commissions and departure for duty. There was no time for formal training but they each got a little pamphlet summing up the "Duties and Responsibilities of Naval Officers."

Van Allen, Dilley, and Peterson received their commissions on November 4, 1942, and they had three days to settle their affairs, get their uniforms, and pack. Parsons told them to bring two blue uniforms, three white ones, about a dozen khaki shirts, and five khaki trousers for everyday wear on the ships. He advised them to bring leather jackets, sweaters, and old uniform caps for night wear at sea. Raincoats would be issued on the ships, he said, but he advised them to bring their own binoculars. "The fleet is very short on these," he noted. Van Allen borrowed $300 from APL to make all the unexpected purchases—he wouldn't have a chance to apply for his uniform reimbursement for nearly a year and that provided only $100. He hastily collected the personal contents of his furnished room into two duffel bags and stowed them away at APL.

Van Allen, Dilley, and Peterson boarded a Baltimore & Ohio train for the first leg of a four-day rail trip to San Francisco to supervise the secret loading of some 5,900 proximity fuzed projectiles in various holds of the USS *Republic,* a troop transport ship docked at nearby Mare Island. There were two projectiles per box and each weighed 120 pounds. The three physicists took turns checking on their secret inventory, now called the VT (variable time) fuzes. A name like radio proximity fuze gave too complete a description of how the fuzes worked, navy brass had decided. On November 19, the three young officers found themselves and their ammunition heading toward an undisclosed location to meet Parsons. Van Allen applied elementary celestial astronomy to keep track of latitude and estimated longitude by the progressive changes in the time of sunrise and sunset compared to the time on his watch.

The ship sailed at only 14 knots without an escort and even the new recruits knew that the ship would be helpless in a Japanese raid. Still, they spent a festive Thanksgiving at sea. The holiday menu included cream of turkey soup, mixed pickles, creamed Tom Turkey, homemade cranberry sauce, mashed sweet potatoes, peas and carrots, bread and butter, a choice of desserts and fruit, and cigars and cigarettes. The prayer on the menu read: "Sharp indeed has been the turn of our destiny since November a year ago."

After two weeks, they arrived at Noumea, New Caledonia, headquarters of the Commander of the South Pacific (COMSOPAC), and they reinventoried the fuzes. "I have to tell you we never got the same number twice when we counted the darn things," Van Allen said. "They were all jumbled in various holds." He adjusted the ammunition count to approximately 5,400 shells after arrival. The three lieutenants met up with Parsons at Noumea, and he introduced them around headquarters. Other officers greased the skids across the fleet to get the physicists time with captains and gunnery officers across the entire fleet. At first, Dilley took cruisers, Peterson took battleships, and Van Allen got the destroyers.

Van Allen couldn't keep a diary of his stops—the navy forbade logbooks lest they fall into enemy hands. The rule wasn't strictly enforced, however; one gunner on the USS *Washington* kept a diary in a pocket-sized notebook marked "Recipes." Van Allen did preserve a thick file of carbon copies of his navy orders that document his movements as he darted across the Pacific to bring the proximity fuze successfully into battle. On one unexpected stop in Hollandia, New Guinea, he saw a sign for an army field hospital and recognized it as the one directed by his brother Maurice. Van Allen greeted Maurice, now a doctor and captain in the Army Medical Corps, on the dock picking up medical supplies. Bravado and good cheer at the brothers' incredible luck of running into each other pushed aside the high stakes of their responsibilities for one night as they had dinner together at the officers' mess.

It was the only time Van Allen ran into anyone he knew from home. Most often, he worked with captains and gunnery officers on each of the ships to explain how the new fuzes worked. "He had a way of speaking that was low-key. He talked to the men rather than lectured them, treating everyone as equals," said Baldwin. Van Allen even-handedly described the superiority and shortcomings of the fuze—duds, early firings, and the dangers of exploding a projectile from another ship. He helped load the fuzed projectiles and stood by as gunnery officers first tried them out on enemy aircraft that circled within range on scouting expeditions.

But Van Allen soon realized that the new invention had a unexpected psychological flaw linked to the innovative design. "The thing about the time fuze was, you'd get a burst every time you shot it. And you could see you were doing something, whereas with the proximity fuze, nothing would happen if you missed the proximity of the plane. That's not a very gratifying experience for a gunnery officer who is told to shoot and shoot and shoot," Van Allen noted.

His films, demonstrations, and test results with drones couldn't always overcome a more general skepticism. Gunnery officers agreed to use the new fuze basically at their discretion. Some opted for a fifty-fifty mix of time fuzes and proximity fuzes. Some agreed to go as high as 75 percent. Some stubbornly refused to bet their lives—and the lives of their men—on any new weapon. At least they knew what to expect with the time fuzes.

Van Allen had to "sell" the proximity fuze ship by ship, gunnery officer by gunnery officer. He took a motor launch loaded with one hundred to two hundred proximity fuzed projectiles to each destroyer and made his case to a cross section of humanity with different fears and divergent levels of respect for physics and physicists. Then Van Allen boarded the USS *Washington* and came under the wing of a dead-shot battle technician, Rear Admiral Willis A. Lee. Lee had sunk the Japanese battleship *Kirishima* at Guadalcanal just a few weeks earlier and laid in wait as Japanese General Shizuchi Tanaka parked four transport ships at the island. Lee destroyed them all, halting construction of the Japanese air base that threatened Allied ships. The arrival of a young scientist who had helped invent a newly minted weapon sparked Lee's curiosity about the projectile. After Van Allen's presentation, Lee wanted to see how it worked.

Van Allen knew the rules about dissembling ammunition on shipboard where it could blow up. But he needed an advocate and offered to cut a fuze open, assuring the admiral that he could do so safely. He unscrewed the fuze from the top of a shell and threw the rest of the projectile overboard. He disconnected the igniter and sawed right through the casing of the fuze. He showed Lee how the miniature radio transmitter and receiver ignited the explosives and explained the rugged vacuum tube spring concept he had designed. Lee was a gunnery expert who knew firsthand that time fuzes often substituted the fallacy of filling the air with explosives for real hits. He ordered all the ships under his command to use the new fuzes immediately and his decision gave the weapon validation.

Van Allen received the first successful battle report on the proximity fuze from Deke Parsons who watched it shoot down an Aichi 99 dive bomber from the bridge of the USS *Helena* as the assault continued to capture Guadalcanal. The *Helena* opened fire on a surprise raid of bombers after their dives into attack position. One came within 7,000 yards of the ship when "a very close burst occurred, obscuring the plane." The Aichi 99 immediately went out of control "with heavy smoke rising as it crashed," Parsons radioed BuORD on January 5, 1943. Antiaircraft guns took down another plane but, with the mix of ammunition, Parsons couldn't credit it to the Mark 32 fuze, which became the official product name. The rest of the bombers withdrew, chased off by U.S. fighters. The report pushed for wider use of the fuze. Fresh supplies of Mark 32s now filled a steady supply line to the South Pacific and most U.S. ships had them in their arsenals after about three months. Dilley and Peterson requested to go back to the lab. Van Allen stayed to continue with the training and documentation of the use of the fuze. In one report, he concluded that it required thirty-two rounds of ammunition on average for a conventional anti-aircraft projectile to bring down a plane, more than ten times the number of rounds required with a Mark 32 fuzed projectile.

With successful initial reports, the navy expanded production of the fuzes and Van Allen soon set up new ammunition depots on Espíritu Santo and other New Hebrides islands. On one trip to Espíritu Santo, Van Allen and his ammunition hitched a ride on a converted tuna fishing boat. The ship provided refrigeration for beef, chicken, cheese, and frozen strawberries. "We ate well on that trip," Van Allen said. But the boat slogged along at 10 knots. Japanese planes cruised high overhead but ignored the small ship until one pilot circled and attacked with a bomb. The bomb fell into the ocean and the crew retaliated with a barrage of machine-gun fire. "It was totally ridiculous—none of the machine guns could reach the plane. It was all short-range stuff but everybody felt better to be doing something, I guess," Van Allen said.

Van Allen continued to move from ship to ship in the Pacific recording results regarding the proximity fuze through early June 1943. Then on June 9, 1943, BuORD ordered him back to Washington, D.C. He got a leave to visit Mount Pleasant and then, back at BuORD, he made a careful analysis of his notes and reports from the Pacific. Then in early fall, reports about the use of the fuze turned suddenly ominous.

The number of duds increased dramatically and the hard-won acceptance of the fuzes began to evaporate. Van Allen requisitioned the return of some of the fuzes from ships where he had so recently convinced gunners to use them. He took several apart. They looked intact. Then he tested the circuit and found that the batteries powering the radio transmitter and amplifier were dead. The batteries had an expected shelf life of ten months but Van Allen determined that the shelf life dwindled to as little as four in the hulls of cargo ships where temperatures reached 120 degrees.

Van Allen felt a special responsibility for the successful use of the fuzes based on his previous tour of duty on combatant ships. He decided the quickest way to get new batteries in the weapons was to return to the South Pacific and oversee the replacement himself. He requested orders to return and Tuve backed his efforts to commandeer enough batteries to replace every single one in nearly 250,000 projectiles scattered among the ships and ammunition depots of the Pacific. COMSOPAC sent a secret dispatch on February 15, 1944, telling ships that the Mark 32 batteries had deteriorated and that the batteries would be replaced locally.

Van Allen and a crew of sailors cobbled together a shop in Noumea, the main city of New Caledonia where the replacement batteries arrived by air. He set up shop in a stifling bamboo storehouse with long tables. Van Allen and gunner's mates laid out each shell on a table, unscrewed the cone holding the VT fuze, popped out the old battery, inserted the new one, and soldered the cone back in place. Van Allen worked side by side with the men. Once a sailor got the hang of it, he could replace a dozen batteries an hour. That meant that eight men working a six-hour shift could replace 576 batteries—576 out of 250,000. And the men had to stop frequently as they worked to brush off the picric acid powder. The poisonous yellow powder used in the explosives spilled out every time they unscrewed the cone. The dust coated their hands and arms and, after an hour of working in the heat and humidity, it stuck like a second skin. After two hours, a fine layer of the powder clung to their hair and faces. It sifted across their uniforms and filtered through the air they breathed. "It washed off, but you sort of had to wait for the skin to grow back after that," Van Allen said.

The men worked day after day to rebattery the fuzes in Noumea. As ships came in for supplies, Van Allen and his crew brought them fresh rounds of rebatteried antiaircraft projectiles and stripped the inventory of proximity fuzed projectiles on board to rebattery them for the next ship. After Noumea, Van Allen moved on to the ammunition depots on Tulagi, Espíritu Santo, and Manus islands. But the biggest job still lay ahead on ammunition barges anchored at the islands of Eniwetok, Kwajalein, and Ulithi. Though the atoll can't be found on many maps, dozens of battleships, cruisers, and destroyers resupplied at the barges where Van Allen completed the rebatterying marathon. It had taken him three months. The allied fleet readied now for an assault on the Mariana Islands scattered like jewels across the Philippine Sea.

The proximity fuzes were ready to explode once again and Van Allen brought them into battle as the assistant staff gunnery officer on the USS *Washington* where Lee brought him and assigned him to work with chief gunnery officer, Commander Raymond Thompson. Van Allen boarded the flagship in early June 1944 as it headed toward the Mariana Islands to capture the Japanese stronghold of Saipan in the Battle of the Philippine Sea. Battleships, cruisers, and destroyers began to bombard Saipan and Guam on June 15—ten days after the D-day invasion at Normandy to liberate France and the same day that the Germans unleashed the first rocket-launched buzz bombs against London. Saipan brought the war home for the Japanese to an island 1,250 miles from Tokyo where Japanese civilians lived and worked. Vice Admiral Jisaburo Ozawa threw submarines into the fight and most of his remaining planes. "From the beginning of the war, the Japanese naval high command had a fixation on bringing the U.S. fleet into one climactic, decisive action," wrote World War II historian Harry Gailey in his book *The War in the Pacific*. "Ozawa still believed in the possibility of a cataclysmic victory similar to the fabled destruction of the Russian fleet at Tsushima in 1905." Now Saipan and the Battle of the Philippine Sea took on this symbolic weight.

The Japanese fought fiercely from the hills and ridges that shielded them from attack and the army convoys and marines suffered heavy casualties as they hit the open beaches. American battleships downed seventy planes at Iwo Jima and Chichi Jima on June 16 and then joined the main fleet that tightened a juggernaut across the Philippine Sea. Fight directors on board each ship used the radar capabilities developed early in the war to detect the first wave of planes that attacked on June 19. As Van Allen watched the first attack of sixty-nine fighter planes swarm through the skies on June 19, he joined Thompson on the bridge as the ship's gunners fired at several bombers headed directly toward the *Washington.* With the help of the proximity fuze, the Americans shot down most of the first wave of planes, but 128 bombers and fighters sped in on the second wave a short while later. Only thirty-one escaped the combined force of the pilots and "Lee's antiaircraft gunners," as Gailey called them. Yet another wave of eighty-two planes attacked. The Hellcats and gunners shot down almost all of them and continued the defense attack.

America finally had a weapon to back up the pilots, a weapon with a projectile that worked, though American battleships continued to shoot a mix of time fuzes, proximity fuzes, and smaller weapons. With clouds of dark smoke from exploding fuselages, with bombs and gunfire pounding the air, it was nearly impossible for Van Allen to differentiate the proximity fuze hits from hits with other weapons. "I could never say with certainty that it was the proximity fuze that did it but we fired a lot of them," Van Allen noted. "The haze of battle is far from a controlled laboratory experiment, and it was difficult to arrive at an objective quantitative summary. Nevertheless, gunnery officers identified many episodes of proximity fuzed rounds being responsible for aircraft hits."

By any measure, the battle of the Philippine Sea decimated Japan's airpower in the most important aircraft carrier battle of World War II. Some 425 aircraft and pilots—75 percent of the planes sent into the battle—were lost, more than at any battle before American ships used the proximity fuze. In a detailed description of the fall of Saipan, Gailey concluded that "The gunners' effectiveness had been significantly improved by the newly developed proximity fuzes." While Americans viewed the 1942 battles at Midway and Guadalcanal as the turning point in the Pacific, the Japanese recognized that they had lost the war at the Battle of the Philippine Sea. "I could say that Japan was defeated by the fuze . . . and the radar," one of the few surviving Japanese pilots of the battle later told Baldwin. "America was rather weak in the beginning of the war but as time went by she strengthened herself. And made those new weapons. In the end, Japan was defeated technologically."

American pilots located Japanese carriers on June 19 and Rear Admiral Marc Mitscher capitalized on the crippled air defenses by sending in the Hellcats to bomb the ships. Mitscher gambled against both time and fuel—the Hellcats had no margin for error if they were to make it back to their carriers. U.S. pilots and torpedo attacks destroyed three ships, damaged several more, downed dozens of Japanese planes, and returned in the darkness with gas gauges fixed on empty. They had no fuel to spare to search for ships maintaining a virtual blackout against nighttime raids. Despite the risks, Mitscher ordered lights on—every available light—to blaze a path to the landing decks.

But carnage lay ahead for both sides. Japanese officers rallied the five thousand troops still on Saipan into a full-scale suicide attack. U.S. soldiers and marines repelled them with heavy casualties and deaths mounted as GIs encountered Japanese soldiers isolated in caves and bunkers. More than three thousand U.S. servicemen died and another eleven thousand were wounded or missing. As the U.S. forces swept in from the south, Japanese military survivors threw themselves off the cliffs on the north side of the island and civilian men, women, and children followed their example.

The navy enforced an ironclad policy against using the fuze over land at Saipan—lest a dud be recovered and copied or used to develop countermeasures such as a radio transmitter that could trigger the fuze prematurely or jam it. But ground troops in the Pacific did use the proximity fuze against planes to hold the Marianas against nightly raids by Japanese pilots on Saipan's neighboring island of Tinian. Casualties mounted with each raid and soldiers fighting from the marshes felt defenseless and abandoned. "Almost every night we would see 3,000 to 4,000 shells fired with only a hit or two in several nights. Then came the VT fuze. On the very first night seventeen Japanese planes were shot down," army veteran Stanley B. Hudler told Baldwin. Hudler credited the fuze with shortening the war and saving his life.

A month-long strike at the Palau Islands north of the Marianas brought U.S. troops back within striking distance of the Philippine Islands. The Americans suffered heavy casualties on the ground once again but the air above the battleships looked ominously quiet as the Japanese air force regrouped.

Van Allen filed regular reports to BuORD on the VT fuze, accompanied by urgent requests for air shipments of fresh batteries, tools, and equipment. Shells with the proximity fuzes continued to pour into the Pacific pipeline, replenishing the arsenal as the USS *Washington* and the Third Fleet positioned for the invasion of the Leyte Gulf in the center of the Philippine chain in October 1944. Now the Japanese hurled a nightmarish weapon at the ships—kamikaze pilots who flew planes loaded with explosives into the decks. The desperation measure made any miss deadly—gunners had to shoot down the planes before they crashed. With the stakes so high, the Third Fleet relied heavily on the battle-tested proximity fuze and successfully defended most of the ships against the kamikaze assaults. Later, U.S. pilots located and hit Japanese ground bases to reduce the kamikaze capabilities, but the kamikazes remained a grave hazard for the rest of the war.

The Battle of Leyte Gulf from October 24–27 marked the greatest naval battle in history, an apocalypse of firing guns from nearly three hundred ships. The American victory destroyed the Japanese navy and eliminated any major threat of a Japanese naval attack. While fighting stormed across the Philippines, the U.S. ships, including the *Washington,* hammered west to liberate Japanese-occupied territory and strike at Japan itself. Nature rebuffed the entire Third Fleet with a typhoon in mid-December before the skies cleared for Christmas. The ships launched B-29 bombers in air strikes against Taiwan and Okinawa in early January; air strikes aimed at Indochina, Canton, and Hong Kong in mid-January; and air strikes on Tokyo itself in mid-February. Major General Curtis LeMay choreographed low-flying nighttime air raids that dropped fire bombs, igniting vast stretches of Japanese cities. Ships fired on the cities as well but, again, never with proximity fuzed weapons. Fear remained that the Japanese could recover a dud.

The noose tightened around Japan as the Allied fleet targeted Iwo Jima, only 750 miles south of Japan, in a bloody campaign. In March, 1945, the island fell and Allied troops moved on to Okinawa, only 350 miles from Tokyo. In March, Manila fell, liberating the Philippines completely. In March, Tokyo itself lay in cinders from fire bombs and millions of homeless civilians roamed the streets and countryside.

Radar and the proximity fuze had helped topple a mighty empire. The Japanese military knew, since the fall of Saipan, that they had lost the war. But they fought on, holding true to ancient codes of honor rather than any hope of victory and surrendering only after the United States military struck Hiroshima and Nagasaki with the third new weapon in its arsenal—the atomic bomb.

Van Allen's job was finally done with the liberation of the Philippines and he returned to BuORD in March 1945. He summarized the technology success of the proximity fuze in several lengthy and detailed reports after his return. Other reports documented successes with the fuze against German bombers. The proximity fuze downed the dreaded buzz bombs bombarding London and Antwerp and ensured victory at the Battle of the Bulge on the German/Belgium border where the Germans planned to split the Allied forces in two.

After his return from the Pacific, Van Allen was appointed assistant group supervisor to work with BuORD and Applied Physics Laboratory in developing new models of the proximity fuze. The weapon was now a staple in the military arsenal. With secrecy no longer a concern, Van Allen even deployed one fuze for a bookend on his desk at APL. He worked as a one-man bureaucracy buster to streamline the approval process for materials and projects. "I had a desk at APL and at the navy [BuORD]. And I would go to APL and discuss with colleagues what we needed there and write a letter, which Tuve would sign. Then I'd carry that letter down to navy headquarters and write a response to it for [the commanding officer] to sign. I had enough clout to pull that off. So I was a one-man liaison between BuORD and APL," Van Allen said. That meant letters could be exchanged and orders executed two days later. He also tested refinements to the proximity fuze in San Juan and with the navy's Atlantic Fleet off the coast of Maine.

Van Allen conservatively assessed the impact of the new weapon in his reports and gave barely a hint of his own dramatic role in ensuring its use. Officers in the field such as Arleigh Burke, the first to use the fuze on January 5, 1943, relied on their own eye-witness observations to conclude that the fuze had shaved years off the war, Baldwin said. Even President Truman credited an unnamed secret weapon as the "determining factor in defensive antiaircraft action by the United States navy" as he awarded Tuve the Medal of Merit in March 1945.

Germany surrendered on May 7, 1945, but the war lingered on in the Pacific. Van Allen knew about the development of the atomic bomb and, with a PhD in nuclear physics, he had been asked to participate in its development. He declined. He was already making headway with the proximity fuze. But he knew several of the scientists involved and shared the dismay of many of them when the atomic bomb was dropped on Hiroshima and Nagasaki in August 1945. He learned later that his friend Deke Parsons, who had opened the initial doors needed to bring the proximity fuze to the Pacific, was the bomber who dropped the atomic bomb on Nagasaki.

With the war over, Van Allen completed his evaluation and continued his liaison role. His senior officers advised him to stay and make a career of the navy. "I was tempted. I really liked the navy. I still do. Being on a battleship is very powerful stuff. I can understand how people feel a deep respect for the power and majesty of the sea. I admired the power of the sea and the way you cope with it on a ship," Van Allen said. His life-and-death responsibilities in the navy, the cross section of people he had met there—from seamen to admirals—and the navy's code of honor deeply touched him.

But Van Allen wanted to get back to research and was racing to build his first cosmic ray detectors for the opportunity to fly them in captured German rockets. By the time he received his discharge as a lieutenant commander from the United States Navy on March 8, 1946, he had taken on a new rank—as husband.

5 Enter Abigail Fithian Halsey

Silver Spring, Maryland—March 1945. Soon after Van Allen returned from the South Pacific, he pulled up to the intersection of Colesville Road and Georgia Avenue and waited for the traffic light to turn green. Suddenly, he felt a bump. Abigail Fithian Halsey, the driver in the car in front of him, accidentally backed into him when the light changed. The contact was negligible. Neither Van Allen nor Halsey bothered to check for damage. He passed her on the right, and gave her one of those looks—male annoyance at a female driver written all over his face.

• • •

Their paths quickly crossed again as they checked through security at the Applied Physics Laboratory. "Who do you think you are throwing dirty looks at me?" she called out. Van Allen said nothing and moved on swiftly. But he was completely charmed. "I thought, this was a gal with real spunk. I liked that," he said, recalling the incident nearly sixty years later. Halsey wasn't hard to find. She worked at APL as a mathematical analyst. Despite her English literature degree from Mount Holyoke College in Massachusetts, she had a good head for math and answered the call when the lab appealed to area colleges for graduates to join the war effort. Overcoming his shyness, Van Allen phoned her several days after their "collision" to invite her to go bicycle riding the following Sunday. She accepted. Her roommates answered the door when he arrived and tried to get acquainted. Van Allen resolutely burrowed behind a magazine until Abbie appeared.

"My roommates thought I was crazy," Abbie recalled. But Abbie Halsey and James Van Allen fell in love. With war fever still in the air and young couples heading into marriage after only a few month's acquaintance, Jim quickly proposed.

Just as quickly, Abbie said yes.

Abigail Halsey grew up in one of America's oldest families. Thomas Halsey and a group of his Puritan friends fled England and came to America to escape religious persecution in the 1630s. He soon owned 100 acres of land in what became the town of Lynn, near Salem, Massachusetts. Persecution in Europe didn't teach the Puritans tolerance, however, and their own inquisition and executions during the infamous Salem witch trials convinced Halsey and others to find a new settlement. An advance party reached Long Island near present-day Manhasset in May 1640, but the Dutch garrison arrested them in defense of New Amsterdam and then sent them back home. Halsey came with a second wave that reached the more isolated eastern end of Long Island and established the village of Southampton. The village anchored the first English settlement in New York State. Halsey helped establish a covenant signed by all the founders that "if he sell his farm he shall not divide it but sell it together."

The covenant didn't stop boundary skirmishes between the Dutch and the New England colonies, however. Native Americans also threatened the new village, and constant raids cost lives on both sides. Halsey's first wife died in a raid and he later married a woman named Phoebe. They had four children. Halsey was a prominent citizen who owned and farmed more land than anyone in the area, helped archive local court records, and served in the court and as a representative of the regional Connecticut Council. Self-reliant and opinionated, Halsey spoke his mind bluntly and sometimes rudely.

"From the town records it is plain that Thomas Halsey was not only an active citizen, but one possessed of independent spirit and a strong will, and not always respectful to his fellow townsmen. He had been well educated and, in accordance with the distinctions of those times, was styled 'gentleman' in the records," wrote his ancestors in their biography of him. They noted that the first recorded instance of his independent spirit was dated March 25, 1643, when he was censured at a town meeting for "irreverent speech" to a magistrate. The house he built on South Main Street still stands, and Jim and Abbie's daughter Cynthia Van Allen Schaffner, a researcher at the Metropolitan Museum of Art, is the curator of the colonial Halsey House.

Seven generations later, on the same street on the north side of town, Henry Halsey applied his trade as a carpenter to build the home at 49 North Main Street. His oldest son, Charles Henry Halsey, took over the house and brought his bride, Melvina Terry Halsey, there. Their children included Harry, Lizbeth, Abigail Fithian (Miss Abbie), and Jesse. Miss Abbie's niece and namesake, Abigail Fithian Halsey II, was born in the family home on August 9, 1922, while her parents, Helen Isham and Jesse, vacationed there.

At the time, Helen and Jesse lived in Cincinnati where he was pastor of the wealthy Seventh Presbyterian Church. Reverend Halsey combined a keen intellect and a Princeton diploma with courtly manners and a vintage New England pedigree. His congregation loved him. So did the city where he was a popular civic speaker, captivating audiences with stories of missionary work among the Inuit and other native peoples in Labrador and of emergency missions to Russia as the Bolsheviks took power. Still, life with a clergyman had been quite a change of pace for the patrician Helen.

Helen grew up in Lake Placid, where her family had a permanent home but lived in Manhattan as well. Her grandfather Henry Lockwood Isham bought land in Lake Placid after establishing and operating the Isham Wagon Company in town, a leading East Coast manufacturer of wagons and sleighs. Her father, Frederick Asher Isham, was a senior partner with the prosperous New York law firm of Isham and Isham. The law practice was the reason he maintained dual residences in Lake Placid, where he served two terms as mayor, and in Manhattan.

The Ishams made appearances in the *New York Times* society pages for their charity events, a European tour, and summer parties at Lake Placid. A society piece in the *Times* noted that "a party of young people making the ascent of Mount Whitney included Miss Phyllis Verity, Miss Anna Cozzena, Miss Alice Dike, Miss Helen Isham, John McGraw, Clarence Dike" and a party of several other fellows.

Helen loved the mountains near Lake Placid and near Saranac Lake where her father bought land and developed homes and home sites for patients of the posh tuberculosis sanitarium. Like those in Europe, the sanitarium offered a combination clinic and resort with hotels and cure cottages where TB sufferers could rest and restore their health in the cool, fresh mountain air. Saranac Lake had a legendary reputation as hallowed ground for healing any case. Dr. Edward Livingston Trudeau had grown up there and thought he was returning home to die in the 1870s when he contracted tuberculosis. But the story of the restoration of his health turned Saranac Lake into a thriving health spa attracting gifted practitioners and fanatic healers alike. The sanitarium was a natural and the new venture offered a fresh start for Frederick after financial setbacks due to the 1890s depression and the loss of his two young sons to diphtheria.

Jesse Halsey made trips to Saranac Lake to visit his sister Lizzie who was taking the cure while he studied theology and attended graduate school from 1907–1909 at Princeton University. As a young Presbyterian minister, he gave inspiring talks to raise money for his mission in Labrador. One speaking venue brought him to the church in Lake Placid where Helen played the organ. They met there.

The reverend's work at Sir Wilfred Grenfell's Labrador Medical Mission appealed to Helen's romanticism and social conscience. The needs of the settlement houses serving the impoverished at home and the missions serving them abroad came up frequently at the club meetings and parties of young ladies of her circle.

Jesse and Helen married in 1910 and, soon after the wedding, they set out for Labrador, a daunting journey to a rugged peninsula of fur trading outposts and fishing villages shared by European settlers and the tribal Inuit, Innu, and Metis peoples. Helen was soon alone at St. Anthony for months at a time while her husband left to replenish supplies and raise funds. Pregnant and tightly holding the hand of toddler Charles Henry, she watched for incoming ships from the cliffs over the harbor, never knowing whether Jesse was alive or dead, never certain of the length of his journey. The couple returned home with Charles, Frederick, and a polar bear for the Boston zoo. They had plans to settle in Boston but the family moved to Cincinnati after Jesse spoke about the mission at the Seventh Presbyterian Church and leaders of the captivated congregation offered Halsey a job on the spot.

Helen gave birth to Helen Augusta, Wilmun, and Abigail. The loss of Wilmun in a car accident, when he was seven and Abbie was five, set the bereaved little girl in pursuit of the sports her brother had loved. She excelled in academics, too.

Abbie attended Mrs. Lotspeich's private school along with the young relatives of both President Franklin Delano Roosevelt and former president William Howard Taft. Through these connections, Mrs. Lotspeich arranged a trip to Washington, D.C., to meet Eleanor Roosevelt and Helen Herron Taft. Mrs.Roosevelt greeted the children in the White House and shook each child's hand.

Mrs. Lotspeich's rigorous demands in mathematics and literature stood Abbie in good stead. She sailed through Hillsdale High School, pursuing drama and musical theater and playing Rosalind in *As You Like It* her junior year. Students elected her president of her class that year and vice president of her class and of the student council during her senior year. Summers in Southampton toned her as an outstanding swimmer and tennis player, as well. But tragedy struck again and the family mourned the loss of Frederick, who died at age twenty-seven, just after Christmas in 1939. Abbie was the only child still at home now, with Charles on his own and Helen studying for a master's degree at Yale University. Still, her heartbroken parents encouraged her to attend her first choice for a college, Mount Holyoke in Massachusetts.

While Abbie was in school, Helen followed in her mother's footsteps and married the Reverend Joseph Haroutunian. They settled in Chicago where

Haroutunian became a renowned theologian at the University of Chicago and taught at the McCormick Theological Seminary. Jesse and Helen Halsey also moved to Chicago when Jesse became a professor at the seminary in 1941, although he took a leave to serve as a navy chaplain during World War II. He had settled back into academia by the time Abbie met Jim.

Even if her roommates teased her about Jim's shyness, colleagues at the lab told Abbie how much they liked and respected her new boyfriend and encouraged the budding romance. Abbie began to appear in Jim's letters home in spring 1945, and his brother George and sister-in-law Winnie took the train to Silver Spring to meet her. Sabina Winifred Martin grew up on a farm near Mount Pleasant and she had married George the year before. The two young couples turned the D.C. visit into a holiday. By summer, when Abbie left for Southampton, she invited Jim to visit the family homestead. "I took the train up and met her father and mother—very nice people—and passed muster somehow," Van Allen said.

Southampton decided their fate. "I have some big news for you," he wrote to his parents on July 31. "Abigail and I have decided to be married. I have never been so happy before. She is a wonderful girl and has fine parents and a brother and sister, all of whom I met and got well acquainted with in Southampton a few weeks ago." He and Abbie had time to become really close while he was there, he confided. "So we decided to make it a lifetime arrangement."

Abbie and Jim's mother, Alma, started corresponding, too. With the wedding date set for October 13, Abbie expressed her hopes that the Van Allens would come to the wedding, a challenging prospect since Alfred Van Allen seemed very weak. "I had thought that we could have [the wedding] in Chicago so that would make it easier for you, however, mother has not been too well" and was recuperating in Southampton, she wrote.

The *New York Times* ran the wedding announcement in a lengthy article on August 28, complete with a picture of Abbie wearing pearls to accent a V-necked lace blouse. "The Rev. Dr. and Mrs. Jesse Halsey of Chicago and Southampton, formerly of Cincinnati, have announced the engagement and approaching marriage of their daughter, Abigail Fithian, to Lieutenant James Alfred Van Allen, USSNR, son of Mr. And Mrs. Alfred M. Van Allen of Mount Pleasant, Iowa," the article reported.

Jim sent his parents a copy of the article in his next letter. The young couple still didn't have a place to live. They spent August in a futile apartment search since wartime growth in Silver Spring, as elsewhere, created a housing shortage. They could plan to drive in style, though, because Alfred Van Allen offered his son a car for a wedding present. Jim gratefully accepted the gift.

"My old Mercury is still going strong—but getting a bit rattly in the joints," Jim wrote his folks on August 30.

But Alfred's weakness made a trip east for the wedding unlikely now so Jim planned a trip home to Mount Pleasant instead. "Am hoping to get three weeks leave for my honeymoon. We don't have it planned fully yet but will come out to Mount Pleasant for a week. I think that you will enjoy Abbie very much. We have wonderful times together," Jim continued hopefully in the August 30 letter.

Then Alfred suffered a severe heart attack. "I regard his condition as dangerous. Wish you could come out for a few days as he is anxious to see you again. I think it very doubtful if he lives until October," George wrote his brother. Jim couldn't leave. He received the letter just a few days before he left to spend much of the month of September on shipboard testing proximity fuzes with the Atlantic Fleet.

Abbie left her job at APL in September. She finalized wedding plans with her mother and wrote Van Allen rapturous letters during his absence and gave him a countdown on their rapidly approaching wedding. "Seventeen more days," she wrote on September 26. "Not hearing from you is awful—gee, it's amazing how much closer we have grown since July 20. I was in love with you then—but as I look back, it was nothing as compared to the way I feel now."

She had continued to hope his parents might yet be at the wedding. Alfred Van Allen never met Abbie, however. He died of heart failure on September 23 at age seventy-six. His death left his son with a crushing choice. "I tried to get a leave but the navy wouldn't give me a few days of compassion leave unless I wanted to sacrifice my honeymoon and I felt I couldn't do any good out there," Van Allen said. In the end, the navy cut short his honeymoon as well, putting off any trip home as his wedding day approached.

• • •

Abigail Halsey wanted to marry in the Seventh Presbyterian Church in Cincinnati where her father had been pastor for twenty-eight years. "I had grown up in Cincinnati and all my friends were there," Abbie said.

But, with her parents living in Chicago now, a homecoming in Southampton made more sense. Abbie's girlfriends converged on Long Island for a wedding with twelve bridesmaids wearing red velvet. Abbie's sister Helen served as matron of honor and Abbie's close friend Caroline Lacey was the maid of honor. Captain and Mrs. Frank Myers, friends from Silver Spring, both attended the bride and groom. Jim's brother Bill was Van Allen's best man and Dilley and Peterson, who had brought the proximity fuzes to the Pacific with him, served among his attendants. The whole party had gathered in Southampton for the noon rehearsal on Friday, followed by Jim's bachelor party and Abbie's bridesmaid luncheon.

The bride walked past columns of white chrysanthemums and autumn foliage decorating the aisles of Southampton's First Presbyterian Church for her 8 P.M. wedding ceremony on Saturday, October 13, 1945. She lit candles at the end of every pew. She wore a white satin gown trimmed in Belgian lace with a fitted bodice, long sleeves, and a train. A tulle veil fell in folds from an heirloom-lace crown worn by her mother and generations of brides in her mother's family. Lieutenant James Van Allen proudly wore his navy uniform and Reverend Halsey officiated at his daughter's wedding ceremony.

"It was a very formal and traditional wedding ceremony," Van Allen recalled. "I remember her father really looking me in the eye and saying, 'Do you take this woman to be your lawfully married wife, to honor and obey, in sickness and in health, till death do you part?' He looked me straight in the eye—he really meant it and I meant it. This was no kidding and the impression is still vivid in my mind."

Abbie's parents hosted the reception in the family home on North Main Street. After the reception, the newlyweds took a ferry to Connecticut and then drove to the New Hampshire coast. There they parked the car on the beach of a small, pristine lake, appropriately near Lovewell Mountain. They packed their gear in a canoe and rowed to their honeymoon cabin on an isolated island in the middle of the lake. It had no electricity and only a pump for water. Golden bursts of the last fall colors surrounded them in the chill of late autumn. They kept a fire burning almost constantly in the cozy cabin, closed off from the rest of the world.

"Marriage is wonderful," Van Allen wrote his mother on October 25. "In fact, I am just deciding that it is the best institution that the human race is capable of," he wrote. Their wedding gifts had included about $600, antique silver treasures, glass, china, pottery, blankets "and even a chair." On a personal note, he asked in the letter if Alma had happened to see the articles on the proximity fuze in *Life* and *Time* magazines.

Returning to Silver Spring, Jim and Abbie moved into a temporary apartment lent to them by friends who were away. With postwar housing shortages, the arrangement gave them time to find a one-bedroom apartment of their own at 529 Dale Drive near the lab in Silver Spring.

In March, Abbie went to Chicago to visit her parents and then made her first trip to Mount Pleasant, Iowa. George and Winnie introduced her to everyone in town and Alma fussed over her pretty daughter-in-law. Staying with Alma in the home where Jim had grown up gave her a profound insight into the lives of his parents. "I imagine life had been very difficult for Jim's mother. But she had a great goodness within her—a warmth," Abbie said. "Jim has this quality—it's an innate quality of altruism, an innate quality of just being a cheerful person." His heart and mind had both attracted Abbie to the naval officer who fell in love with her so quickly and so completely.

With summer weather in 1946, the newlyweds bought a sailboat and christened it the *Mesotron,* an arcane name for the subatomic particle now called a meson. They sailed on Peconic Bay and Chesapeake Bay that summer of 1946, and Jim helped his father-in-law build a bay-front cottage called "the camp."

Alma visited Southampton that summer as well. She met Abbie's whole family but the two school teachers—Alma and Miss Abbie (Babbie)—hit it off from the start.

"Every morning when I wake up, I think of something pleasant downstairs, breakfast with you, and then I realize you are not here. I hope you had a safe and pleasant journey home. We enjoyed your visit so very much, all of us, and I hope you come again next summer," Babbie wrote Alma after the visit. "You know what I think of your Jimmy. With love to you—and keep me in your heart, [signed] Abigail (Babbie)." The indefatigable Babbie, a legendary producer of historic pageants in the area, died in her sleep later that summer.

The Southampton home and the family "camp," a cabin seven miles away on Peconic Bay, became Abbie Van Allen's base for lengthy expeditions that kept her connected to her family and her heritage. Abbie taught Jim to swim there the summer after they married and it became the center of their family life every summer to come. From Long Island, Abbie took her five children to New York to museums, stores, and concerts. Abbie inculcated a love of the ocean and the arts along with an attachment to her family's East Coast heritage. "It helped me survive during the long months when Jim was away," Abbie said. The Van Allens eventually bought the family home in town.

"We followed my mother's interest to a large extent. She was interested in art and she exposed us to the museums. She took me to Europe when I was thirteen and dragged me through all of the museums," said Peter Van Allen, the youngest of the children. As adults, all of them returned to the East Coast.

That first summer of their married life in Southampton, Abbie became pregnant. As she borrowed a baby carriage from her sister Helen and a bassinet from a cousin to set up a nursery area in their apartment, she adjusted to her husband's frequent research trips. Now, at a remote desert military base in New Mexico, James Van Allen opened the door to outer space with a small cadre of American scientists. They worked with German rocketeers brought to America under house arrest.

6 The Dawn of Space Exploration

White Sands Proving Ground, New Mexico—April 16, 1946. Van Allen had a cosmic ray experiment ready to fly on the first V-2 rocket launch, scheduled for April 16, 1946, at White Sands Proving Ground (now White Sands Missile Range) in New Mexico. The APL group headed there by train and loaded instruments in the baggage compartment where sensitive electronics rattled across 3,000 miles of track to El Paso, Texas. An army truck picked up the group and took them on another bumpy ride through the desert to the primitive base at White Sands. Van Allen bunked with the other scientists in hastily constructed barracks and worked in the makeshift rocket assembly building. The spartan accommodations, long hours, and pitiless heat indoors wilted spirits but couldn't dull the excitement and the promise of new discoveries.

The V-2 rocket gleamed four-stories high in the sun on April 16, a surreal monolith dwarfing the ridges of the distant San Andreas Mountains. Van Allen climbed a ladder hydraulically raised against the sides of the rocket while German physicist Ernst Stuhlinger scaled a second ladder and unscrewed the door to the nose cone. They rechecked the instrument wiring and battery connections. All the other rocket systems had to be rechecked as well and then launch preparations hit full gear. A favorable weather report cleared the rocket for fueling with kerosene and liquid oxygen. Plumes of gas clouded the rocket as trickles of oxygen escaped to the air. Fueling took ninety minutes and nearly tripled the 9-ton weight of the rocket and payload.

At about 2 P.M., forty-seven minutes prior to launch, the German team, the scientists, and army personnel headed into a cement block bunker with 10-foot thick walls and a pyramid-shaped roof designed to withstand a direct hit by a wayward rocket. The scientists gathered at two slits in the walls to watch

"The Earth from 65 Miles Up," title of the first photo of Earth taken from space with an APL 35mm motion-picture camera developed by APL engineer Clyde Holliday and launched in a V-2 rocket on October 24, 1946. Courtesy of the Applied Physics Laboratory of Johns Hopkins University.

the launch 350 feet away. At 2:15 P.M., base personnel closed off the proving ground with roadblocks and cleared roads and parking areas of vehicles. Ten minutes before firing, a red flair that could be seen throughout the area gave warning of the launch. Two minutes before firing, the post siren blared a final alarm. Then Van Allen heard the countdown from behind him in the blockhouse and, at 2:47 P.M., German electrician Ian Werner Rosinski fired the rocket. The V-2 blasted into the air, rising to a maximum speed of 5,600 feet per second with a fire bolt of exhaust cutting through the sky. Inside the nose cone, APL's counters streaked toward a new scientific horizon.

Then, just seconds later, the rocket veered off course on a trajectory that targeted it beyond the proving ground. Wernher von Braun, mastermind of the V-2, aborted the flight from the bunker with a radio-triggered fuel cutoff. Nineteen seconds after takeoff, the rocket tumbled to the ground, exploding into a crater of crushed metal, melted instruments, and blazing fuel. Von Braun and his team had seen the misfires before—dozens of times. But the forced crash was a devastating blow for the APL physicists who watched months of work go up in smoke. They had built a new generation of experiments hearty enough to stand up to a rocket launch. Nothing they constructed could survive the crash. Most physicists and astronomers steered clear of such unpredictable operations but Van Allen and the handful of mavericks at White Sands shrugged off the disappointments and just built more instruments to take the first steps into space exploration.

• • •

"Most people say space exploration began on October 4, 1957, with the launch of *Sputnik* but it's much earlier, from my point of view. It started at White Sands with the V-2s," Van Allen often said. Little more than a year after the German V-2 rockets terrorized London and Antwerp, the U.S. Army transplanted them to New Mexico for reassembly and testing as part of America's own fledgling missile program. The army's top secret Operation Paperclip brought the Nazi rocket engineers to the United States as well and settled them at Fort Bliss in Texas. The military testing program undertaken at nearby White Sands didn't require warheads in the nose cones, so the army invited scientists to fly cosmic ray detectors and other equipment in them instead.

Scientists from institutions interested in this new mobile laboratory gathered at the Naval Research Laboratory in Washington, D.C., on January 16, 1946, and formed the V-2 Panel, later called the Upper Atmosphere Rocket Research Panel, and later still, the Rocket and Satellite Research Panel. By any name, the all-volunteer panel ran the only civilian space program in the country until Congress created the National Aeronautics and Space Administration (NASA) in 1958. Panel members met monthly to allocate payload space for scientific equipment on the sixty-seven V-2 launches made from 1946 through 1952 and continued the job with subsequent rocket programs.

Van Allen drew directly on his research in nuclear physics, his geophysics training at the Department of Terrestrial Magnetism and his ballistics expertise from World War II to design instruments with his spring-loaded vacuum tube that could stand up to a rocket launch. He designed the simplest and most direct instrument to measure cosmic rays—a single Geiger counter pushed as far forward as possible in the 44-cubic-foot nose cone and instrument compartment of the V-2. The counting rates steadily increased with altitude as cosmic rays splintered in the air and left an increasingly dense trail of debris—the trail of debris that geophysicists had been tracking for years with instruments flown by balloons.

Balloons could soar some 20 miles into the air; the V-2 rocket could top altitudes of 100 miles. Somewhere in that 100 miles, Van Allen hoped to find the primary stream of cosmic rays that hurled toward Earth across billions of miles of space before they disintegrate in the earth's atmosphere.

Rockets date back nearly two thousand years. The Chinese used them in fireworks first until the Grand Khan combined rockets with an earlier invention—gunpowder—to make "fire arrows" in the thirteenth century. But three pioneers—a Russian school teacher, an American inventor, and a German physicist—laid the foundation for modern rocketry, even as improvements in guns and other weapons temporarily pitched the rocket into the military dustbin by the close of World War I.

Rocketry and rocket clubs flourished after the war, although they were often linked to the fantasy of space travel. Russian Konstantin Edvardovich Tsiolkovsky envisioned a way beyond fantasy to achieve space travel, however, even though scarlet fever robbed him of his hearing at age ten and deprived him of any further education at the schools in the tiny village of Ijevskoe where he grew up. He read his way through his father's library and every other book he could find until his parents sent him to Moscow for tutoring at age seventeen. Tsiolkovsky passed the teaching exam and found a position as a high school math teacher in the town of Borovsk in 1880. There he married the daughter of the local preacher, started his family of seven children, and settled into a small home where, inspired by Jules Verne, he sat at his desk each night to spin tales about space travel. Soon the tales evolved into rudimentary equations and drawings until Tsiolkovsky applied the concept of propulsion to space travel in a groundbreaking 1883 manuscript, "Modification of the Force of Gravity."

Tsiolkovsky made the first major advance in rocketry in six hundred years by substituting liquid fuels for gunpowder in hypothetical designs that retooled the rocket as a vehicle instead of merely an explosive device. He published his mathematical plan for rockets in Russia's *Scientific Review* in May 1903, seven months before the Wright brothers successfully flew an airplane at Kitty Hawk. Later, he proposed multistage vehicles in which a huge booster could launch a smaller vehicle into orbit. Tsiolkovsky never had the means to develop a rocket as he continued to create designs now on display in his home in Kaluga, where he lived and taught later in life. Few people knew of his work. But to the Russians who built *Sputnik*, he was a hero. They targeted September 17, 1957, the hundredth anniversary of Tsiolkovsky's birth, for the launch date for *Sputnik*, missing it by 17 days with the actual October 4 launch.

Unaware of Tsiolkovsky and others theorizing on spaceflight, American physicist Robert Goddard began experimenting with liquid-propelled rocket designs in 1911. Students at Clark University in Worcester, Massachusetts, often found their reclusive professor in his office sketching engineering designs for rockets. Goddard spelled out the design for a rocket that could reach the moon in his classic 1919 paper, "A Method of Reaching Extreme Altitudes." He, too, combined practical engineering with visionary dreams, such as the migration of human beings to other solar systems to survive the death of our sun. Newspapers and scientists ridiculed his ideas, aviators hounded him for permission to pilot his future spaceships to the moon, and the state of Massachusetts declared his experiments a public nuisance.

The attention drove the shy and introverted Goddard to later spurn legitimate offers of help and many of his brilliant, patented innovations languished as a result. He found relative isolation to experiment with his rockets at his cousin Effie Ward's farm in Auburn, Massachusetts. In the snow-covered stubble of the fields, he built a simple wooden launch tower and a plywood windshield. He secured a 10-pound liquid-fueled rocket in the tower on the morning of March 16, 1926, ignited it, and watched it soar upward 41 feet in 2.5 seconds, proving the feasibility of spaceflight. But Goddard retrieved the rocket from the vegetable garden 184 feet away and then kept the whole achievement a secret for nearly ten years, always suspicious that others would steal his ideas. Only his wife, his two assistants, and two representatives of the Smithsonian Institution, which provided funding, knew about the brief rocket voyage.

Goddard decided to pursue rocket development full time in the solitary confines of Roswell, New Mexico. The lack of government support of science between the world wars and Goddard's own mistrust of government involvement made progress slow and financial backing precarious. He spurned Millikan's invitation to work cooperatively with Caltech's Guggenheim Aeronautical Laboratory (GALCIT), soon to become the Jet Propulsion Laboratory. By 1940, Goddard had perfected a 10-foot rocket with liquid oxygen–gasoline fuel that generated 900 pounds of thrust and reached altitudes of 9,000 feet. Rockets attracted serious interest as the military sought new weapons for the war effort, and Goddard finally accepted government support in 1942 when the navy made him director of research for rockets within the Bureau of Aeronautics. He continued rocket tests in Roswell and Annapolis, Maryland.

The third pioneer of modern rocketry was German physicist Hermann Oberth, a fan of Verne's science fiction and Goddard's principles of rocketry. He transplanted Goddard's work to Germany at a pivotal time. He provided a blueprint for modern rocket design and spurred German rocket development with his 1923 book, *The Rocket into Planetary Space.* Oberth tackled everything from multistage rockets for human flight to alcohol fuel mixes that burned hot and maximized thrust. He summed up Goddard's groundbreaking 1919 paper in the appendix to his book. German rocket societies mushroomed based on Oberth's book, including the *Verein für Raumschiffahrt* (Society for Space Travel), or VfR. With Oberth's book providing the gospel, the engineers and amateurs in the VfR designed and tested "minimum rockets," the Miraks.

Like Tsiolkovsky, Oberth never developed a rocket but he directly inspired the twentieth century's key crusader of high-performance rockets for spaceflight: Wernher von Braun. Von Braun, the son of a baron, first read Oberth's book when he was twelve years old. He decided he wanted to explore space and never turned back. He joined the *Verein für Raumschiffahrt* in 1929 as an eighteen-year-old engineering student at Berlin's Charlottenberg Institute of Technology. Von Braun's technical flair, good looks, and aristocratic manner caught the attention of everybody within the rocket society, including Oberth, who teamed him up with the core group of experimenters.

Von Braun's growing acumen singled him out once again after he graduated with a degree in engineering in 1932. That summer, Army Captain Walter Dornberger visited the VfR's *Raketenflugplatz,* or "rocketport," near Berlin and spotted von Braun. He offered the young engineer a job as a research assistant to develop military rockets at Kummersdorf. Von Braun accepted, sensing a well-funded route to develop rockets for spaceflight. "Successful firing brought increased military support, and increasingly longer flights required a larger testing range. In 1937, the team moved to a large new facility at Peenemünde, on the Baltic Coast," wrote Susan Enscore in a brief history of the V-2s.

German military interest in the rocket was clear-cut. While the Versailles Treaty, which ended World War I, banned nearly all weapons development in Germany, the ban didn't cover rockets. As the Nazis cemented their grip in Germany, von Braun was ordered to join the Nazi Party in 1937. Then in 1940, an SS colonel visited and asked von Braun, now the director of rocket development, to join the SS. "I told him I was so busy with my rocket work that I had no time to spare for any political activity," von Braun later explained. But Dornberger insisted he join because the SS was trying to get their "fingers in the pie" of the rocket program. Von Braun is known to have worn his SS uniform in public exactly once, though he received honorary advancements to the rank of captain. While von Braun joined the military rocket program because it provided the means to continue his work on spaceflight, he advanced the rocket as a terrifying threat. During World War II, U.S. physicist Lloyd Berkner headed an intelligence team to investigate the possibility that the German rockets could shoot death rays.

Hitler pegged his hopes on the rockets, especially later in the war as the Nazi regime turned rockets into propaganda to bolster Germany's morale as the prospects for victory dimmed. German propaganda minister Joseph Goebbels even renamed the rockets *vergeltungswaffen*, or weapons of vengeance—also known as V-1 and V-2 rockets. The German army built them at the great rocket factory at Peenemünde until the British Royal Air Force targeted the factory along with seventy-three of ninety-six launching sites in 1943. The near destruction of Peenemünde opened the way for blasting a catacomb of tunnels in the mines near Nordhausen to create the underground rocket factory operated by the government-run company Mittelwerk.

Thousands of political prisoners and prisoners of war worked long hours in horrific conditions to carve out the caverns and set up the assembly lines where the 46.5-foot V-2 rockets took shape. An estimated sixty thousand prisoners passed through the Dora complex of concentration camps that served the rocket works. A staggering twenty thousand prisoners died of illness, starvation, and maltreatment during the eighteen-month life of Mittelwerk; about ten thousand deaths were associated with the V-2 production. The death toll meant the V-2 rockets killed twice as many people at Nordhausen where they were made than in London and Antwerp where the majority of them exploded.

The slave labor camps came under renewed scrutiny in the 1980s when the Office of Special Investigations (OSI) initiated a dragnet to find possible Nazi war criminals. On the rocket team, the OSI singled out Arthur Rudolph, Mittelwerk's civilian V-2 production manager and America's project director for the Saturn V rocket program that sent the *Apollo* crews to the moon. Rudolph was a Nazi Party member early on and was investigated by the OSI for alledgedly exploiting slave labor for the wartime rocket program. Rudolph maintained his innocence for the rest of his life but voluntarily gave up U.S. citizenship and returned to Germany rather than face proceedings to strip him of American citizenship by the U.S. Immigration and Naturalization Service. Von Braun had no control over the slave labor system and tried to point out to the SS that improving the health conditions at Mittelwerk would improve the quality of work, according to Stuhlinger. The creation of a virtual army of rocket makers at Nordhausen—half of them paid civilians—explains how Germany could have built so many rockets in so short a time. Attacks on London with eight thousand of the V-1s began in June 1944, keeping Londoners running for shelter as the low-pitched whistle of the rockets cutting through the air won them the nickname "buzz bombs." The slow speed of the V-1 made it an easy target, especially to proximity-fuzed weapons. Only a quarter of the V-1s exploded and most zigzagged hopelessly off target. Missed targets and premature blowups plagued the V-2 as well. But the V-2 evaded ground weapons with terrifying speed of up to 3,800 miles per hour. As the war effort became more desperate, Hitler pinned the redemption of the Reich on the rockets. The SS sought von Braun's support in snatching the rocket program from army supervision. Von Braun resisted and the SS retaliated. Gestapo officers arrested him, imprisoned him, and charged him with sabotage.

"He was not really interested in war rockets, but was working on space exploration; he was opposed to the use of V-2s against England and he was about to escape to Britain in a small plane," von Braun and coauthor, Frederick Ordway III, wrote of the SS charges against him. Dornberger, pleading that the whole rocket program would die without von Braun, got an order from the Führer's headquarters for his friend's release. In the end, however, von Braun's lukewarm commitment to the military application of rockets early on cost the Nazis dearly.

"Had the V-1 and its even more formidable successor, the V-2 liquid-fuel rocket, been developed earlier in the war, the balance of power might have been fatally tilted against the Allies," concludes Doris Kearns Goodwin in her Pulitzer Prize–winning history, *No Ordinary Time.* Von Braun's most ardent critics, as well as his equally ardent supporters, agree he was apolitical, determined to build rockets to explore outer space even if it meant selling the program based on its potential as a weapon.

As the Allies tightened the noose around Germany in early spring 1945, Americans came secretly to the rocket works a few weeks prior to the fall of the Reich and offered asylum to the rocket specialists. With the fall imminent, von Braun gave his men a choice. They could surrender to the Russians closing in on the rocket complex from the east or be "captured" by the Americans.

Von Braun sorted through conflicting orders from Berlin telling him to stay put at Nordhausen and to evacuate by turns. He picked out the latest evacuation orders and labeled a caravan of trucks and cars for an imaginary rocket mission to bring his rocket scientists and engineers, their families, and a vast archive of records to the West in the dead of night. When the Americans moved in on the rocket plant later, they found nearly one hundred rockets gleaming in various stages of completion beneath bright lights with tools still on the ground in a surreal ghost town. The Americans packed up the rockets and parts, located a hidden cache of documents in the mines, and commandeered every vehicle they could find to retreat west just a few days ahead of the

Russians in an area designated for Russian occupation. The rocket engineers who had stayed behind headed to Russia as consultants. Now, in the first volley of the Cold War, both sides had the same rocket and members of the same rocket program in the race to develop ballistic missiles.

Richard Porter of General Electric, contracted by the army to develop guided missiles, nuclear physicist Ernst Krause of the Naval Research Laboratory, and Army Ordnance officers pieced together Operation Paperclip, the code name for a program that brought nearly 650 German and Austrian scientists to the United States. The program included the 118 rocket scientists initially transferred to Fort Bliss, Texas, from November 1945 to February 1946. The army promised protection for their families in Germany as sixteen liberty ships carried the remnants of the dreaded "weapons of vengeance" to Boston with tons of engineering and other documents.

The cache of rockets and documents from Germany filled three hundred freight cars for transfer to White Sands. The Allied troops had shipped captured V-2 rocket components to the United States earlier in the year, and the military turned to Goddard to analyze them. He realized that his own patents covered many of the components he found in the rockets. The Germans didn't have the patents but they did have Goddard's seminal thinking from Oberth's book. One German engineer questioned in 1945 reportedly said, "You have a man in your country who knows all about rockets, and from whom we got our ideas—Robert H. Goddard." With almost Shakespearian staging, the V-2 rockets came to a site just 100 miles from Goddard's Roswell station and arrived there within a few weeks of his death on August 10, 1945.

The Germans arrived shortly after the rockets and the army confined the "Peenemünders" under house arrest to a 6-acre area of Fort Bliss. The Germans converted an area of the officers' quarters into a clubhouse with a bar, reading room, and game room. They even gave it a little touch of Europe, tapping a water main running through the area to cultivate gardens and vines across a trellis at the front the clubhouse, Stuhlinger said. They constructed their own workshops at the Fort Bliss camp to conduct rocket research in the primitive facility.

The fort opened the bowling alley, pool, and movie theater to the German rocket group and the drill grounds around the barracks provided space for evening soccer games. But they remained isolated from outsiders most of the time. "We had no one to practice English with except each other, so most of us had lifelong accents," Stuhlinger said. The army did arrange Saturday bus outings to El Paso for shopping and movies and to tourist stops such as the Billy the Kid bar and museum in Lincoln, New Mexico. Major James Hamill, who coordinated the rocket specialists, looked the other way when they utilized a hole in the fence to take long walks in the desert around White Sands. Best of all, after years of short rations, the first reliable supply of food made Fort Bliss seem like paradise, Stuhlinger recalled. Texas steaks became an instant favorite.

• • •

Work got underway almost immediately at the newly established White Sands Proving Ground to reconstruct and test the V-2s for America's fledgling Hermes missile program. The army recovered portions of one hundred rockets but only two could be assembled completely from parts on hand. General Electric was contracted to engineer new parts for the rest. Meanwhile at NRL, Krause moved forward with a crazy idea he had as he helped confiscate the V-2s. He proposed filling the empty warheads with scientific instruments, a practical alternative to filling them with sandbags to stabilize the rockets for testing. Colonel (later General) Holger N. Toftoy and Army Ordnance backed the idea. NRL's Milt Rosen designed a new nose cone to enclose 1,000 pounds of scientific payloads, more than fifty times the payload capacity of the early American satellites, and Krause spread the word about the rocket program to colleagues such as Henry Porter, assistant director at the Applied Physics Laboratory.

Porter, an affable administrator with a business background in Milwaukee real estate, sensed a great opportunity as he advised Tuve to join the new program. Tuve offered the idea to his staff but few people shared Porter's enthusiasm. Only Van Allen and a few others expressed interest and Tuve gave them a free hand in undertaking the project. Van Allen headed the team—fondly nicknamed the "five percenters" because five percent of APL's budget went to the rocket program. Guided missile development now dominated APL, and few APL scientists wanted to abandon this influential new domain to research the upper atmosphere.

Porter's insider tip sealed Van Allen's fate, joining him with a handful of scientists from across the country who crossed the boundary through the earth's upper atmosphere into space. Krause greeted them all at NRL headquarters for the January 1946 meeting where the V-2 Panel was created. It had no official status, budget, or staff, yet "every agency in the United States taking part in upper atmosphere rocket research [was soon] either directly or indirectly represented on the panel," wrote NRL's Homer Newell, a panel member. The scientists met monthly at rotating sites where the panel screened experiments, scheduled them, and shared ideas for making instruments.

Van Allen described the panel as a "closed fraternity" of scientists willing to trade away the security of earthbound laboratories for altitude—high altitude. The panel offered a freewheeling flow of ideas, contacts, and tips for making instruments. Oftentimes the laboratory selected to provide the payload for the flight was the laboratory with instruments closest to completion, as the scientists worked frantically to keep pace with the launch dates.

"I think the panel came to feel that the V-2s belonged to them, as far as scientific use was concerned. All of us had miscellaneous professional colleagues around the country, and the whole matter was conducted on a handshake basis. If somebody who was a friend of yours wanted to do something, and it sounded to you like a good idea, we would work it in," said Van Allen.

The panel met for the first time on February 27, 1946, at Princeton University. The founding members in attendance included Van Allen, still a naval reserve officer, Krause, Charles Green, K. H. Kingdon and George Megerian of General Electric, Myron Nichols of Princeton, Fred Whipple of the Harvard College Observatory, William Dow of the University of Michigan, Marcel J. Golay of the Army Signal Corps Engineering Laboratory, Marcus O'Day of the Air Force Cambridge Research Laboratory, and Newbern Smith of the National Bureau of Standards. The group elected Krause as chairman. Megerian volunteered as secretary and took the minutes.

The real question the panel confronted at that first meeting was who could have an instrument up and running for the first planned launch of a V-2 in mid-April, less than two months away. Van Allen reported that APL had an instrument underway. He and a handful of colleagues cobbled together techniques from nuclear physics labs and previous balloon and ballistics experiments to configure detectors, circuitry, and battery-power supplies to fit the warhead. The scientists included Lorence Fraser, Harold Clearman, Russell Ostrander, Clyde Holliday, John Hopfield, and James Jenkins.

Van Allen also drew inspiration from the top researchers at the University of Chicago and University of Minnesota who were using innovative instruments in a continuing exploration of cosmic rays with balloons. But Geiger tube detectors, developed for a graceful ride in a balloon, provided only rudimentary models for devices Van Allen would need to withstand a blast into space. He designed the sturdier, tubular instruments that he would refine and miniaturize for years to come. They consisted of Geiger counters to detect cosmic rays wired to a battery-powered circuitry of receivers, relays, filters, pulse generators, and a transmitter to radio the data to the ground station.

• • •

The April crash gave the scientists a new and sobering perspective on the price tag set for discoveries in space. Van Allen returned to APL, built new sets of instruments and headed back to White Sands again and again as the research program settled into a routine. After the jolting train rides, Van Allen tuned up his experiments with tests for operation, calibration, and voltage. He placed the instrument assembly on the shake table where vibrations simulated those of a launch. After days of shakedown and testing, the rocket team attached the nose cone full of instruments to the rocket.

The scientists shared a lab in the hangar where the German rocket engineers worked with General Electric engineers to assemble the V-2s and make new parts under GE's contract. NRL set up a makeshift machine shop in a trailer. Von Braun supervised operations, showing up at the hangar with army officers and three or four members of his own team. He clipped his orders to a few words in German. Charismatic and authoritarian, he moved among his army captors with an air of complete command. The American scientists knew him by reputation but kept their distance from him and his specialists. The Germans did the same, both sides uncomfortable with the awkward partnership.

"At the end of the day, they would go back to their quarters [at Fort Bliss] and we would do whatever it was we needed to do," Van Allen recalled. Mostly, the scientists worked on their instruments, with a poker game or occasional trip to a nearby cantina breaking the routine.

After the April mishap, the rockets performed flawlessly for two launches staged for the press in May. Some 50,000 pounds of thrust drove each rocket upward and out of sight to a maximum speed of about 3,800 miles per hour and an altitude of about 100 miles. The rockets won feature articles in *Popular Science, Aviation,* and other publications across the country.

"Weekly trips to higher altitudes than men have ever ascended are being made now by German V-2s from the Army's White Sands Proving Ground in New Mexico. The 46-foot, 26,000-pound rockets are being fired to bring America's knowledge of long-range, guided flaming arrows up to date and to answer the physicists'questions about the unexplored outer fringes of the earth's atmosphere," *Popular Science* reported, with step-by-step photos of the launch.

Applied Physics Laboratory got another opportunity to fly instruments in May but now that the rockets worked, the instruments failed. All the rocket team could do was build more and remain philosophical. "The immense opportunity for finally being able to make scientific observations through and above the atmosphere of the earth drove us to heroic measures and into a new style of research, very different from the laboratory type in which many of us had been trained," Van Allen wrote, reflecting back.

Naval Research Laboratory, APL, the University of Michigan, and a handful of other institutions continued to perfect the experimental style that beat the odds in space. Princeton dropped out after a series of mishaps with flight experiments in 1946. The instruments had to work, the battery-power systems had to work, the rockets had to work—all in perfect harmony. NRL perfected the radio telemetering instrumentation that allowed direct communication with the rocket and the ground. But anything scientists needed returned, such as film, required protective armor that could survive the ultimate fall of the rocket back to the ground.

The possibilities for disaster sometimes appeared overwhelming. A hasty rewiring of a faulty connection just before launch resulted in a botched switch control that turned all the instruments on board off instead of on in one case. "A post-flight review showed that there were several ways in which the switch could have been connected to do the job intended and only one way in which it could fail. The one and only wrong way had been chosen, an important object lesson concerning hasty, last minute changes," Newell wrote.

Another time, a physicist failed to remove the lens cap from a recording camera on a near flawless flight. It's not that such oversights didn't occur in experimentation on the ground. Newell pointed to these examples to underscore the unforgiving nature of rocket research and the fortitude it took to swallow hard—and wait for the next flight. In addition to scheduling missions, the panel members offered each other sympathetic ears and a clearinghouse for dos and don'ts as members laid the foundation for space science and competed to make groundbreaking discoveries.

Then, as the panel hit high gear late in 1946, Ernst Krause announced he was leaving NRL to take a job in private industry at Ford Aeronautics in California. His departure meant the panel needed a new chairman. Even in the friendly ranks of the panel, some politics now came into play. Newell wanted the chairman's post but that meant handing another wheel of control to NRL, already a powerful presence at White Sands, where the lab ran the telemetry and command systems. Panel members preferred a chairman from a more neutral institution and recruited Van Allen by unanimous consent. Van Allen had earned the respect of everybody on the panel with a low-key management style and a high-key instinct for designing instrumentation that worked in a rocket. Van Allen led the panel for the next eleven years, until the formation of NASA. He took over in January 1947, as all the pieces finally fell into place for the V-2 experimenters. The telemetry system hit full operation that January and the rocket engineers improved instrument retrieval by blowing off the nose cones with radio-ignited grenades at approximately 330 seconds into the flight. The rocket was blown up as well so that it fell back to Earth in scattered bits rather than as a smoldering carcass that might collide with the instruments.

Panel members were inventing instrumentation and procedures as they went along. They borrowed from each other to develop experiments that measured cosmic rays, the sun's ultraviolet light, and the earth's magnetism and they competed with each other to fly the first instruments that might make a new discovery. Naval Research Laboratory won the prize where solar research was concerned. Fifty-one observatories across the country were engaged in solar research, but NRL identified the spectrum of ultraviolet light for the first time in October 1946, with a V-2.

Van Allen made his first discovery in space with the thirtieth flight of a V-2 rocket on July 29, 1947, and a perfect choreography of the flight, instrumentation, and data retrieval. He searched for the incoming stream of cosmic rays before they pierced the ceiling in the earth's atmosphere where they began to disintegrate. He pushed a specially designed Geiger tube into a custom-built, spiked extension of the tip of the nose cone to avoid contamination by showers of particles produced by the heavy boiler-plate body of the V-2. A bundle of other Geiger tubes, configured lower in the nose cone, checked for particles coming off the body, and still more bundles measured cosmic rays from different angles.

The rocket soared past 100,000 feet, quickly eclipsing Millikan's balloon altitude record. The rocket rose to 200,000 feet, then 300,000. Van Allen could hear the changing chorus of beeps—radio transmissions from the rocket that confirmed a range of data collection. The arm of a bar recorder rose up and down with the pitch of the beeps and transcribed the data on paper tapes. The counts rose rapidly and then began to drop off. He could tell that an odd dip in the burst of currents from APL's ionization chambers matched the dip in counts from his Geiger counter on the flight. While the Geiger counter detected particles, the ionization chambers generated an electrical current from the particles found. The intensity of the current steadily increased and decreased with the counts of the particles and then hit the same plateau.

"Dip at about 100 seconds has appeared before in a counter? May be real!" Van Allen wrote in his logbook for the flight. Van Allen took the data back to APL to begin the arduous task of reducing the troughs and slopes on the paper tapes into particle counts. He spent weeks charting the hand-calculated numerical entries and hand-drawn graphs to record the counts. When he completed the work, the pattern was clear. The counts rose rapidly, confirming Millikan's maximum intensity at 12.6 miles as cosmic rays disintegrated in the atmosphere into a shower of secondary particles. But the counts fell off after that until the rocket touched about 31 miles above the earth. Here, the counts confirmed the plateau.

He realized he had encountered a phenomenon that showed up on several data records. The plateau was indeed real. It marked the ceiling of the atmosphere for incoming cosmic rays. Above it, the particles hurled toward Earth in a continuous stream of energy. Just below it, the counts rose with bursts of subatomic reactions in the first volatile collisions with the earth's atmosphere. Van Allen had pierced the uppermost shield of the atmosphere before it splintered cosmic rays into subatomic debris. His achievement proved with direct evidence what scientists had surmised for decades—that Earth orbited through a storm of cosmic rays. The tumultuous Milky Way galaxy hurled them across billions of miles of space but 31 miles of sky successfully scrubbed away all but a trickle of them.

In addition, Van Allen validated previous findings over the composition of cosmic rays. The dramatic bursts of particles that bombarded the atmosphere left the fingerprints of nuclear reactions picked up by Van Allen's detectors. Nuclear reactions meant that protons, not electrons, made up the primary stream of cosmic rays.

Balloon research continued and, within the year, a team of scientists with pioneering instruments on an unmanned balloon discovered more components of primary cosmic rays at about 100,000 feet. They identified the nuclei of atoms of helium and heavier elements that included carbon, calcium, and silver. The trail the cosmic rays left through stacks of special photographic plates allowed scientists from the University of Rochester and the University of Minnesota to make the discovery.

But rockets had established their research potential. Van Allen's discovery of the plateau and follow-up discoveries about cosmic rays made by the NRL team reflected the new scientific horizons trailblazed by the first space scientists on the sixty-six V-2s launched from 1946–1951. The rockets reached altitudes of more than 100 miles above the earth. In all, the V-2s carried 223 experiments into space including thirty-eight that measured cosmic rays, thirty-two that made solar observations, twenty-six that surveyed the ionosphere, twenty-five that studied the temperature of the upper atmosphere, twenty-five that measured atmospheric pressure, and nineteen that studied the composition of the atmosphere.

Scientists reaped new information about weather currents and the atmosphere from the flights as well. Everyone knew that layers of the ionosphere mirrored back radio waves, knowledge that had led to the invention of long-distance radio transmission. But now, scientists reached and identified these layers for the first time. They confirmed the hypotheses about a protective ozone layer of the earth that absorbed ultraviolet light. They developed the first round of instruments and telemetering technology critical to the fledgling space programs. Clyde Holliday of APL loaded a modified and heavily armored movie camera, adapted from a navy reconnaissance camera, into the V-2 and created a sensation with the first black-and-white photographs that showed Earth as seen from space. Newspapers and television stations across the country brought the images to millions of readers and viewers.

"The news we had last night about the rocket that photographed the earth from a height of 65 miles is followed tonight by the actual pictures," reported Lowell Thomas on *NBC News* the evening of November 21, 1946. The pictures captured 40,000 square miles of land beneath wisps of clouds—a perspective that conveyed a magnificent yet humbling firsthand sense of Earth as a fragile planet orbiting in the vastness of space. *Life* magazine showed a map of the area, from Los Angeles past Kansas City, encompassed by the amazing photographs. "The camera, encased in one-inch walls of armor steel, is the first to survive a V-2 rocket ride," shooting 50 feet of film for four-and-a-half minutes, the *Washington Post* reported. Despite fears about wayward V-2s and threats they might pose to civilians, only one rocket escaped the proving ground. It crashed near a cemetery in the small Mexican village of Juárez, just over the border. Army officials raced to the cemetery, ready to offer reparations and divert an international incident. Instead, they found local entrepreneurs busy selling pieces of the wreckage as souvenirs by the time they arrived.

The army's initial plan called for the firing of twenty-five V-2s over the spring and summer of 1946—before the initial six-month contract with the German rocket team expired. As the flights continued, it was obvious the team should stay on. The army extended the contract, first for another six months and then for another five years. Life changed rapidly for the German personnel with the contract extensions. They moved into converted wards of the William Beaumont General Hospital Annex at Fort Bliss in October 1946. The annex offered more spacious living quarters and improved lab and office space. In November, the army lifted the curtain of secrecy surrounding the Germans and capitalized on the public relations potential of the team instead.

Still, the social divide persisted between the American scientists and the German rocket team. The scientists barely talked to von Braun, even on a professional level, but the courtly and ever helpful Stuhlinger crossed the divide. "He had a kindred, professional, scientific interest and was extremely pleasant and helpful," Van Allen said. "He was almost our ombudsman between the investigators and the rocket engineers there. The rocket engineers thought we were a terrible pain in the neck to have to cope with and Stuhlinger was one of those who smoothed the way."

"I knew about some of Dr. Van Allen's work. We could get past our differences with the common language of physics and the common interest in cosmic rays," said Stuhlinger, who had helped develop the Geiger counters for cosmic ray research particles as a student of Hans Geiger in the 1930s. "Everybody whom Professor Geiger took under his wing at the time worked on cosmic rays. When Geiger was very sick [he died in 1945], I visited him in the hospital. He was in great pain but I told him about the rockets and our plans to take them into outer space. 'Do you think you could put a Geiger counter in a rocket and take it into space?' he asked. I promised him we would do that." Stuhlinger never got the chance with the war production of the V-2s. Now Van Allen took him one step closer to fulfilling his promise.

The divide existed as well between the dedicated corps of scientists on the V-2 Panel and other researchers. The hazards of losing equipment in a failed launch, the short flight times for gathering data, and the unpredictable altitudes made respected balloon scientists such as Bruno Rossi of the Massachusetts Institute of Technology (MIT) skeptical of rocket research. Few astronomers expressed any interest in the rocket program and, with little experience loading instruments into rockets, one-time trials often turned novice experimenters into critics.

Astronomer Jesse Greenstein of the University of Chicago's Yerkes Observatory made a camera with a rotating shutter to photograph the ultraviolet spectrum of the sun. Greenstein proposed the instrument to APL and they contracted with him to develop it and fly it in one of the APL-assigned V-2 flights. The film never advanced in the holder due to a mechanical failure during a V-2 flight in February 1947. The security clearances, the secrecy, and the lack of institutional support for an "outsider" at White Sands only added to Greenstein's disappointment and frustration with the fraternity on the panel. Van Allen apologized for not helping him more and urged him on in a letter, assuming he would learn from the hard knocks inherent in rocket research and try again. Greenstein continued as a consultant on APL instruments but got back to his "real" work after the failure and didn't pursue further research with rockets.

Greenstein wasn't the only detractor. J. Allen Hynek, later famous as the UFO expert at Northwestern University, blamed his former buddies on the APL team for mishandling and fogging his film when he added his experiment to an APL payload. University of Chicago physicist Mel Gottlieb, who later came to Iowa to work with Van Allen, complained that the army's attitude at White Sands resulted in inequities and blatant hostilities.

In general, Van Allen had only a few requests by scientists outside the panel to load instruments in a V-2. Part of the reason, he admitted, was that the inner circle of the rocket panel considered the V-2s their own. The panel members gained the experience to overcome the obstacles of rocket research and get results unobtainable in any other way. The results won more proponents than critics, and the panel provided a clearinghouse for ideas to promote research rockets and rocket design.

The V-2s upstaged Jet Propulsion Laboratory's Corporal rockets that had a similar size and functioned as a surface-to-surface missile. Despite flight failures of the rocket overall, many of the innovations in safe, reliable rocket engines and in precision guidance systems can be traced back to the early rocket program at the Jet Propulsion Laboratory. The V-2 also pushed aside a research rocket called the WAC-Corporal. JPL rocket engineers Frank Malina and Homer Joe Stewart designed the WAC under an army contract and launched this rocket from White Sands as well. WAC stood for Women's Auxiliary Corps, a name suited to the fact that the WAC was considered the "little sister" to the larger Corporal military rocket. Officially, WAC stood for "Without Altitude Control," meaning without control of the angular trajectory of the flight pattern. Development of it lapsed soon after the higher-flying V-2s arrived on the scene. But the army retooled the WAC as the upper stage of a two-stage rocket using the V-2 as a booster and proceeded with the first successful tests of multistage rocketry.

More importantly, the WAC provided a starting point for the design for a lower-cost, high-powered research rocket called the Aerobee, Van Allen's brainchild. These were the "good old days" of space science, when projects took shape quickly with a wartime can-do spirit and little administrative bureaucracy, according to Van Allen.

7 The Mighty Little Aerobee

White Sands Proving Ground, New Mexico—May 1947. The U.S. Navy leased space at White Sands and began construction of a 150-foot launching tower that May for a streamlined new research rocket, the Aerobee. The rocket stood 19 feet high, compared to the 46.5-foot V-2. It carried 150 pounds of instruments instead of more than 1,000 pounds. It was Van Allen's baby, "the realization of the dream that Professor Robert H. Goddard had when he began his pioneering work on rockets," praised Homer Newell.

* * *

But White Sands Commanding Officer Colonel Harold Turner halted construction of the Aerobee tower on June 19 and called in Van Allen for a stormy session on rocket safety.

Unlike the V-2, the Aerobee had no guidance system. How did the U.S. Navy plan to keep it on the range? Turner wanted to know, especially in the wake of the misfire that sent a V-2 crashing into a cemetery in Juárez, Mexico.

"It's a fin-stabilized rocket," said Van Allen, going over the design that featured four wide fins strutting like propellers from the base of the rocket.

"It's a hazard to the neighboring communities," Turner countered. And he ordered the tower to come down. Van Allen stalled the teardown by garnering Turner's permission to get another opinion about the danger. He turned to the Aberdeen Proving Ground and Bob Kent, a ballistics engineer colleague who had helped test the proximity fuze. Kent looked over the Aerobee design and offered a simple solution.

"He said that all you had to do was have an adjustable tower to counteract the wind on the direction of flight. He worked with me on a set of wind tables based on the prevailing winds that blow at White Sands," Van Allen said. Those winds blew from the southwest, so Kent and Van Allen rigged a three-legged tower with a hinge and a jack-screw that would tilt the rocket into the wind. The harder the wind blew, the more the rocket would tilt into it, essentially allowing the wind to push the rocket upright upon launch. Van Allen also designed a sky screen with a wire grid that could be used to visually track the rocket's course from points to the south and east of the launch site. If the Aerobee veered off course, a radio command would destroy it. The simple approach promised to provide the necessary safeguard and Turner gave permission for the construction of the tower to proceed.

While the Navy Bureau of Ordnance funded the compact Aerobee, the Naval Research Laboratory developed the Viking to carry on research in a rocket near in size to the V-2. Both the Viking and Aerobee research programs tooled up in 1947. It cost $450,000 to build and launch a single Viking rocket compared to $25,000 for the compact Aerobee. By 1950, the army, the navy, and the air force were all using Van Allen's little "work horse" rocket. It spawned several variants, such as the Aerobee-Hi with a 168-mile altitude peak. The Aerobee achieved more than a thousand research flights during several decades of service.

In pressing for the new research rocket early in 1946, Van Allen exhibited a hallmark trait of his involvement in the space program: he planned ahead for the next program almost as soon as the current one got off the ground. As the V-2 hit its stride with an average of two launches a month in 1946 and 1947, APL and the other laboratory teams worked feverishly to complete instrument packages one step ahead of the next firing date. But just beyond the frantic pace, Van Allen saw the quick demise of the fledgling space science program while it depended on the limited supply of V-2s and the cost of shooting them.

Van Allen approached Tuve with the idea of making a small, affordable rocket just for research, and Tuve suggested he survey the branches of the military and universities about interest in sustaining upper atmosphere research. Van Allen fielded his network of rocket panel contacts to garner support for his project. Military contacts also expressed interest, viewing the compact Aerobee as a way to develop liquid-fuel propulsion technology for artillery weapons. As a result, the Navy Bureau of Ordnance contracted APL to supervise development of a research rocket that could deliver 150 pounds of payload to an altitude of 76 miles. Van Allen headed the group at APL that identified the relatively compact size, thrust, and other parameters for performance. The navy awarded the design and production contracts to Aerojet Engineering in Azusa, California, and Douglas Aircraft Company in Santa Monica in May 1946. Now, the rocket needed a name and Van Allen suggested "Aerobee," a tip of the hat to both Aerojet and APL's Bumblebee program developing surface-to-air missiles.

Weathered design diagrams and blueprints in Box Number 12 of the Van Allen Archive at the University of Iowa's Special Collections show the rocket taking shape over a mere eighteen months. The original Aerobee measured 19 feet and weighed only 1,100 pounds, an enlarged and improved version of the WAC-Corporal. Though only three feet longer than the WAC, the new rocket packed six times more payload and launched to altitudes of more than 75 miles, rather than 44. The Aerobee's secret? Lots of extra thrust packed in a two-stage rocket design. Viewed immediately after launch, the Aerobee and its booster looked like a rocket with two sets of fins. More than 21,000 pounds of solid propellant thrust gave an initial 0.2-mile boost to the rocket. Then a liquid-engine propellant of red-fuming nitric acid, aniline, and furfuryl alcohol sustained power. The mixture sounds potent, and it was—a fuel that gave the Aerobee a maximum velocity of 2,800 miles per hour.

Van Allen also customized the Aerobee to correct research problems the White Sands scientists encountered. The heavy-steel warhead of the V-2 made it a "dirty" rocket, littering cosmic ray data with maddening showers of particles from the body of the rocket itself. The Aerobee body was made of stainless steel but the nose cone was crafted of light-weight aluminum, a "clean" metal in terms of gathering magnetic and cosmic ray data. Van Allen also made the decision to simplify manufacture and cost of the Aerobee by eliminating any guidance system to keep it in a stable, vertical trajectory during flight. Douglas Aircraft designed stability into the rocket with four wide, squat aluminum fins.

The navy began construction of the launching tower at White Sands as assembly of the first Aerobees neared completion. Then construction halted in the tug-of-war between the army and the navy over rocket stability.

"The stop order caused a substantial delay in the program, but its over-all effect was very beneficial since it forced the development of full-fledged techniques for the reliable control of the trajectories of Aerobees and for the reliable pre-firing prediction of their impact. [I] immediately prepared a detailed draft of a range-safety doctrine for the firing of Aerobees," Van Allen stated in an early guide to research rockets.

Van Allen redesigned the tower and designed the safe-harbor sky screens mounted south and east of the launch site. A radio-triggered fuel cutoff, similar to the ones used for the V-2s, was added to the rocket to terminate the flight if the screens showed it veering beyond the safety range. The three-pronged safety system of the fins, the tower, and the sky screen convinced army brass to allow the tower construction to resume.

Aeronautics Physics Laboratory and Aerojet tested the first Aerobee rocket at White Sands in September 1947. The rocket launched without a payload for this test and two subsequent test flights in October. Van Allen began loading cosmic ray detectors into the rockets on November 24, 1947, reinforcing the results he had recorded with the July V-2 flight. The rocket reached 34.8 miles in altitude, just above the ceiling of the cosmic ray plateau, before the command fuel cutoff aborted the flight as it headed off course. Aerobees reached summits of over 75 miles on subsequent flights, reliably measuring primary cosmic rays, solar radiation, and magnetic fields. The rocket was a resounding success. "The least controversial and most coveted vehicle for upper atmospheric research was the Aerobee," stated space historian David De Vorkin. "The Aerobee was the sounding rocket of choice."

The Aerobee testing dovetailed with what Van Allen called the most-important discovery of his life—fatherhood. Abbie Van Allen gave birth to a baby girl on January 28, 1947. Cynthia Van Allen weighted 6 pounds, 6 ounces. The warmth and love of that moment of holding his baby daughter for the first time radiated through his voice fifty-five years later. "That little life, those tiny fingers curling around my hand—I'm not a particularly religious man but a baby really makes you think about the miracle of life," Van Allen said.

With a child to consider, the couple bought their first home with a low-interest VA loan. Van Allen planted maple trees to shade the two-story brick Georgian house with a white picket fence that still stands at 1105 Meurilee Lane in Silver Spring, Maryland, near APL. Abbie decorated and refurbished the house, room by room.

Jim continued the frequent trips for launches at White Sands. The young family split summers between Abbie's family in Southampton and Jim's family in Mount Pleasant. George and Winnie and their growing family lived next door to Alma, who enjoyed cooking big Sunday dinners for everyone. The families joined together for picnics and outings. The heat and isolation of the small Iowa town oppressed Abbie, especially when Jim was away at White Sands. But one pleasure included conspiring with Winnie to take their mother-in-law on shopping trips. "We urged her to buy a few luxuries she had never allowed herself," Abbie Van Allen said.

• • •

As the V-2 launches slowed down to fewer than one a month in 1949, Van Allen shifted more experiments to the Aerobee, locating compact parts and streamlining electronics to fit in a smaller rocket. All the research teams participating in the V-2 launches supplemented their testing with the Aerobee flights by then, and the V-2 Rocket Panel, now the Upper Atmosphere Rocket Research Panel (UARRP), scheduled experiments for both sets of launches. But the workhorse research rocket doubled as a military test vehicle where different mixes of liquid fuels could be tried and evaluated. Both Van Allen and APL capitalized on the dual role the Aerobee offered. For Van Allen, it was a lesson well learned. The opportunities of dual-role space missions provided a critical and continuing network of support when he returned to the University of Iowa.

Van Allen also learned the art of salesmanship, since the ballistics researchers at APL often dismissed the rocket programs involving space science. Van Allen and APL physicist Fred Singer, a new member of the rocket team with an instinct for public relations, promoted the group's work tenaciously at Wednesday morning staff seminars. Singer emphasized the pioneering work in learning to instrument rockets. Van Allen stressed the importance of rockets in reaching the "huge nuclear physics laboratory" offered by primary cosmic rays—a laboratory that even the latest particle accelerators on earth couldn't come close to matching.

Tuve made personal efforts to validate the rocket team as well. In a research report for 1946, Tuve justified APL's high altitude research as "frankly intended to be one of the navy's contributions to pioneering physics, without regard to practical end results. However, we are keeping our eye on the matter of very energetic particles carrying energies much greater than those involved in the fission of uranium." But the "five percenters" couldn't resist poking fun at some of the efforts to connect scientific research in the upper atmosphere to nuclear energy. They printed a mock newspaper with a bold headline and subhead that read: "V-2 Helps Army Swell Stockpiles of Cosmic Rays: Rocket Wrests Rays from Skies, U.S. Soon May Possess Enough for Cosmic Bomb."

Debates about the scientific data itself posed a bigger concern. Not everyone accepted the discovery of the cosmic ray plateau. Van Allen decided to take the Aerobees on the road or, more precisely, on the high seas, to prove his point with shipboard launches. He would revisit Millikan's cosmic ray latitude surveys made in the 1930s with instruments carried in his balloons. Millikan's survey "left the nagging question" of whether cosmic rays would show the same pattern of rising intensity at altitudes higher still. Van Allen hoped that by locating the cosmic ray plateau at a consistent ceiling at different places above the earth, he would rebuff the skeptics, including MIT physicist Bruno Rossi. As one of the first scientists to apply Geiger counters to cosmic ray measurements in the 1930s, Rossi's word carried a lot of weight. Rossi referred to the misleading readings that could be obtained with Geiger counters as "a kind of witchcraft" and suggested that Van Allen's data was compromised further by particle readings off the body of the V-2 rocket. However, leading cosmic ray researcher Marcel Schein at the University of

Chicago was among those who embraced Van Allen's work and gave active encouragement and advice.

Van Allen's navy contacts got him and the Aerobees passage on the USS *Norton Sound,* a navy seaplane tender that became the first ship outfitted after the war to fire military missiles from a broad afterdeck. Rear Admiral A. G. Noble supported the Aerobee launches for further study of the upper atmosphere but also with the ulterior motive of additional training for his crew in handling and launching "rocket-type missiles." The navy had set the precedent for shipboard launches with a V-2 fired from the carrier *Midway* in September 1947. Then it began testing new military missiles from shipboard, since the wide stretches of open sea provided an even better safety zone than a proving ground.

The *Norton Sound* departed from Port Hueneme, California, on March 2, 1949, with Van Allen and Singer among the APL staff members, along with enough Aerobees, detectors, and replacement parts for three flights. The entourage included the destroyers USS *Richard B. Anderson* and USS *Agerholm.* Even Cold War paranoia didn't necessitate such protection but the destroyers had the telemetering apparatus to receive transmissions of data from the detectors. Five technicians from the New Mexico College of Agriculture and Mechanic Arts signed on to operate the telemetering equipment. Aerojet engineers came along to assist with the launches, and representatives of the Naval Ordnance Laboratory and the army's Signal Corps joined the excursion as observers.

The *Norton Sound* sailed along the Pacific coastline of South America. Aerobee Flight 10 launched on March 17 and Flight 11 launched on March 22. Van Allen's Geiger tubes successfully measured cosmic rays and the earth's magnetic field at low latitudes south of the equator. A third flight misfired.

"If any small detail of the rocket vehicle or of the instrumentation fails, the participants can only shrug—and hope to try again in six months or a year. But if the flight is successful, a new chapter in the physics of the upper atmosphere may be opened. A long-standing hypothesis may be verified or disproved. Or a new phenomenon may be observed. Rocket experimentation appeals to the most adventuresome of physicists," Van Allen wrote in 1950 in the "The Rockets Report," an article for the *Johns Hopkins Magazine.* Shipboard launches added new elements of high adventure to the report.

As was the case at White Sands, the spirit of improvisation prevailed. Singer adapted magnetic equipment the navy used to detect underwater mines when Van Allen asked him to examine the earth's magnetic field off the South American coastline. The data revealed that the electrojet, powerful electrical currents associated with the earth's magnetic field, were lower and more intense than scientists had expected near the equator.

Van Allen often compared sending experiments into space to dropping a fishing line in the water and waiting to catch something. Singer and Van Allen had to wait to get back to the lab to figure out what phenomena their data had "caught." Answers had to be gleaned from hundreds of feet of paper tapes. Data reduction in the space days before computers meant taking a ruler to manually measure the peaks and troughs on the paper tapes. The measurements had to be statistically analyzed and converted to a graph—a process that took weeks for data collected from the few minutes of each rocket flight.

Van Allen was happy for the effort because it kept him close to home in June as he and Abbie awaited the birth of their second child. Margot Isham arrived June 10, 1949, and weighed six pounds, six ounces, just like her older sister. Abbie's mother, Helen, came to help with Cynthia and the new baby. Once she left, Abbie took charge of caring for the two children and running the household. Van Allen came home every night for dinner, but the demands for travel increased with the new potential of the Aerobee rocket for research.

In December 1949, Van Allen won his first "space" award, a medal from the American Rocket Society (ARS) that honored his upper atmosphere research and his push to develop the Aerobee rocket. The ARS also honored Robert Goddard posthumously, presenting his wife, Esther, with a medal named in honor of her husband.

Van Allen, Singer (now at the University of Maryland), and the APL team boarded the *Norton Sound* once again to sail to the Gulf of Alaska in early 1950. Temperatures remained in the 40s for the trip north but stormy swells prevented a launch until after January 10. "The North Atlantic much like the North Pacific" with "arctic air masses causing one-to-three storms per week," Van Allen wrote in his field log. The ship tuned into Japanese, Liberian, and even Chinese weather reports to obtain warnings of weather fronts heading their way. Van Allen spent the time below deck testing equipment in quarters set aside for a makeshift lab. The weather finally cleared on January 15, 1950, and the APL team started an 11 1/2 hour prep for the launch: 0400—commence radar search for other ships or crafts in nearby waters; 0700—load the batteries in the rockets; 0800—take the rocket from the hangar; 0930—portable tower erected; 0930—telemetry check in the tower; 1100—start fueling; 1230—complete fueling; 1300—clear flight deck and search for aircraft in the area; 1330—blast-off.

The launch went flawlessly as did a second flight launched three days later and a few hundred miles south. With the pressure of the flights over, Van Allen, Singer, and the ship's officers Lieutenant Lee Lewis and Lieutenant Commander George Halvorson discussed other options for high altitude research. Lewis piped up that he'd heard about an inexpensive way to get a small rocket into the upper atmosphere by hitching it to a balloon and firing it by radio control when the balloon reached its maximum altitude of about 20 miles above the earth. A balloon-launched rocket sounded interesting to the budget-minded Van Allen and he made a mental note of the idea.

Back at APL, Van Allen reduced the data from the northern sojourn and combined it with the readings from off the coast of South America near the equator. His rocket data confirmed at much higher altitudes the same pattern Millikan's balloon data had demonstrated years before: radiation levels resulting from cosmic rays rose dramatically with distance from the equator. His readings off the coast of Peru correlated with Millikan's balloon data from the mountains of India; his readings from White Sands to Millikan's at San Antonio and his counts from the Gulf of Alaska to Millikan's from Saskatoon, Canada. Van Allen found that four times the intensity of cosmic rays collected in the northern skies as compared to the regions just south of the equator where most cosmic rays are deflected by the earth's magnetic field. Three times more collected at White Sands than at the equator. But he found the cosmic ray plateau at about 31 miles above the earth at every location.

The earth's magnetic field arches upward from the poles of the planet, a natural deflector shield that reaches its highest point at the equator where Van Allen found the light intensity of cosmic rays. At the same altitudes, cosmic ray intensity increased with distance from the equator as the earth's protective magnetic field arched downward toward the poles. But the atmosphere wraps the earth uniformly at all latitudes. Van Allen's discovery of the plateau located the ceiling in the atmosphere where primary cosmic rays begin to splinter into secondary particles.

At a conference in Echo Lake, Colorado, in June 1948, Van Allen reported on the exciting results in discovering the incoming stream of cosmic rays in the atmosphere. He told his audience how his work had been inspired by Millikan's balloon soundings in the late 1930s. At the close of his presentation, a tall, distinguished white-haired scientist wearing a signature bow tie approached him. It was Robert Millikan. "He sought me out and congratulated me warmly. I shall never forget that," Van Allen told an Iowa audience in a June 27 presentation exactly fifty years later. The 1998 event, hosted by the Jackson County Historical Society in Maquoketa, Iowa, celebrated the seventy-fifth anniversary of the award of Millikan's Nobel Prize for the discovery of the charge of the electron.

Van Allen's latitude data quieted the skeptics such as Rossi and his trips proved that the Aerobee was a manageable traveling rocket. "The equipment used for launching Aerobees from the *Norton Sound* is portable with relative ease and could be set up at any convenient land location. Thus the possibility exists for setting up at moderate expense an upper air rocket research site at some land location other than White Sands," recommended Homer Newell. Van Allen was already lobbying for a new rocket range at Fort Churchill, a Canadian base on the Hudson Bay.

Research with rockets slowly began to expand beyond the small circle of pioneer experimenters. The cost and reliability of the Aerobee helped encourage that early expansion. When the air force began shooting Aerobees in 1949, new players took places in the roster of experimenters. The University of Colorado, the University of Denver, Boston University, and the University of Rhode Island began loading experiments in the Aerobee starting in 1950.

Public infatuation with the rocket program escalated rapidly. The rocket scientists and the American Rocket Society became missionaries for a new technology and a new field of science as they attended conferences and other events across the country. The rocket societies organized many of the meetings and promoted the potential of rockets. Van Allen calmed fears that cosmic rays could mow people down like the cosmic rays of science fiction.

"Overall, the potential biological and medical hazards of cosmic radiation are not particularly alarming—especially for flights of such brief duration as may be achieved in the near future," Van Allen said. But even as scientific research with rockets and the futuristic possibilities it represented fired the public imagination, the rockets wrought deep divisions in the V-2 Panel. As war loomed with Korea, Van Allen and others recognized that scientific research remained an opportunity linked to military rocket development. He and others of the rocket research panel openly looked to military development of the next generation of research rockets, but the harmony of the group splintered in 1949 and early 1950 as Pickering, Newell, and Marcus O'Day of the Air Force Cambridge Research Laboratory sought panel endorsement for competing military rockets. Pickering wanted backing of the army's Corporal, Newell wanted support for NRL's Viking and Marcus O'Day promoted the air force's MX. As the only proven system, the UARRP endorsed the Viking for research but Van Allen deftly recommended that the panel produce a report sought by the Defense Department on the value of all three of the competing military systems and assess them in terms of a dual role for research and ballistics testing.

The report, written by panel founding members Charles Green of General Electric and O'Day, ultimately put aside rivalries to stress the wealth of collective experiences of panel members and rocket programs that could be applied to a national emergency. They linked the scientific research launches and the training of military personnel to use guided missiles. The report essentially applied the V-2 and Aerobee agenda to more rocket programs and promised the support and expertise gained by the rocket scientists in any national emergency.

The pragmatic link Van Allen and the panel forged between space science and military research gave universities a seminal foothold in the space program, including the University of Iowa where Van Allen had just been recruited in 1950 to return as head of the physics department. Giving his change of jobs as his reason, Van Allen attempted to resign from the UARRP as well but panel members pressed him to stay. "Discussion of your resignation has led to the unanimous opinion that the Panel is keenly interested in retaining you both as Chairman and Member," stated a December 15 letter signed by Newell, Pickering, O'Day, Whipple, Megerian, and several other members. "It is felt that the Panel has functioned efficiently and harmoniously during your tenure as Chairman for the past three years. This has been attributable, in large measure, to your competence, fairness, impartial judgment and intimate knowledge of all concerned agencies and persons." They asked Van Allen to remain during this "most critical period."

Van Allen thanked them but held firm in his reply on January 10, 1951, shortly after his return to Iowa. His determined friends simply responded with another laudatory letter and appeal. "In response to the very kind letter from members of the panel and since there does not appear to be ready agreement on a new chairman, I shall be pleased to continue as chairman," he wrote on February 10. But he had suddenly become the only member of the rocket research panel without financial access to a rocket, a reason he offered in his January letter to support his panel resignation.

Still, the university offered him a free hand to develop research programs while APL pressured him to step back into weapons development. After World War II, scientists who had mobilized their talents to produce the atomic bomb, radar, and the proximity fuze divided into two camps. Many, like Tuve, felt scientists should return to basic research independent of government contracts and military agendas. Others, including physicist Lloyd Berkner, wanted to see a continuing partnership in a new political world where they felt science and national security should march hand in hand.

Berkner, whose research opened the door to radio transmission, participated in policy boards that outlined how academics could assist national security with advances such as new radio research to prevent jamming of the *Voice of America* behind the Iron Curtain. His efforts promoted formation of the national laboratories and Berkner took over as the first director of Associated Universities Incorporated, a government-funded consortium organized to coordinate university participation in research programs at the labs. Brookhaven National Laboratory, located on Long Island, opened in 1950. The laboratory network already included Los Alamos National Laboratory in New Mexico, established in 1943 to develop the atomic bomb, and Argonne National Laboratory near Chicago, chartered in 1946.

At the Applied Physics Laboratory, the focus continued on weapons development. The flamboyant and frequently hotheaded Tuve found the research and the politics increasingly disheartening and in early 1950 he decided to return to civilian science at DTM. His departure set Van Allen's rocket team adrift. Tuve had protected "the five percenters" and the opportunity to pursue pure science with the rocket tests. His replacement, Ralph Gibson, gave only feeble support to the rocket scientists and reassigned Van Allen to work on a new generation of proximity fuzes.

"I had given my all to developing the fuze and bringing it to the Pacific," Van Allen said. "But I had no interest in returning to weapons per se. I could see the writing on the wall and I was more of the Tuvian line of thought." Discouraged by his reassignment, Van Allen quietly began to look for another position in 1950 when the perfect job came to his attention—the head of the physics department at the University of Iowa.

Just prior to World War II, the university took a headlong leap into theoretical and nuclear physics. Stewart, who turned sixty-five in 1941, pushed in this direction and secured funds to buy a Van de Graaff accelerator, which could accelerate particles to more than 1 MV (million volts) as compared to 400 kv (thousand volts) with the cranky Cockroft-Walton. The more energy, the more interesting the subatomic particles scientists could expect to find as accelerated protons smashed into targets. He wanted to set the course for the future of the department in anticipation of his retirement. Stewart trenched beneath the lawn between Schaeffer and MacLean Halls to create the elongated room for the Van de Graaff. The underground location and the zigzag, cement-lined entry shielded students from any encounter with stray particles. But the hectic war years put actual assembly of the Van de Graaff in moth balls.

Stewart stayed on to provide a continuity of leadership through the war years and then retired in 1946. His departure meant the first change of leadership in the department since he took the helm in 1909. Stewart, still determined to accomplish his dream of establishing a strong nuclear physics department through his successor, led the charge to convince Princeton nuclear physicist Louis A. Turner to come to University of Iowa.

Turner spent the 1930s researching nuclear fission at Princeton, his alma mater, and wrote the first comprehensive summary of both the military and peacetime potential of nuclear fission in 1939, publishing it in the *Review of Modern Physics*. During the war itself, he headed a radar research and development team at MIT. The job at Iowa appealed to Turner's streak of empire building—he could set his own stamp on the direction of physics research at a major university. The course was already set at Princeton by titans such as theoretical physicist John Wheeler.

Turner accepted the job at Iowa with the understanding that he could quickly bring on board another theoretician. Hans Bethe at Cornell helped him recruit thirty-one-year-old Josef Jauch. The Swiss-born Jauch was young enough for the $6,000 a year salary to be attractive, even though his credentials already included important theoretical work in developing the atomic bomb. Jauch, in turn, hired Fritz Coester, a twenty-six-year-old wizard of mathematical physics who earned his PhD at the University of Zurich in 1944 and came to Iowa in 1947.

Turner championed further development of the Van de Graaff and, by 1948, the accelerator was a state-of-the-art device at a total investment of $100,000. Then the money ran out and Turner, beleaguered by administrative tasks he hated, hit a brick wall in terms of expanding the nuclear physics programs he loved as the Iowa state legislature tightened the purse strings on the university overall. Meanwhile, the research budget was soaring at Argonne, 40 miles west of Chicago, as the lab equipped itself for both applied and experimental particle research. Turner jumped at the chance when Argonne offered him a position and submitted his resignation to the University of Iowa in January 1950.

Among the candidates to replace him, Van Allen stood out as a rising star "His former mentors in the physics department had followed his developing career with interest and a large measure of pride," noted James Wells, in a department history written in 1980. Van Allen's high altitude physics, spurned at APL, sparked lots of interest when he lectured on the topic in Iowa in 1948 at the annual department colloquium.

"We wanted to hire someone younger who was active in research," said Coester, who later followed Turner to Argonne but continued to teach at Iowa and retained an office at the university. "One field which wasn't represented here was cosmic ray research. That was an important field since cosmic rays were high energy and accelerators at the time were relatively low energy. Many particles later found with accelerators were found with cosmic rays during that period," Coester said. Edward Tyndall, Van Allen's master's thesis adviser, headed the search committee and approached his former student about applying for the job in March 1950.

"Tyndall was very frank about it. They had two or three other people under consideration and he said he would let me know in due course," Van Allen said. Van Allen was the front-runner as the faculty discussed the candidates late in 1950, but then someone pointed out, "You have a candidate with no teaching experience," Coester recalled. That stymied the group for a minute until they arrived at a consensus based on Van Allen's well-received talk on cosmic rays at an Iowa colloquium. "We said that, given the superb pedagogical presentation at the colloquium, we don't have any doubts about his ability to teach," Coester recalled. Van Allen was the unanimous choice of the search team.

When Van Allen submitted his resignation, Gibson tried to convince him to stay on at APL, promising over a quiet dinner that he could turn things around on the space science front. Van Allen saw little hope of such change. He headed home to build a rocket program from scratch.

8 It's a Rocket! It's a Balloon! It's a Rockoon!

U.S. Coast Guard Cutter *Eastwind*, Thule, Greenland—July 29, 1952. Van Allen boarded the USCGC *Eastwind* with an unlikely crew of space explorers—two graduate students, a university technician, balloon experts from General Mills Corporation, and a lieutenant from the Office of Naval Research. Ship's Captain O. A. Peterson had invited him to join the ice cutter's trip toward the North Pole. No one had measured cosmic rays so near the poles before and Van Allen approached the task with rockoons, the catchy name he adopted for his balloon-rocket hybrid. This was the ultimate economy model in rocketry since Van Allen could no longer afford to launch the Aerobee he had helped develop.

The University of Iowa gave Van Allen carte blanche to do any research he wanted with just one catch—no money.

"How will I get funding," Van Allen asked Tyndall.

"That will be up to you, young man," Tyndall stated firmly.

Van Allen dusted off the idea of the balloon-launched rockets that he had discussed during the shipboard launches of the Aerobees. He estimated that he could piece together a rockoon and launch it for a total cost of $1,800 compared to $25,000 for an Aerobee and $450,000 for the Viking. He sent off a one-page application to the Office of Naval Research (ONR) for a grant to fund his bargain-basement space program. Office of Naval Research approved the funding, giving him the money to purchase weather balloons from General Mills for $200 each. As for rockets, he picked up surplus Deacons for free from his friend and fellow UARRP board member Bill Pickering at JPL.

"Jim's idea of flying these things from a balloon was a wonderful idea," said Pickering, who had built the Deacons for the air force. "We were delighted because the air force didn't want the rockets and we had quite a few of them."

The military dismissed the Deacon because, fired from the ground, it reached an altitude of only 12 miles due to drag from the air. But carried by balloon and fired from the stratosphere, it promised to reach altitudes of more than 70 miles, an altitude on a par with those attained by the Aerobees.

Now Van Allen needed a launch site. A ship was the obvious choice since it offered the opportunity to take cosmic ray measurements across a wide span of latitudes. Van Allen's navy contacts suggested he hitch a ride on a coast guard ice cutter to approach high latitudes but he was asking a lot of any captain who hosted his trip. He needed quarters for himself and his team, a makeshift lab and storage space for rockets, balloons, and other equipment. Then he needed an open deck for every launch. And that was the easy part. He also needed a captain who could understand the value of sailing off course at times for a man in search of cosmic rays with balloons carrying rockets. But the arduous trip to and from the Arctic offered few diversions and Peterson found the rockoons a fascinating project to engage his crew. The ship embarked the same day Van Allen boarded it.

* * *

The rockoons put Van Allen back in the rocket business. He parlayed his bare-bones budget and surplus rockets into an international mecca for space exploration within two years of his unceremonious return home on a frigid day in January 1951. He drove into Iowa City with Abbie, Cynthia, and Margot wedged between their luggage in the family station wagon. A trailer carried most of their household possessions. "We plowed through the snow to move into a 'barracks apartment,' one of a cluster of small metal-sheathed buildings which had been erected during the war as temporary quarters for naval cadets and other personnel," Van Allen wrote in a lighthearted autobiographical sketch. The source of heat was a single cast-iron stove that ran on fuel oil from an external 55-gallon drum. "The small living room could be made comfortably warm but the remainder of the apartment presented a challenging problem in heat transfer. However, the monthly rent was only $35," Van Allen wrote.

Within a few months, the Van Allens bought a home at 130 Ferson Avenue in Iowa City. Van Allen became a familiar figure in the neighborhood on brisk walks with the family dog, Domino. Abbie and Jim joined the South Ferson Avenue Improvement Society, dedicated to halting any improvements that might mushroom out of the era's ever-encroaching urban renewal programs. Margot, already in love with animals and the outdoors, rode her tricycle relentlessly with Domino as her constant sidekick. Cynthia, the social sister, preferred playing with the other kids on the block.

Sarah Van Allen, born in Iowa City on January 7, 1953, fondly recalled Christmases on Ferson Avenue. "Dad always read *The Night Before Christmas,* the last thing we always did before we went to bed on Christmas Eve. He's a wonderful reader when he reads out loud. The next morning, we all lined up at the top of the stairs, and went down. The oldest one went down first—we always did that. Dad would be building a fire. He's a very meticulous fire-builder," she said.

However hectic his schedule, he continued to come home every night at six for dinner when he was in town and everyone joined in for the "boiler factory." "That's what Mother called it," Sarah said. "We'd sit down for a family dinner and we'd go around the table and everyone was asked, 'What happened in your boiler factory today?' Mother really pressed that."

At the physics building, Van Allen threw himself into the departmental chores of teaching and administration from his "official" office in the center of the second floor of the physics building where the secretary's entry alcove and a conference room completed the suite for the department head. His favorite office soon became a small, sequestered nook with a west-facing view just beyond the conference room.

Before the rockoons came on the scene, he continued his cosmic ray research with counters carried by arrays of weather balloons, still the norm for most geophysical research. Stewart, who still cast a long shadow across the physics department he had guided for so long, provided a grant for the balloon research through the privately operated Research Corporation he now directed.

Van Allen tested the particle counters and circuitry for a new set of experiments in the nuclear physics lab at Iowa and also tested a new battery for the counters, the EverReady Minimax, which substantially increased the research life of instrumentation carried by the balloons. As the balloon program took shape, Van Allen spent the summer of 1951 at Brookhaven National Laboratory to salvage three months of a yearlong Guggenheim Fellowship he had been awarded and wrote a paper based on cosmic ray research. They were 90 percent protons stripped from hydrogen atoms, 9 percent helium nuclei, and 1 percent nuclei from heavy elements accelerated by the cataclysmic explosions of dying stars—the stars scientists called supernovas.

Writing the paper stimulated his determination to continue some form of rocket research in the fall despite his "zero budget." Frugality was nothing unusual to Van Allen and while he didn't have funding, he had a well-supplied machine shop where instrument maker Joe Sentinella could help him build just about any gadget he could imagine with the help of his machinists and a stockpile of parts from obsolete electrical gear. Most importantly, Van Allen began to assemble a team of faculty and graduate students to carry out the research.

University of Chicago physicist Melvin Gottlieb, whom Van Allen had known at White Sands, joined him on the faculty that first spring. Ernest Ray and Joe Kasper, already in graduate studies when Van Allen arrived, began to work with him as well. The knack of applying electronics meant for home gadgetry to cosmic ray experiments became the hallmark of Van Allen's recruits as he enlisted new students for the research team.

Tall, gangly Les Meredith was the first to unsuspectingly walk through Van Allen's doorway and straight into the fledgling space program he hadn't known existed ten minutes earlier. He knew nothing about astronomy either, but he was skilled at rewiring communications gear from training in the Army Signal Corps during the war. That made him a perfect candidate for a master's degree. Van Allen quickly involved him in building and testing the Geiger counters sent up in the weather balloons to measure secondary cosmic rays.

"From my perspective, this was being at the right place at the right time. I was looking for something you could get your hands on instead of theory," said Meredith. Their relationship set a pattern Van Allen relied on for the next thirty-five years. Always adept at making the most of resources at hand, he entrusted his graduate students with all the responsibility they could handle. The highly motivated ones got full partnership status in Van Allen's research efforts and parlayed their own new ideas into equipment or experiments that quickly could be harnessed to a space mission in the frontier years of space exploration. The bargain meant that Van Allen got economical and talented labor and the grad students took pioneering roles in space science as they earned PhDs.

Meredith grew up in Iowa City where his father headed the Child Welfare Research Station at the University of Iowa. He was an undergraduate senior majoring in physics and looking for a master's thesis project when he met Van Allen and joined his team sending the weather balloons into the upper atmosphere.

Meredith's roommate, Bob Ellis, who started his PhD in 1951, wanted nothing to do with the high altitude balloon research, however. "He told me, 'I'm not doing that. The objective of graduate school is to get out of here and you don't know if you'll get data from those balloons,'" Meredith said. Van Allen won Ellis over and Ellis was among the first African Americans in the country to earn a PhD in physics.

Ellis grew up in a solidly middle-class background and earned an undergraduate degree from Tennessee Agriculture and Industry, one of the few colleges open to blacks in his home state at the time. The University of Iowa accepted black students but, in many ways, Iowa City of the 1950s did not. Ellis had to go to a black barber in Cedar Rapids to get a haircut because the white barbers near campus wouldn't seat him, recalls Meredith, who shared an office and work space with Ellis in the basement of MacLean Hall. Traveling with Van Allen and Meredith on a research trip to White Sands, Ellis was denied service in the train station restaurants. And when the physics department made hotel reservations for Van Allen, Ellis, and other researchers at a physics conference in Tennessee, Ellis insisted on staying with relatives. "You don't understand the chromatic situation in the South," he told Van Allen.

But people who wouldn't cut his hair or serve him a meal had no qualms at all about joining the crowds gathered at the Iowa City airport for the unusual balloon launches orchestrated by Van Allen, Gottlieb, Meredith, and Ellis on sunny, windless mornings in 1951. Van Allen could lift payloads of up to 27 pounds into the upper atmosphere with an array of nine Darex J 1600 weather balloons filled with hydrogen. The balloons couldn't reach into the primary cosmic ray stream but the team gained expertise in building counters, circuitry, and telemetry receivers. Meredith based his master's thesis on the research. Then the ONR grant came through to build the rockoons.

"That started a whole new line of research," said Van Allen. He hopped back into the rocket business with one of the oddest, cheapest, and most effective tools available for space exploration in the early 1950s. Gottlieb, Ray, and Kasper pitched in to configure the balloon-rocket system. The design was simple. A Skyhook balloon hoisted a rocket loaded with instruments to an altitude of about 10 miles. There, the timer and switch in a firing box ignited the rocket. Van Allen's new rocket team designed the firing box that dangled from the rocket's tail. The team also replaced the slender fins meant to improve the speed of the Deacon, a rocket designed for jet-assisted takeoff of military planes on short runways. Iowa's machine shop customized a set of fins nearly four times larger than the original design to stabilize and maximize the rocket's ascent. The fin assembly weighed 11 pounds 12 ounces, and the rocket overall weighed a mere 78 pounds. As for the cost, Homer Newell broke it down this way in his book *Sounding Rockets*: surplus solid-propellant Deacon rocket: $900, a cost waived for Van Allen; balloons, $200; firing box, $50; helium $70. That added up to $320 for the entire rocket assembly out of pocket, though the value was $1,220 plus approximately $600 in costs for freight, travel, and expenses involved in shooting the rockets.

The group made three trips to White Sands in June and July of 1952 for shakedown launches to test the whole system. The balloon with the rocket dangling from it by a long rope resembled a strange apparition as it rose in the sky. The balloon carried the rocket to an altitude of about 10 miles, as planned, and then the timed fuse ignited black powder and sparked the rocket's ignition. Van Allen's team couldn't see the launch but knew it had occurred from the data transmissions they began to receive. The rocket soared to a maximum altitude of about 60 miles, a near match for the Aerobee at a small fraction of the cost, though the Aerobee could carry larger payloads of instruments.

Elated with their success, Van Allen, Gottlieb, and Ray packed up their rockets, balloons, and detectors and all the backup gear and shipped it via train to the *Eastwind*. The rockoon team that left Iowa for the Arctic voyage included Van Allen, Meredith, Kasper, lab technician Lee F. Blodgett, and Navy Lieutenant Malcolm S. Jones (representing ONR). They flew on a military air transport from Dover, Delaware, to Thule, Greenland, on July 29, 1952, and met up with technicians from General Mills who arrived to help with the balloons. To get support for the trip, Van Allen had suggested that the Office of Naval Research network with other university research teams in need of an opportunity to fly balloon missions, and scientists from the New York University and the University of Chicago won passage on the *Eastwind* too. The Iowa group organized gear in a cramped, makeshift laboratory, with the rockets and the payloads lined up along the hull of the ship.

The *Eastwind* departed Thule at 1800 hours (6 P.M.) with the ship's band playing the catchy tune, "So long—it's been good to know you." It headed north, carving a lifeline of open water through the ice to bring supplies to the weather station at Alert, Canada, on Ellesmere Island in the Arctic Sea.

The Iowa team settled into a routine of spending half the time below deck checking instruments, wiring gear, and waiting for launch times, and the other half as eager tourists. The arctic adventure covered ominously beautiful landscapes of ice castles populated by seals, walruses, auks, and gulls diving for fish around the ship amid a landscape of icebergs. Meredith took photographs, documenting life on ship and recording the various launches. He also practiced skeet shooting with the crew from the deck. Meredith and Van Allen grew beards, a habit Van Allen repeated on subsequent field expeditions. It was a practical move aboard the pitching ice breakers. But it also linked him to polar explorers of the past, who all—including his mentor Poulter—grew beards on their sojourns.

The ship sailed at 10 knots when ice floes didn't trap it completely. Despite the landscape, temperatures typically remained near the thirties at first in the endless daylight of arctic summer. But even when the sea remained calm, the ship rocked precariously.

"The thing about an ice breaker is that it has no keel to stabilize the ship—it has a perfectly round bottom. And we'd be down below deck in the lab and the ship is rolling 10 to 20 degrees on each side. And you think, God what kind of storm is going on. So you go up to the deck and there are (just) a few swells. The rolling of the ship really got to me. I lived off Dramamine," Meredith said.

The rolling was a necessity, not a flaw, however. The hull of this hulking steel ship resembled a giant canoe, a design essential for the heeling system that enabled the crew to rock the ship free as it pushed aside floes of ice. The bow of the ship was designed to sail right up onto the 4-foot thick floor of arctic ice and crash through it with the weight of the vessel. The floes seen at the surface often topped huge floating freighters of ice, 10 to 30-feet thick. The ice didn't yield to the weight of the vessel, didn't yield to dynamite, and often didn't yield to 40-pound projectiles of explosives. Sailors throughout the ages liked to sail with the wind at their back. But not the sailors on the ice breakers. When the *Eastwind* headed north into the thickest and densest ice, a brisk wind at its back pushed floes in behind it and engulfed the ship. It drifted for miles like a small toy in the destructive grip of the ice. When floes collided, they heaved up ice and water that froze and sealed the floes together in thick, mounded seams of ice. The crew on the *Eastwind* stayed away from those fault lines and the civilians on board quickly learned a healthy respect for the merciless rule of the ice. The ice snapped half of the two-ton blade off one of the ship's propellers as if breaking a match stick. The ship vibrated as well as rocked after that.

"Already I can tell by looking at a floe of ice about how hard it will be for the ship to go through. If it's just flat and white, it will not be bad. If it's a bright light blue (covered with white) or has blue puddles in it, it will be hard," Meredith wrote in his journal. Even a light snow on the ice "blinded" the crew because they could no longer differentiate between white ice and the dreaded blue. Van Allen accepted the rigors of launching rockets in this unforgiving environment for one simple reason—it was the only way to get cosmic ray readings at such high latitudes near the North Pole where no one had taken them before. His team had no set schedule for launches but relied on weather conditions and the goodwill of the ship's officers. Frequently, the officers and the scientists dined together. Van Allen swapped World War II stories with Captain Peterson and patiently but persistently pressed for launches despite the delays and hazardous sailing conditions that often enveloped them.

"This noon we got stopped. We weren't really stopped but were only gaining ground by inches," Meredith wrote on August 11 with a full account of a futile attempt to explode their way through the blue ice blocking them. "First, they put two 20-pound charges on the floe. These blew holes about 3 [feet] deep and a foot across. Then they put about six pounds of regular explosives in each hole. This just enlarged the holes but didn't hurt the floe. Then they tried a 40-pound shaped charge. This blew a hole probably all the way through but the hole filled up so they could only fill the top 6 feet with explosive. Then they put 52 pounds of explosive in the three holes. This made the crater larger but the floe remained. We ended up going around it. I would estimate it was about 30-feet thick."

The ship finally reached the harbor at Alert on August 12. The outpost with several buildings painted bright orange was maintained by a staff of two who served as weathermen and radio operators. Steep brown cliffs rose up to 1,500 feet out of the ocean and icebergs dotted the inlet home of auks, gulls, and penguins. Sailors unloaded food, medical supplies, mail, clothing, and machinery in shifts and Meredith went ashore to mail letters from Alert, which boasted the post office closest to the North Pole. The letters were postmarked and loaded back on the *Eastwind* for delivery via Thule. The *Eastwind* inched north again the next day, making two miles in five hours, then stopped entirely for a whole day, then pulled forward amid thick gray glaciers along the coast. It was a momentous occasion on August 16 when the ship reached a record-breaking 384 miles south of the North Pole, nearer than any ship had ever reached under its own steam. In honor of the feat, everyone on board ship became "Companions of the Polar Key," Meredith reported in his journal. He dryly adding that "key, of course, is derived from the Latin saying, 'key, key, Christ (pronounced key-rist) it's cold.'"

After nearly two weeks of rough going, the ship finally reversed course for the return to Thule and sailed into the Kane Basin close to the shore of Greenland on August 20. Not one rocket had been launched on the whole arduous journey as yet, launches critical to Meredith's planned PhD thesis. Van Allen was as anxious as Meredith to take the rockets to latitudes where cosmic ray measurements had never been made before. He had three points on the latitude map showing that the intensities of primary cosmic rays rise with distance from the equator. His points also showed that, regardless of the energy levels of these particles, they pierce a ceiling in the atmosphere at about 31 miles up—the altitude of the cosmic ray plateau—and then start to splinter into secondaries. But more points on the map would make the pattern clearer.

The Iowa team actually began to prepare for a launch on the second day of the trip. Then the wind whipped up to 17 knots and the ice breaker couldn't outrun it. The New York University balloon scientists pressed for a launch on a seemingly windless night two days later and got the okay from the captain, even though the ship was locked in the ice. Their gear wasn't ready to go so they urged Van Allen to take their place. He reluctantly agreed and crew members got up to help.

"Then the wind picked up. And they had to cancel it. Being about midnight, some of the crew was quite disgusted. Nobody was more disgusted than Van Allen and myself," Meredith wrote. Van Allen was predictably silent on the point in his own field notebook but he never again okayed a launch when a ship was locked in the ice. When the ship was mobile, it could sail downwind at a speed that could cancel out the impact of the wind on his launches. An icebound ship couldn't move.

With the weather finally cooperating in late August, the captain set a whole series of launches on short notice, with plans announced at midnight for a 3 or 4 A.M. rockoon flight. He abruptly scheduled the first of these launches at 1:30 A.M. August 21. Van Allen, the Iowa group, and the General Mills team quickly mobilized the hour-long setup across the entire helicopter deck at the bow of the ship and the crew moved the helicopters, used to scout the safest path through the ice, to other locations. The mammoth Skyhook balloon encircled the entire deck but Van Allen kept the rocket warm below deck for most of this period and then sailors brought up the rocket body and laid it between two sawhorses so that the fins and nose cone could be attached. Only now, as the group prepared for the first launch, did Meredith discover that the holes to bolt the new fins to the body of the rocket were slightly too small. The bolts had to be hurriedly filed down to fit. Then they loaded the payload, capped it in place with the nose cone, and plugged the firing circuit into the ignition nozzle at the base of the rocket. The other end of the firing circuit attached to the boxlike gondola with the timer that ignited black powder and sparked ignition of the rocket. Connecting the firing system was Van Allen's responsibility—coast guard orders. Premature firing of a rocket on shipboard would be the same as the explosion of a missile and the coast guard brass decided that the leader of the mission would have to hook it up.

Shortly before the launch, Van Allen attached the rocket to the balloon with rope tied to a hook on the side of the rocket. A mild wind blew that night and Captain Peterson sailed with it to create zero wind conditions aboard ship. The balloon lifted off at 1:30 A.M. as planned with the rocket and gondola dangling from it. It sped upward at 900 feet per minute, almost in slow motion compared to the speed of a rocket. It took sixty-two minutes for the balloon to reach the ignition altitude of just over 10 miles. The counters loaded in the rocket were devoted to Meredith's PhD research on the composition of low-energy cosmic rays. They ticked off particle counts but then lapsed into irregular transmissions after about twenty minutes. Still, exhilaration mounted as the balloon neared the ignition point at 2:32 A.M. Nothing happened. Minute by minute the Iowa team waited for the rocket to fire. After seventy minutes, they knew it wouldn't.

Deeply disappointed by the unidentifiable malfunction, they continued to monitor the spotty data they were obtaining from the stratosphere until about 7:30 A.M. What happened? The counters continued to work, proving that the rocket itself was intact and still aloft. Meredith heard the ship's horn blow at a stubborn seal that wouldn't budge from an ice floe dead ahead. Then everyone went to bed with plans to try another launch the next day as the ship headed south along the jagged Greenland border of icebergs. But ice once again locked the ship in its tracks on August 22 and slashed a 9-foot rip in a fuel tank, luckily leaving the main hull of the ship intact.

The Iowa team used the period to work feverishly below the deck, rewiring and rechecking the ignition system of every rocket. Meredith repaired the counters to correct the problem of the intermittent transmission of data. The team launched the second rocket at 10:30 P.M., Saturday, August 23. This time the instrumentation worked flawlessly with the telemetry system recording the rising count of charged particles. Van Allen wrote down the counts in his field notebook, counts so well known to geophysicists that they could quickly map the ascent of the balloon from them—about twenty counts of particles per second at 30,000 feet, about forty at 45,000 feet, and about fifty counts per second at 55,000 feet, the ignition zone. The team watched the data counts expectantly, looking for evidence that the rocket had fired. Again, it didn't. The counters worked perfectly as the balloon continued to slowly rise, but hope quickly evaporated into fears over the failure of the entire trip. "We double did everything on this flight. Now we're up a tree," Meredith wrote in his journal.

The Iowa group met in the makeshift lab and decided that the rocket circuit was freezing during the ascent of the balloon. Van Allen requisitioned two 32-ounce cans of juice from the ship's mess for the next launch on August 28. He wrapped the rocket in a plastic sleeve to insulate it and heated the cans of juice. He placed them in the firing box that dangled from the rocket. The balloon lifted into a clear sky at 8 p.m. on August 28. The Iowa team monitored the ascent with a good case of anxiety rather than excitement this time around. Other than record the counts, there was little to do once the balloon was aloft. Meredith and Van Allen checked the counts again and again to pass the time during the agonizing hour-long wait to ignition. Then suddenly, at sixty-two minutes, the counts leaped, suggesting an altitude range that only a rocket could reach. The Deacon had fired and the group monitored Ellis's experiment until the rocket plummeted into the Arctic Sea and cut-off the signal some four-and-a-half minutes after firing. Meredith collected his friend's data, elated. Iowa prepared for another launch, this time with Meredith's counter. The balloon lifted at 2:30 A.M. Friday, August 29, followed by another successful launch that same day after lunch. The navy ice breaker *Atka* docked about a mile from the *Eastwind* and took pictures of the sixth rockoon launch on Sunday, August 31, after two days of gale-speed winds. The rocket worked this time but the counter didn't. Regardless, the *Atka* got great shots of the rockoon ascent with the entire *Eastwind* in view. It also recovered film plates flown as part of the University of Chicago's research on fallout from nuclear weapons testing in the upper atmosphere.

Van Allen lobbied for as many launches as possible in the time remaining. With the additional launches, Meredith had a wealth of data to analyze back at the university. The Iowa team hopped a military air transport back to Delaware on September 6, thirty-nine days after they had left there. Still sporting their beards on their way back to Iowa, "several people asked us to which religious order we belonged," Meredith joked in notes added to his journal later.

Everyone who participated in the first rockoon adventure agreed it was a success. Five of the seven rockets had ignited, yielding high latitude cosmic ray data from altitudes of sixty miles above the earth, altitudes comparable to those reached with the much more expensive Aerobee research rocket. In addition, the Geiger tubes broke new ground in the research of the composition of cosmic rays, establishing the absence of low energy protons at the plateau, more evidence that cosmic rays are high-energy particles darting through interstellar space.

The 1952 expedition launched research articles and PhDs as well as rockets. Local newspapers and radio stations quickly picked up on the escapades of the University of Iowa's arctic explorers. George M. Ludwig hosted a popular morning radio show called *Breakfast at Ludwigheim,* broadcast from his farm a few miles west of Iowa City. Ludwig invited Van Allen and Meredith to breakfast and interviewed them on the show. Afterwards, he introduced them to his son George, an air force pilot with a flair for electronics who had recently completed his military service. He enlisted right out of high school, closing shop on his high school electronics repair business, where he fixed everything from radios to vacuum cleaners for families in the small town of Tiffin, Iowa. Van Allen suggested George Ludwig stop by and put some of his expertise to use in the physics department. He offered him a job for 75¢ an hour. Ludwig accepted, started taking college classes early in 1953 and quickly became involved in building instruments for Van Allen's rockoons.

With a research program established, Van Allen quickly lined up support for a second rockoon mission from ONR, General Mills, and JPL, where Bill Pickering promised a fresh supply of surplus Deacon rockets. "I was very good at switching my full attention to whatever I was doing—meeting with people in the shop or in the lab, meeting with graduate students. I was juggling grants and contract support, lining up rockets to make sure they were shipped in time, getting the balloons from the General Mills people," Van Allen said. In the meantime, he continued the Iowa City balloon launches, ran the physics department with a trademark open door policy, wrote scientific papers, continued to chair the UARRP, taught undergraduate classes, and mentored his graduate students.

Van Allen received help from a vigilant secretary who warded off traffic at his office. "I remember one day while I was waiting to see [Van Allen] that the Dean of the College of Liberal Arts stopped by and asked to see [him]. Agnes (Costello) McLaughlin replied, 'Dean who? Please wait your turn after the rest of the students,'" recalled space scientist Louis Frank, another early graduate student. McLaughlin, the feisty chief secretary, managed the office with the style of a drill sergeant.

• • •

Preparations went forward for the 1953 rockoon mission to the Arctic, this time aboard the naval ice breaker *Staten Island*. But behind the scenes, a secret government project to build a thermonuclear fusion reactor was underway at Princeton University. Physicist Edward Teller had pressed for development of the first thermonuclear bomb, made at Los Alamos and successfully tested in 1952. Unlike fission bombs that released energy by splitting atoms apart, the H-bomb vastly increased explosive power by fusing deuterium (heavy hydrogen) atoms into helium. Van Allen's graduate work with deuterium and follow-up research with Breit at Department of Terrestrial Magnetism ten years earlier made him a prime candidate for the secret Project Matterhorn that his colleague, astrophysicist Lyman Spitzer, initiated in 1953. The goal was to build a fusion reactor to generate energy as the sun does. Scientists believed—and hoped—that fusion meant a cheap, near limitless source of energy to promote a peaceful world.

Jim and Abbie Van Allen moved to Princeton with Cynthia, Margot, and Sarah in May 1953 and rented a house at 106 Broadmead Road for a sixteen-month stay. During the previous year, Van Allen's PhD student Ernest Ray had worked under Spitzer's guidance to build and operate a tabletop model of Spitzer's concept for a fusion reaction. This was a figure-eight "racetrack" made of glass tubing and equipped with a system of current-carrying coils to spark and magnetically confine the super-hot, high-pressure plasma needed to achieve fusion. The machine embodied the basic design of what Spitzer called the Model A "Stellarator," a step toward development of a power source akin to the sun and the stars. The device produced evidence of fusion reactions but with greater input than output of energy. The need was clear for a much larger machine to push development. Van Allen immediately undertook the design and construction of a room-sized Model B Stellarator, with an assembly of much larger coils and a massive bank of condensers for storing energy to provide impulse. He added diagnostic equipment as well.

Meanwhile, Gottlieb took over the 1953 rockoon mission on the *Staten Island*. Ellis joined the excursion this time and he and Meredith flew their Geiger counters and ionization chambers. Van Allen received frantic calls in Princeton as some of the rocket igniters failed to work again, although seven of the sixteen rockoon flights launched perfectly. The team reinforced their 1952 findings and made an unexpected discovery of high radiation levels in the aurora borealis, the donut-shaped oval of charged particles above the North Pole.

"The 1953 expedition yielded a remarkable new finding, namely the first direct detection of the electrons which, we surmised, were the primaries for producing auroral luminosity," Van Allen noted. The scientists referred to their discovery as "soft radiation" because it was easily absorbed by the upper atmosphere as it endlessly rained in. Composed of low-energy electrons and other particles, the soft radiation followed the flux lines of the earth's magnetic field as it arches sharply downward at the poles toward the center of the earth. The particles pour into the aurora at about 60 miles above the earth and then collide with nitrogen, oxygen, and other gases to create the colorful showers and bolts of the Northern Lights and Southern Lights that human beings have beheld in wonder for untold thousands of years.

Back in Iowa City, Frank McDonald, a PhD from the University of Minnesota, joined the physics department for post-graduate study in September 1953. McDonald started developing innovations in counters to identify the different atomic nuclei found in cosmic rays. An accomplished high altitude researcher, he had learned the technique flying Skyhook balloons in graduate school. He and Gottlieb joined the 1954 rockoon mission that focused on the aurora that summer.

That same summer, Van Allen's Model B Stellerator was producing the brilliant gaseous discharges from fusion reactions for durations of one-millionth of a second. But even with this leap forward, Van Allen judged that development of a practical fusion reactor would require much more time than he had envisioned. "I was optimistic [about fusion] when I first got there," Van Allen said. But as one year at Princeton dragged into two with no end in sight, he felt he could no longer justify running Iowa's physics department long distance. And he wanted to take a more active role in Iowa's growing rockoon program as plans heated up for a year of international collaborative research involving rockets. Van Allen returned to Iowa for the start of the fall term in 1954. To keep the fusion project fully staffed, he asked Gottlieb to take a stint with the research. Van Allen returned to a rapidly changing team of people.

Gottlieb replaced Van Allen at Princeton and never left, becoming director of the Plasma Physics Laboratory that evolved from Project Matterhorn. Ellis and Meredith, PhDs complete, found jobs. Meredith went to the Naval Research Laboratory and Ellis returned to teach at his alma mater, Tennessee A & I. Ellis later joined Gottlieb at Princeton, where he worked for the rest of his career. George Ludwig was completing accelerated undergraduate studies and working nearly full-time in the new Cosmic Ray Laboratory established in the basement of the physics building. Several other graduate students had projects well underway.

Shortly after his return, Van Allen called a meeting in his office with the latest graduate recruits to get acquainted and divvy up the new round of research projects. Larry Cahill arrived with an undergraduate science degree from the U.S. Military Academy and additional study at the University of Chicago. He undertook the task of fabricating a magnetometer to measure magnetic and electrical currents in the upper atmosphere. "I can almost see Van Allen writing on the blackboard. There were all these different things to do. I still have my notes—just scribbles," admits Carl McIlwain, a flute player from North Texas State College, who came to the University of Iowa to continue his studies and teach flute in the music department.

"When the job [I came for] disappeared, I decided, yes, science is very interesting and signed up as a physics [graduate] student," McIlwain said. "Dr. Van Allen probably didn't know that this student was fresh out of music school." Now McIlwain piped up that he would take on the 1955 rockoon latitude survey of cosmic rays using a different rocket—the streamlined Loki. That meant miniaturizing instruments to fit the 3-inch diameter Lokis that would be used in addition to the six-inch diameter Deacons for the 1955 trip back to Greenland. "We were starting from scratch. Nobody ever built instrumentation so small before, and batteries. Oh yeah, I had to have batteries and the batteries on hand couldn't possibly fit. So I took a trip up to Madison and talked to the engineers at Rayovac. I said, 'Can you package me a battery that will fit in here?' And they said, 'Okay.' It's incredible in retrospect how much [Van Allen] let us make our own decisions," McIlwain said. Van Allen also got McIlwain involved in the exciting idea of launching two-stage rockets from the balloons—a Loki saddled atop a Deacon. Ludwig gave McIlwain a crash course in instrument design as the flutist built an assembly line of counters and igniters on a lab bench in the northwest corner of the basement.

Kasper helped McIlwain keep pace with classes he missed as the Iowa rockoon team departed for the 1955 expedition on September 10. McIlwain, Kasper, Ludwig, and McDonald flew to Norfolk to board the USS *Ashland* in Chesapeake Bay in mid-September. Van Allen stayed home as Abbie recovered from a calcified disk in her spine and from childbirth. After three daughters, Thomas Halsey Van Allen was born at 7:12 P.M. on September 1, weighing 6 pounds, 8 ounces.

McIlwain and other members of the team got upper bunks on the ship and quickly learned how to wedge themselves in with pillows lest they be tossed to the floor as the ice breaker rolled. Not that there was much time for sleep. McIlwain and Ludwig improvised their way out of several crises before the ship ever left the harbor. Ludwig completed and repaired experimental instrumentation in a marathon effort after boarding and then found that their Geiger tubes weren't functioning. They called Van Allen to commandeer forty backup Geiger tubes. Then, as they unloaded the Lokis, McIlwain and McDonald realized they still had the stubby 1-inch "factory" fins from JPL instead of the broader "high altitude" fins. "Frank McDonald and I ran into a Norfolk hardware store for some sheet aluminum, came back, cut some fins and bolted them on," McIlwain reminisced in his lighthearted article "Music and the Magnetosphere."

When the first Loki launched on September 23 with McIlwain's experiment, dead silence followed. After a harrowing twenty seconds, McIlwain located the data signal on another channel. He successfully launched the rocket and described himself as an "ex-music student" from this point on, a man who was now receiving a symphony of cosmic ray counts.

The team pinned down new details about the particles in the aurora and probably reached an altitude record with a two-stage Loki-Deacon combination, undocumented because the nose cone melted in the tremendous heat of acceleration generated by the two-stage launch. During the two-stage assembly, Frank McDonald "bravely stood on the NRL trailer to steady the tall pair of rockets during the balloon launch. He shudders now when he sees pictures of it," McIlwain wrote.

Van Allen later speculated that this inexpensive two-stage rocket, given more heat resistant shielding, could have discovered the "horns" of the outer radiation belt that arch toward the poles.

The only mishap with a rockoon occurred on the 1955 mission. The radio transmitter on the Loki accidentally fired the rocket as ship's Lieutenant Commander Gus Ebel checked the igniter box. Ebel was badly burned on his face, scalp, and shoulder as the rocket accelerated toward the stern of the ship, slicing through everything in its path, including the cable of a phone held by a sailor who managed to leap from danger. The rocket crashed into several empty helium bottles and then exploded. Parts and red-hot propellant rained down, setting fires. The burning balloon had to be doused and tossed overboard. "The report we filed on the incident strengthened precautions against the accidental ignition of civilian and military rockets with radio transmitters," McIlwain said.

Ebel recovered and the Iowa team stayed in contact with him, but the 1955 mission was suspended after the accident. By then, Iowa had successfully launched eight deacons, four Lokis, and one of two two-stage rockets. NRL had adopted the rockoon method of research as well and made ten successful launches on the same trip. There was no return trip in 1956, as the Iowa team prepared for two lengthy rockoon missions for the summer of 1957 and the start of the International Geophysical Year (IGY). Van Allen had multiple IGY programs in the works, including an experiment for the Cold War competition to launch a satellite into orbit around the earth.

9 *Sputnik* and the Space Race

Iowa City, Iowa—November 16, 1956. Van Allen picked up the phone and recognized the courtly but urgent voice of Ernst Stuhlinger even before the physicist identified himself. Stuhlinger cut straight to the point. A launch of a Jupiter C rocket in September could have sent a satellite into orbit but, under direct orders of the defense department, it fired with a dummy payload instead of the real thing.

"The Vanguard program won't deliver on time," he told Van Allen, referring to America's official contender in the race to launch a satellite into orbit—and to get it there first. Both scientists agreed that the choice of the Vanguard had been a big mistake. Now Stuhlinger asked Van Allen to develop a hush-hush cosmic ray experiment to load on the undercover satellite being built for the Jupiter C.

Van Allen gambled. His cosmic ray experiment already held one of the top four priority spots planned for the Vanguard satellite launches. But he decided to hedge his bets and configure instruments to fly on either the Vanguard or the Jupiter C. "Presently, [von Braun's scientists] have the proven capability of projecting 18.5 pounds into orbit . . . might have about two pounds available for us later in the year," Van Allen wrote in his journal.

• • •

Project Vanguard entered the first lap of the space race under the auspices of the International Geophysical Year (IGY), a year of worldwide scientific collaboration planned for 1957–1958. The IGY eventually involved sixty-six countries, some 60,000 scientists and hundreds of research programs. But all eyes were fixed on the satellites and the two players building them: the United States and the Soviet Union. The Vanguard incorporated retooled renditions of the Viking launcher with a high-flying version of the Aerobee for the upper stages—seemingly tried-and-true technology. Yet little more than a year into the Vanguard program, it was clear to Van Allen that America had bet its money on the wrong rocket and left the ready contender in the dust.

That contender was the Jupiter C rocket and satellite being developed by von Braun and the rocket team at the Army Ballistic Missile Agency in Huntsville, Alabama. Rejected for the IGY mission and dropped into limbo as an army missile project, the Huntsville team developed the Jupiter C officially as a military test vehicle but hid one rocket away as a satellite launcher. General John Medaris funded the satellite development out of unrestricted budget allocations, and Pickering kicked in some funding from JPL.

Von Braun's colleagues often moonlighted on the project on their own time. Stuhlinger built the hand-operated detonator to fire the upper stages of the rocket in his garage, buying and making parts himself. He even designed a simple apex calculator to predict when the Jupiter C had reached its highest point, the split-second point where upper stages should be fired. Josef Boehm worked at the arsenal hour after hour on a satellite, perfecting a design he and von Braun had discussed for years. Notations about the "Boehm-Van Allen line-up" appeared in von Braun's daybooks soon after Stuhlinger's call to Iowa. The Jupiter C's success and escalating delays in the Vanguard program brought increased urgency to build the secret satellite. They had found a way "to bootleg a program to launch satellites," said Herbert York, an Eisenhower science advisor.

A top secret report completed by Douglas Aircraft Company engineers in May 1946 laid out the feasible steps necessary to develop an earth-orbiting satellite and determined it could be designed, built, and launched within five years at a cost of $150 million.

The report, "Preliminary Design of an Experimental World-Circling Spaceship," envisioned the satellite as the "baby" of a four-stage rocket that had no fins and looked like a giant artillery shell. The Douglas engineers conceded that such a rocket would be "bulky, expensive and inefficient" but justified the project as a potent military and research tool to "inflame the imagination of the world." The report was commissioned by the air force from Project Rand, the research and development think tank of Douglas Aircraft.

"In our [rocket] panel, we expected satellites to be launched as soon as propulsion capability existed. It was implicit in our discussions and most of us were acquainted with the military projects," Van Allen said.

You didn't need classified reports to find out about satellites, however. "The orbit of a space rocket would assume this form. The trajectory would be an ellipse with the center of the earth in one of its focal points," noted science writer Willy Ley in his 1944 book *Rockets: The Future of Travel Beyond the Stratosphere.* He discussed the need for a two-stage rocket to achieve orbit. *Popular Science* reported in March 1947 about plans to send into orbit fragments of an explosive launched by a V-2 rocket. "We first throw a little something into the skies. Then a little more, then a shipload of instruments—then ourselves," said Caltech physicist Fritz Zwicky in the article. Van Allen predicted in the article that America would send a rocket to the moon ["one way, no crew"] within fifteen years. "A conservative estimate," he said.

Yet reporters scoffed when Van Allen confidently noted at a symposium in 1948 that people could brace themselves for satellites in space within the next ten years. Satellites did go into space by then and an impromptu after-dinner conversation in Van Allen's own living room in Silver Spring set the stage for the race to put them there.

British geophysicist Sydney Chapman came to the dinner at Van Allen's home on April 5, 1950, along with Berkner, Singer, and APL geophysicist Wally Joyce. Chapman had cemented his reputation with the cornerstone concepts of terrestrial magnetism in the early twentieth century, tracing the earth's magnetic field to electric currents deep in its interior. He combined a brilliant and ever-curious mind with chivalry, warmth, and a sense of humor—qualities that made him not just a respected scientist but a beloved one. "Chapman came to APL to look at high altitude measurements from the equator. He was a very forward-looking guy and very supportive of rocket research," Van Allan said.

He readily accepted the invitation to dinner that Van Allen extended during one of their conversations. Van Allen made a quick call to Abbie and then enlarged the party to include the other scientists. Abbie hastily cleaned the house, got the kids off to bed early, and cooked a dinner complete with her famous seven-layer chocolate cake. Van Allen credits the cake for the high spirits and a freewheeling "bull session" that followed.

"As we were all sipping brandy in the living room, Berkner turned to Chapman and said, 'Sydney, don't you think that it is about time for another international polar year?' And Sydney said, 'Yes, I do. In fact, I've been thinking along the same lines myself.' And Berkner said, 'Why don't we do it?' We all kicked it around. It was a good time because of new rocket techniques and greater participation in geophysics around the world."

Pulling together the scientists, experiments, and communication network for an international collaboration would take time, and the group decided 1957 would be the perfect year for the kickoff. It was the seventy-fifth anniversary of the start of the First Polar Year and the twenty-fifth anniversary of the second, both periods of international collaboration motivated by polar research. But the new effort expanded into the International Geophysical Year, a research program encompassing the entire earth and the regions around it that could be reached by a rocket.

Attitudes were changing about rockets. "The success of the rocket program [at White Sands] was one main element, the fact that we showed you could do good scientific work," Van Allen said. "But the principal driver was the development of high-performance rockets within the military, the so-called intercontinental ballistic missiles." Van Allen recognized the opportunity to rapidly expand the scientific applications of rockets and upper-atmosphere research with an international collaboration. Chapman's international reputation and contacts made him the obvious ambassador to field the idea abroad and he agreed to the task. The universal language of science could perhaps ease the tensions of a jittery world—the Russians had successfully tested an atomic bomb in September 1949.

• • •

The specter of a Russian nuclear arsenal heightened U.S. urgency to build ballistic missiles by 1950. The army began looking for sites to pursue nuclear weapons development and reactivated the Redstone Arsenal in Alabama near the small, languid cotton town of Huntsville. Von Braun and members of his rocket team had opted to stay in America and now they moved as a group to Huntsville. With the reopening of the Redstone Arsenal, Huntsville rapidly grew from the watercress capital of the South to the "Rocket City" of the United States.

After living in a gated compound under close watch, the German scientists loved the freedom of Huntsville. The verdant, hilly landscape reminded them of Europe and the avid hikers in the group felt at home. Ernst and Irmgard Stuhlinger found a hilltop south of town and built a home there with a yard and garden that rolled to the edge of a scenic, wooded ravine. "Others followed us and so this area became known as 'Kraut Hill,'" Stuhlinger said. A glimmering needle of light just visible from the yard today is the towering Saturn V rocket that Stuhlinger helped design to launch American astronauts to the moon.

But in the early years at the Redstone arsenal, von Braun and the German rocket team designed the surface-to-surface Redstone missile, a huge piece of boiler-plate fashioned after the V-2. The Redstone evolved into a more powerful intermediate-range ballistic missile called the "Jupiter," and provided the booster for a four-stage test rocket, the Jupiter C. Recycled railroad tank cars trenched into earthen berms to serve as observation bunkers still stand testament to the missile testing program at the site that became known as the Army

Ballistic Missile Agency in 1956 (and NASA's Marshall Space Flight Center in 1960). Even with the focus on military missiles, von Braun once again found opportunities to pursue his dream of developing rockets for space travel.

• • •

Public awareness over the likelihood of satellites changed quickly in the early 1950s with events such as the First Annual Symposium on Space Travel, hosted by the American Museum of Natural History in New York City to honor Columbus Day in 1951. Reporters who had scoffed at predictions about satellites shifted to a tone of boosterism and started to attend technical seminars just to see what those scientists were dreaming up next. The scientists didn't disappoint at a historic four-day symposium in San Antonio on the physics and medicine of space that year. Van Allen delivered preliminary conclusions on the impact cosmic rays could have on space travel. "Cosmic radiation does not represent an overwhelming obstacle to high-altitude and space flight by manned vehicles" and no hazard at all to passengers in aircraft traveling below 60,000 feet, he advised.

In a roundtable that wasn't part of the official sessions, several scientists, including von Braun, spun futuristic fantasies. "Connie, go to San Antonio and find out what these nutty Germans are doing," said *Collier's* editor Louis Ruppel, an ex-marine, to *Collier's* associate editor Cornelius Ryan, a former war correspondent. Ryan's report spurred an eight-part series of articles in *Collier's* in 1952 on the future of space travel. Kaplan, Whipple, and others wrote for the series but it was von Braun's descriptions of journeys to the moon and Mars that ignited space fever. He introduced the idea of a space station that would orbit the earth, and artists Chesley Bonestell, Fred Freeman, and Rolf Klep painted dramatic cosmic renderings to go with the articles. After that, Walt Disney hired von Braun to help turn space into popular culture in television shows and with the design for Tomorrowland at a new amusement park—Disneyland. For von Braun, the International Geophysical Year meant one thing—the opportunity of fulfilling his dream of launching a satellite.

Scientific societies worldwide immediately bolstered Chapman's proposal for the International Geophysical Year (IGY). He, Berkner, and Hugh Odishaw, assistant to the director of the National Bureau of Standards, began the daunting task of forming an international network of committees to orchestrate a heavy schedule of research projects. They invited projects in polar geography and oceanographic studies but mainly focused on balloons and rockets. The IGY would map cosmic rays, ozone, and magnetic field readings from around the world, providing a one-time global snapshot of the upper atmosphere. Van Allen and a handful of other researchers slowly and painstakingly mapped the upper atmosphere point by point and trip by trip, sending rockets into the skies over years of time. Then they connected the dots to map the influences of solar flares and cosmic rays above the earth. An international collaboration promised to fill in thousands of dots simultaneously, and IGY organizers worked out a schedule of three World Days a month to target dates for intensive observations.

Chapman helped organize the Comité Spécial de l'Année Géophysique Internationale (CSAGI) in 1952 to coordinate research efforts for the IGY and was elected president of the group. Berkner became vice president. The National Academy of Sciences orchestrated U.S. participation and formed the U.S. National Committee (USNC) for the IGY late in 1952, appointing UCLA physicist Joseph Kaplan as chairman and Odishaw as executive director. The new committee set up fourteen technical panels in 1954–1955 to divide up research programs. Van Allen participated in the panels on rocketry, cosmic rays, and aurora as he planned thirty-six Aerobee flights and fifty rockoon launches among the projects for Iowa alone. The original panel lineup omitted one notable committee where he played a key role, however—the Technical Panel on the Earth Satellite Program.

Organizers hadn't considered satellites as part of the planning for the IGY at first. In one of the ironies of modern science, the IGY focused on rocket and balloon research but shepherded a tool that eclipsed them both. And though satellites seemed to be an afterthought for the international planners, behind the scenes, insiders seized the IGY as a catalyst for their development. After a meeting at the Jet Propulsion Laboratory early in 1954, Pickering, JPL's rocket designer Homer Joe Stewart, Van Allen, and Newell turned to a discussion of "long-playing rockets" that could potentially establish an earth orbit. LPR (an acronym for 33 1/3 long-playing record) became a handy code name in preliminary deliberations about U.S. satellites. It eventually became a standard code name for early satellite proposals, including a joint army-navy venture.

The Office of Naval Research (ONR) made an overture for this joint project in February 1954. Fred Durant, an operative of the Central Intelligence Agency (CIA) and a former president of the American Rocket Society, knew exactly where to find designs for a satellite in the U.S. military. In the best tradition of a spy thriller, Durant approached the Huntsville team on behalf of ONR Commander George Hoover and laid the groundwork for a collaborative army/navy venture that laid the foundation for *Explorer I,* America's first satellite. Hoover, Whipple (a member of the rocket panel), Singer, Durant, and a few others met with von Braun and his team. Von Braun proposed building and launching a satellite with a Redstone topped by upper-stage clusters of

JPL's Lokis, the same rockets used for the rockoons. The program, dubbed Project Slug, incorporated Singer's idea for a Minimum Orbital Unmanned Satellite of the Earth (MOUSE), a small, scientifically instrumented, five-pound concept satellite that he had introduced at a meeting of the International Astronautics Federation in Zurich, Switzerland. The MOUSE, in turn, was based on a British satellite proposal using a V-2 rocket booster. The V-2 plan thus came full circle back to its creator.

Singer presented the MOUSE to the rocket panel Van Allen chaired in April. By May, Van Allen had a much bigger stake in the unfolding satellite saga. Stuhlinger first approached Van Allen about an experiment for Project Slug in May 1954 while Van Allen was at Princeton still working on fusion research. Abbie invited Stuhlinger to dinner and, after dinner, the two men settled on the sofa in Van Allen's study. Van Allen lit his pipe. He and Stuhlinger discussed the Iowa successes in cosmic ray research with the rockoon excursions. Then Stuhlinger filled him in on the plans afoot to launch a small scientific satellite and sought Van Allen's participation in building a cosmic ray experiment to carry into space.

"Van Allen sat [on] his sofa without motion. The only sign of life was the vivid smoke production of his pipe," Stuhlinger recalled. "He showed no signs of interest, let alone excitement. All he said was, 'Thanks for telling me all this. Keep me posted on your progress, will you?'" Stuhlinger left Princeton thinking he had failed in his mission. He had only failed in reading through Van Allen's caution.

"I thought it was a legit idea and I have the highest respect for him. My reaction was one of keen interest but [the idea] was also pretty speculative. I decided to keep in touch and I did," Van Allen said. "It was obvious that only a few days of satellite observations would be equivalent to years of work by the rocket technique. Also, the necessary instrumentation [Iowa had] was compatible with the limited weight, electrical power, and telemetry capacity envisioned by von Braun."

Van Allen said nothing of the collaboration, but interest in satellites heated up as the rocket panel sent out notices that summer about a satellite briefing scheduled for the September 8 meeting. The agenda indicated that satellite proposals by the various branches of the military would be made. A flurry of special requests came to Van Allen by scientists, military officers, and corporate leaders seeking permission to attend the meeting. Van Allen opened the way for attendance as long as guests could meet security clearance requirements.

Early in September, the von Braun team proposed Project Slug to army authorities at a joint army/navy meeting and got preliminary approval to proceed as long as the work didn't interfere with missile development. Hoover prepared to present the project at the September meeting of the rocket panel.

The secret meeting is reported in a handwritten set of minutes recorded by panel secretary George Megerian on twenty-five tightly filled pages of five-hole ring-binder graph paper. Megerian typically sent minutes to Van Allen, who edited them and had them typed and distributed. In this case, there is no evidence of editing with Van Allen's red pen and no typed version in files kept by both Newell and Van Allen. Megerian's yellowing hand-written sheets may be the only record of the nation's first meeting on a satellite program.

Van Allen greeted the fifty-six scientists, military officers, and corporate executives who filled the NRL auditorium that day, swelling the attendance at the usual fourteen-member panel meeting. "The purpose of the meeting was to try to assess the future prospects of high altitude and satellite vehicles in an effort to assist the panel in planning its programs from the standpoint of time scale, capabilities of the vehicles and the amount of free payload," Megerian wrote, summarizing Van Allen's comments.

Whipple spoke next, outlining development steps for "sending objects into space," Megerian reported. "A) An observable object such as a MOUSE (church species)—employ radar techniques for observations. B) Unmanned physical laboratories—perhaps 50 pounds in weight" with a TV, telescope and remote control robotics. And what about that "church species" satellite? "Jim, I couldn't resist this," Megerian added in a circled note to Van Allen.

Then came the moment everyone had been waiting for. Hoover detailed plans for a satellite. "ONR also is engaged in Project Slug, which contemplates keeping a research vehicle aloft for 10 days. Cosmic rays, meteors and solar rays will be studied. Project Slug will be a joint operation of the services. The project has been approved tentatively and ONR is preceding with the various stages involved." He indicated Singer would probably do the orbital studies and Aerojet (with its close association with JPL) would design the multistage rocket using the Redstone booster. "Construction of satellite vehicle—unassigned," the minutes note, referring to an actual construction contract. Both Megerian's minutes and Van Allen's handwritten notes indicate that the satellite design for Project Slug was to be the minimum-weight "von Braun satellite" projected to reach altitudes up to 200 miles.

That was merely phase one. "At the termination of Project Slug, Project Mouse will be initiated. Mouse will also have an orbit at 200 miles altitude but will remain aloft for a month," Hoover stated, predicting a 1957 launch for this larger, 100-pound satellite.

And then Hoover issued a warning: "The technological prestige in being first with a satellite vehicle is of tremendous importance to the defense establishment. The Russians appear to have a 10-year program pretty well laid out, which means that we in the United States must act NOW if we want to be first. The Russians MUST NOT be underestimated."

Gerhard Heller from the Redstone Arsenal followed Hoover and filled in the panel and guests on missile development, the Redstone launcher and plans for a satellite program in Huntsville. That was it for satellites. Other speakers addressed development of supersonic jets, new rockets, and the IGY. The rocket program had a projected budget of nearly $1.4 million for 1955 alone for rockets, rockoons, and associated costs.

But after the September 8 meeting, rocket panel members joined Singer and an upsurge of other scientists urging Berkner to include satellites in the International Geophysical Year. "Everyone assumed that Project Slug would be our satellite," said Van Allen. Almost everyone—Milt Rosen, the father of the Viking rocket, thought he and the NRL where he worked could do better than a five-pound satellite. He was just completing a two-year study on space exploration as part of an ad hoc committee for the American Rocket Society. Rosen, the committee chair, gave high priority to satellites in his committee report. But the committee advised that a scheme to create a satellite as just a stunt wouldn't justify the cost. Rosen began to plan a far larger research satellite to be launched using a Viking rocket. Suddenly, the Huntsville satellite program, now called Project Orbiter, had a formidable rival.

Berkner, Singer, and Newell prospected for international support for an IGY satellite program. On the evening of October 3, 1954, the three of them joined most members of the U.S. National Committee and brainstormed into the night about a satellite mission statement they presented to the CSAGI for a vote the next day in Rome. They essentially presented the concept behind the MOUSE—a small research satellite that could be launched with existing technology and with just enough room for instrumentation to give the project scientific merit. The CSAGI adopted the proposal with little discussion. The announcement by the Russian Embassy in Rome that the Soviet Union had decided to participate in the IGY stole the show. It was an interesting turn of events since only two countries—Russia and the United States—had any hope of launching a satellite. The CSAGI, with its mission of international collaboration, had just fired the opening shot in the space race.

The satellite objective spurred Van Allen into action. "I hadn't had a chance to give much consideration to Stuhlinger's proposal earlier in the year," Van Allen said. Now, he pushed aside the rockoon data, academic responsibilities, graduate theses, and article deadlines for a few days and sketched basic plans for a new, more compact cosmic ray detector that relied on newly developed transistor technology. The V-2 had offered more than 1,000 pounds of payload space for experiments, the Aerobee offered 150 pounds, and the rockoons accommodated up to 35 pounds.

Van Allen struggled this time to scale down an instrument package for a satellite with an expected cargo space of five to ten pounds. On November 1, he wrote a four-page physics department memorandum "Outline of a Proposed Cosmic Ray Experiment for Use in a Satellite (Preliminary)," his earliest document of plans for such an instrument. Ludwig began piecing together a prototype on a table in the northwest corner lab of the basement of the physics building, already cramped with experimental instruments in various stages of completion.

On November 15, 1954, the army received authority from the Department of Defense to develop the Jupiter, an intermediate-range ballistic missile using a modified Redstone rocket. The Huntsville rocket team began developing it to do double duty as the launch vehicle for the proposed satellite. The Jupiter was fitted with upper-stage clusters of solid-propellant rockets. To reach a target, nose cones had to stand up to the friction of reentry into the atmosphere without overheating or causing missiles inside to fire prematurely. Von Braun's team began to test the nose cones and the reentry test vehicles became known as the Jupiter C. Stuhlinger kept Van Allen in the loop on progress regarding the satellite application. "With tongue in cheek, von Braun decided that one of the Jupiter C vehicles should be set aside and carefully subjected to a 'long-term storages test'—it was quietly understood that this vehicle represented a potential satellite launch rocket," Stuhlinger noted.

With the IGY ready to put satellites on public spectacle, advisors to President Eisenhower alerted him to the strategic military potential of satellites for reconnaissance and the question of overflight. Satellites posed a tricky political boundary involving "freedom of space." At what altitude would a nation's protected air space become international outer space so that "overflight" would be tolerated? Edwin Land, founder of Polaroid Corporation and a technical advisor to the administration, recommended a seemingly painless solution: let a small, innocuous scientific satellite establish the legality of "freedom of space," smoothing the way for military reconnaissance satellites later.

The IGY initiative offered the Eisenhower administration the perfect cover for launching a scientific satellite and establishing "freedom of space." Science adviser Herb York and the Department of Defense brass backed the idea. The administration stipulated, however, that satellite development for IGY should not interfere with military development of ballistic missiles.

"In the early 50s the U.S. was not as advanced as the Soviets in the development of long range rockets. So we effectively started a crash program and when the question of the satellite came up, the government said let's not get this mixed up with this crash program," Pickering said.

The U.S. National Committee for the IGY made a formal call on March 5, 1955, for a U.S. satellite project to participate in the IGY. A month later, Moscow radio announced Russia's intention to participate as well. Van Allen and his colleagues now expected Project Slug to win designation as the U.S. satellite entry. But, as the Huntsville scientists pushed ahead, Rosen announced the competing project he had spearheaded at NRL. He proposed an adaptation of the Viking rocket with single, upper-stage Aerobee-Hi rockets, a new but untested higher-flying Aerobee under joint development by the navy and air force. Rosen sent Kaplan his proposal for the Viking-Aerobee hybrid to launch a satellite with 40 pounds of payload for experiments. At the suggestion of his wife, he called the NRL candidate the Vanguard.

With the sudden competition, the U.S. National Committee for the IGY directed Kaplan, as chairman, to select the satellite program. He initially chose the Vanguard in a letter to Alan Waterman, director of the National Science Foundation. He pointed out that Vanguard's Viking and Aerobee components established "the civilian character of the endeavor." He referred to the Huntsville project only as the "German V-2 developments," suggesting the connection to war and old wounds.

Kaplan turned the tables on the army project with his open emphasis on a civilian satellite. The position—fraught with the assumption that the Vanguard, built under the auspices of the U.S. Navy, would be viewed as a civilian project in Russia—triggered an immediate military feud. Army brass challenged the summary dismissal of their program for the upstart Vanguard proposal.

The National Security Council stepped in and muddied the waters further by directing that no rocket launcher under development for defense purposes should be used for the IGY satellite. The directive seemed to eliminate any debate, since Jupiter C development fell clearly within the ballistic missile program. Still, it was a test rocket rather than an actual ballistics launcher, it had a proven track record, and it held the likeliest promise of getting a satellite into space quickly.

Don Quarles, assistant secretary of defense for research and development, formed The Ad Hoc Advisory Group on Special Capabilities to settle the matter and select the official IGY rocket and satellite from three contenders: the Air Force Atlas, the ONR/army program now called Project Orbiter, and the NRL's Vanguard. The original June 27, 1955, committee charter called for a nine-member body, but only eight served. Quarles picked two of the eight members and the army, navy, and air force each picked two. Kaplan was named to the committee in his capacity as the chairman of the U.S. National Committee for IGY. His national position makes it likely that Quarles selected him, an open Vanguard supporter. As to the appointment of Orbiter advocate Homer Joe Stewart of JPL as chairman of the committee, the declassified committee papers make it clear that it was Quarles's job to name the chairman and he named the man who had personally helped develop the upper-stage rockets in the Jupiter C design.

The ad hoc committee quickly became known as the Stewart Committee. Before it ever met, Homer Joe Stewart endorsed the Orbiter for the official IGY satellite in a letter to Paul A. Smith, secretary of the General Sciences Staff of Quarles's office and one of two Defense Department staff members assigned as nonvoting members to the Stewart Committee. As for the National Security Council directive regarding military launchers, Stewart discounted that in his letter as nonbinding and advisory only.

"What I disliked was that almost none of the [Vanguard] existed even on paper yet," Stewart said in one of his last interviews, taped for the University of Iowa production *Flights of Discovery,* a television biography about Van Allen. "Things like that are more apt to take five or six years than two or three. The Redstone had a good operational history."

Stewart had six weeks to make that case with the committee as it met in early July in Washington, D.C., and later in the month, in Pasadena at JPL headquarters. The committee convened with two key members—Stewart and Kaplan—supporting rival programs and a series of fuzzy distinctions between what constituted military and civilian projects. All the rockets to launch the satellites would be built under defense contracts, after all. The committee eliminated the Air Force Atlas as an entry almost immediately because its development for the intercontinental ballistic missile program clearly crossed the line into interference with weapons development.

Stewart said that committee members should settle next which American rocket could launch a satellite into orbit first, or, more precisely, ahead of the Russians. Project Orbiter, entered the scene as the frontrunner given that consideration. The Orbiter relied on existing rockets for launch and Stewart backed it as the only feasible choice if the U.S. intended to win the race for the first satellite in space. But Rosen's Vanguard impressed the committee with superior payload space for scientific experiments and its proposed international picket fence of ground tracking stations known as the Minitrack network. And it retained Kaplan's influential support. When the votes were cast, there were two immediate votes for Orbiter (and a later proxy vote) and five for Vanguard. The committee that bore Stewart's name selected the program he opposed, but many critics down the years gave him all the credit for picking the wrong rocket.

"Word got around that the Stewart Committee had made this decision so it was sort of imputed that Stewart himself was in favor of it. But that's contradicted by the evidence," said Van Allen, viewing copies of the papers from the Stewart Committee for the first time nearly fifty years later. "I think very few people knew the true story so it was sort of blamed on Homer Joe Stewart because he was chairman of the committee and the committee carried his name. They were the rascals who made that decision."

The Stewart Committee issued its recommendation in favor of the Vanguard on August 3, 1955, in a classified majority report, focusing on optimism about the Vanguard and on the overflight issues. Stewart challenged key positions in his minority report. The majority report focused on the anticipated scientific superiority of the Vanguard. The design offered 10–40 pounds of payload, as compared to 5–10 pounds with the Orbiter. The NRL schedule for Vanguard, based on estimates from Viking rocket contractor the Glenn L. Martin Company, convinced the Stewart Committee that the Vanguard could be ready in eighteen months, the same amount of time needed to fully develop and test the Orbiter.

Rosen, as technical director of the Vanguard, originally estimated that the project needed thirty months for development, he told Constance McLaughlin Green as she researched the first full-length history of the Vanguard. But Rosen also told her he felt "pressured" to support the more optimistic eighteen-month timetable. With all things seeming equal in terms of scheduling, committee members opted for the promise of a superior scientific satellite, according to the majority report.

"Unanimity prevailed on one point: the superiority of the NRL tracking system and indeed all the NRL satellite instrumentation," Green reported. Stewart and scientist Clifford Furnas, chancellor of the University of Buffalo, challenged other aspects of the Vanguard program, including the timetable in the majority report. They reiterated that the Orbiter components already existed while Vanguard was purely a design on paper. Committee member Robert McMath, University of Michigan astronomer, ill and unable to participate in the committee meetings, also voted for the Orbiter by proxy.

"I think they were optimistic about the Vanguard propulsion development coming along speedily and I wasn't. I think that was the essential problem," Stewart said. In the end, his committee bowed to the political expediency of backing a civilian satellite that relegated that satellite to low priority for funding and development. Overflight, interservice rivalries, and prejudice against von Braun played key roles in that decision.

"Overflight," the sticky problem of satellites traveling over other countries, presented a major Cold War political dilemma for the Eisenhower administration. Eisenhower's advisors recommended that he establish "freedom of space" with an innocuous IGY civilian satellite and then use satellites for reconnaissance. Walter McDougall's Pulitzer Prize–winning 1985 book, *The Heavens and the Earth, A Political History of the Space Age* argued that the administration's "political insistence" on a civilian satellite program pressured the committee to back the Vanguard.

Quarles himself addressed the committee on the importance of a civilian image and the majority report indicated that the May 20, 1955, National Security Council directive regarding military launchers influenced their recommendation of the Vanguard. Yet Quarles chose Stewart to head the committee even though Stewart's correspondence make obvious his critical and dismissive attitude toward the administration's position. His expertise in rocketry certainly gave him the credentials to lead the committee.

But behind the scenes, military rivalries carried even more potential weight. Michael Neufeld of the National Air and Space Museum scrutinized affiliations of members of the committee and concluded this was a more important factor than the overflight issue. Kaplan and committee members with air force or navy affiliations tended to vote for the Vanguard. The support gave the Vanguard five votes compared to three votes (including army supporters Furnas and Stewart, with his JPL connections to Huntsville) for the Orbiter. Furnas, who voted with Stewart in favor of the Orbiter, later stated in a *Life* magazine article that the Stewart Committee would have liked to combine the proven track record of the Jupiter C with the superior instrumentation and tracking potential offered by the Vanguard but didn't dare go there. "We finally decided," he wrote, "that breaking the space barrier would be an easier task than breaking the interservice barrier."

In addition, the Orbiter encountered prejudice from those who felt that America's IGY satellite should not be developed by the German rocket team in Huntsville, according to Green and McDougall. Fred Whipple and Ernst Stuhlinger expressed the opinion that this issue was the *deciding* factor in the selection of the Vanguard even though the members of the Huntsville rocket team had become American citizens by then.

The deep-seated resentment of von Braun had already surfaced more publicly in a 1952 *Time* magazine article responding to his wildly popular portrait of the future in space, published in *Collier's* magazine earlier in the year. *Time* asked insiders to appraise the realities of von Braun's vision for a satellite space station with shuttles commuting to and from Earth to ensure preemptive U.S.

military domination of space. In the article, Rosen, who spearheaded the NRL's Vanguard project, offered his assessment that the space station had little military value. But an unidentified missile expert assessed von Braun instead: "Look at this von Braun. He is the man who lost the war for Hitler. His V-2 was a great engineering achievement but it had almost no military effect and it drained German brains and materials from more practical weapons. He was thinking of space flight, not weapons, when he sold the V-2s to Hitler. He says so himself. He is still thinking of space flight, not weapons, and he is trying to sell us a space flight project disguised as a means of dominating the world."

• • •

Eisenhower's press secretary James Hagerty announced the American satellite program at a White House press briefing on July 28, 1955, and, a few days later, the Soviets announced they would launch a satellite into orbit during the IGY as well. At an IGY meeting in Copenhagen in August, Soviet scientists announced their satellite would be far bigger than anything the Americans were planning. The branches of the U.S. military discounted Russia's claims and continued to wrangle over the IGY satellite program. The army once again adamantly objected to the Stewart Committee decision and asked for a rehearing, which the committee granted on August 16. Quarles offered a compromise involving all branches of the military in the Orbiter program. The committee rejected the compromise and stood by its original recommendation in a final report issued August 23.

With the Stewart Committee decision made, Secretary of Defense Charles Wilson ordered the Huntsville team to abandon work on a satellite. The team retained permission to continue reentry testing with the Jupiter C. That gave Huntsville adequate "cover" to sidestep orders and proceed with satellite development. Stewart and Pickering visited Huntsville and encouraged the rocket team to continue working in secret on the Orbiter program as a backup to Vanguard. Pickering offered special funds he had available for general research to supplement the funds Medaris bled from general operating sources to help support the effort.

Van Allen made only a terse reference to the selection of the Vanguard in his journal: "recent public announcement" that the navy will coordinate satellite vehicle development for the IGY. He, like Stewart, considered the Vanguard a "paper proposal." Never one to dwell on defeats, he and Ludwig began adapting their cosmic ray counter designs to the Vanguard without waiting for payload specs. They assumed that their traditional tubular shape for the instrument would be fitted into a spherical satellite and Van Allen estimated the cost at about

$66,000 for three years of developing test payloads and actual flight payloads. He sent the IGY's first satellite instrument proposal to Kaplan on September 28, 1955. The straightforward plan reinforced Van Allen's reputation in Washington as a "backyard scientist" who believed in keeping things simple. A week after it arrived, Kaplan called Van Allen to ask him to serve on the newly formed IGY Technical Panel on the Earth Satellite Program, the TPESP.

TPESP met for the first time on October 20. Kaplan selected as chairman Richard Porter, the General Electric engineer who had voted with the majority on the Stewart Committee. The TPESP established four working groups. As chairman of the rocket panel, the organization with more collective experience than anyone in loading experiments into rockets, Van Allen headed the Working Group in Internal Instrumentation. The subcommittee had a clear agenda: select the experiments with top priority to go into orbit in the satellite.

The rocket panel celebrated its tenth anniversary soon after by hosting the historic symposium on the "Scientific Uses of Earth Satellites," in January at the University of Michigan at Ann Arbor. The symposium opened January 26, 1956, and offered the first comprehensive overview of work contemplated in a new arena for research. But even now, with a decade of rocket research experience available, only fifty scientists attended, almost the entire American population of scientists with the expertise to develop a satellite instrument. Van Allen edited the thirty-three papers presented at the symposium and published them as a book, *Scientific Uses of Earth Satellites.*

Meanwhile, Whipple enlisted an army of amateur stargazers around the world and institutional observatories to provide tracking of the satellites. Whipple's working group on tracking also enlisted ham radio operators who could fill in the "picket fence" of worldwide radio receiving stations being outfitted by NRL for the telemetry network to record data from the satellites.

• • •

The initial $18 million estimated cost for the proposed satellite program—a small percentage of the ultimate cost—nearly stalled the program at the start. The budget amounted to two-thirds of the IGY funding sought from Congress. Congressmen bombarded Kaplan with questions about why satellite data would be shared with other countries at U.S. taxpayer expense, what information could be found to share, and how scientists would collect it.

"We literally talk to the satellite, using radio waves, which is a technique known as telemetering," Kaplan said. His patient and unflappable responses to the inquisition sold the satellite. In addition, Congress had to authorize $26 million for the Defense Department to build the rockets that would launch the satellites. As costs escalated with a revised estimate of $63 million, Eisenhower pared back the launches but the NRL already needed an extra $6 million just to keep the Vanguard construction afloat. The national committee for the IGY reluctantly kicked in a share of its money. Hand-to-mouth funding added to the technical difficulties that soon slowed development of the Vanguard.

• • •

Van Allen anticipated "substantial budgetary support" of $390,000 from various sources during the IGY period for Iowa's programs with balloons, rockets, rockoons, and the satellite instruments. It was quite a transition from the lack of any research funds a few years earlier but still a tight budget, and Van Allen watched costs closely. For one 1956 trip to Washington, D.C., for a meeting, he spent $107 on airfare and cabs and a total of $15 for a hotel, meals, and all other costs. Van Allen and the Iowa space team parlayed every area of research at their disposal into the IGY programs.

McIlwain targeted studies of the aurora, launching Aerobees and Nike-Cajun rockets at Fort Churchill in Canada. Van Allen and Cahill mapped out summer and fall rockoon expeditions to the Arctic and Antarctic, fabricating new instruments to measure electrical currents and cosmic rays. Ludwig spent hundreds of hours working on the satellite instrument with Van Allen advising. Ludwig also began to develop an ingenious tape recorder for the instrument so that Iowa could have a continuous read of data rather than just data captured by the radio receivers along the picket fence. "I located and tested subminiature gears, ball bearings, switches, recording and playback heads, magnets, solenoid, tape and other components," he noted in the memoir he is writing. His neighbor, Iowa City watchmaker Eugene Boley, helped him locate watch parts and tools.

Then Ludwig and Van Allen confronted another major retooling of the instrument. Despite all the promises for payload space, Vanguard engineers now shaved back their design to a spherical satellite that would accommodate 2.2 pounds of experiments instead of 10–40 pounds.

Van Allen's Working Group on Internal Instrumentation issued a formal invitation for the submission of satellite instrumentation proposals. They received twenty-five and met June 1, 1956, in Washington, D.C., to recommend Priority A flight status for the top proposals based on scientific importance, technical feasibility, the field record of the participants, and the importance of global data. The Iowa cosmic ray detector made the short list that gave four experiments a seat on the "first few" Vanguard launches along with NRL experiments to identify temperature and the general environment of space.

James Van Allen, University of Iowa graduation photo. Courtesy of James Van Allen.

Lieutenant James Van Allen, U.S. Navy Reserve, 1943. Courtesy of James Van Allen.

Navy gunner's mates replaced thousands of deteriorated batteries in the proximity fuzes for antiaircraft shells at stations such as this one at Tulagi in the Pacific during 1944. Van Allen led the rebattery effort as part of his mission to bring the fuze he helped develop into battle. Courtesy of the Applied Physics Laboratory of Johns Hopkins University.

Lieutenant James Van Allen in New Guinea, 1943. Courtesy of James Van Allen.

Van Allen marries Abigail Fithian Halsey on October 13, 1945. Courtesy of Cynthia Van Allen.

Physicist Ernst Stuhlinger in his quarters at Fort Bliss, Texas, where the German rocket team lived as they helped rebuild and launch captured V-2 rockets at the White Sands Proving Ground in New Mexico, circa 1946. Courtesy of Ernst Stuhlinger.

V-2 science package and nose cone at the Applied Physics Laboratory. Van Allen, standing at the far right, headed the lab's rocket team that launched cosmic ray detectors and other experiments in the V-2 rockets starting in 1946. Courtesy of the Applied Physics Laboratory of Johns Hopkins University.

V-2 rocket with launcher and ladder at White Sands Proving Ground, 1947. Courtesy of the Applied Physics Laboratory of Johns Hopkins University.

While on a research trip to the Antarctic in support of the Apollo *moon mission, Wernher von Braun and members of his rocket team step outside for a quick, cold portrait, 1966. One of the "regulars" at the polar station indicated this was a tradition. Courtesy of Ernst Stuhlinger.*

Van Allen helped design the Aerobee research rocket. He launched it from the USS Norton Sound *to measure cosmic rays in 1949 and 1950. "These expeditions pioneered the extension of scientific rocketry to a wide range of geographic locations," Van Allen wrote. Courtesy of the Applied Physics Laboratory of Johns Hopkins University.*

Van Allen returned to the University of Iowa to head the physics department in 1951. Department faculty, front row, left to right: James Jacobs, John Eldridge, Joseph Jauch, James Van Allen, Edward Tyndall, and Richard Carlson; back row, left to right: S. Bashkin, Melvin Gottlieb, Edward Nelson, Fritz Coester, and George Stewart. Courtesy of James Van Allen.

University of Iowa physics students launch scientific equipment with balloons, 1952. Courtesy of James Van Allen.

Van Allen and Frank McDonald prepare for a launch of cosmic ray detectors in the gondola of a high altitude balloon at the Iowa City airport, March 13, 1956. Courtesy of James Van Allen.

Van Allen and University of Iowa faculty and graduate students launched rockoons (rocket-launched balloons) from various military ships to study cosmic rays from 1952–1958. U.S. Navy photograph from James Van Allen's collection.

Van Allen, graduate student Leslie Meredith (standing), and Lee Blodgett assemble cosmic ray detectors for a rockoon launch, working in their makeshift lab in the afterhold of the U.S. Coast Guard Cutter Eastwind *in 1952. U.S. Navy photograph from James Van Allen's collection.*

Van Allen and Lieutenant Malcolm S. Jones inspect and prepare detectors for a rockoon launch from the USCGC Eastwind. *In the area of science research rockets, a rockoon cost about $1,800 to build and launch compared to about $450,000 for a Viking in the same era. U.S. Navy photograph from James Van Allen's collection.*

(left) The USCGC Eastwind *navigates glaciers and ice floes near Greenland, 1952. U.S. Navy photograph from James Van Allen's collection.*
(right) The USCGC Eastwind *maneuvers through the ice off the coast of Greenland, 1952. U.S. Navy photograph from James Van Allen's collection.*

Left to right: Gunner Mate Gordon, Iowa graduate student Lawrence Cahill, Van Allen, and Lieutenant Stephen Wilson prepare a Loki II rocket for a rockoon flight from the USS Glacier *in 1958. U.S. Navy photograph from James Van Allen's collection.*

Liftoff of a rockoon. U.S. Navy photograph from James Van Allen's collection.

Rockoon launch. U.S. Navy photograph from James Van Allen's collection.

Left to right: William Pickering of Jet Propulsion Laboratory, James Van Allen, and Wernher von Braun hold aloft a full-scale model of Explorer I *at a press conference in the Great Hall of the National Academy of Sciences, February 1, 1958. Courtesy of James Van Allen.*

SATELLITE EXTRA

The Huntsville Times

SATELLITE EXTRA

Jupiter-C Puts Up Moon

Eisenhower Officially Announces Huntsville Satellite Circles Globe

Wail Of Sirens Brings In Era On Space Here

Thousands Gather On The Square For Noisy Success Demonstration

9 Labs Here Aided Project Of Launching

Weather Change Sped Launching

AUGUSTA, Ga. Feb. 1 (AP)—President Eisenhower announced early today America's first satellite is in orbit around the earth.

The President's dramatic announcement was issued at his vacation headquarters a few minutes before ... a.m. EST by White House press secretary James C. Hagerty.

The satellite was launched at Cape Canaveral, Fla. at 10:48 p.m. EST last night.

Army Reveals Second Moon Is Scheduled

70-Foot Carrier Roars Into Starry Night At Cape Canaveral

Explorer I *made headlines across the world, but people celebrated in the streets in von Braun's home base of Huntsville, Alabama, where the* Huntsville Times *identified the satellite as a "second moon." Von Braun's rocket team developed the Jupiter C satellite launcher at the Army Ballistic Missile Agency in Huntsville.*

Abigail Van Allen models her dress borrowed for the state dinner at the White House on February 4, 1958, in honor of Explorer I. *Courtesy of Abigail Van Allen.*

Abigail and James Van Allen leave for dinner at the White House to celebrate the Explorer I *victory. Courtesy of James Van Allen.*

Walter Cronkite comes to Iowa City to interview Van Allen about satellites and radiation belts, 1958. Courtesy of James Van Allen.

Van Allen and faculty member Ernest Ray, a former Iowa graduate student, examine the orbit of Explorer I *in 1958. Courtesy of James Van Allen.*

Left to right: Sarah, Cynthia, Tom, Abbie, and Margot Van Allen, circa 1958. Courtesy of Cynthia Van Allen.

Van Allen and the discovery of the radiation belts make the cover of Time *magazine on May 4, 1959.*

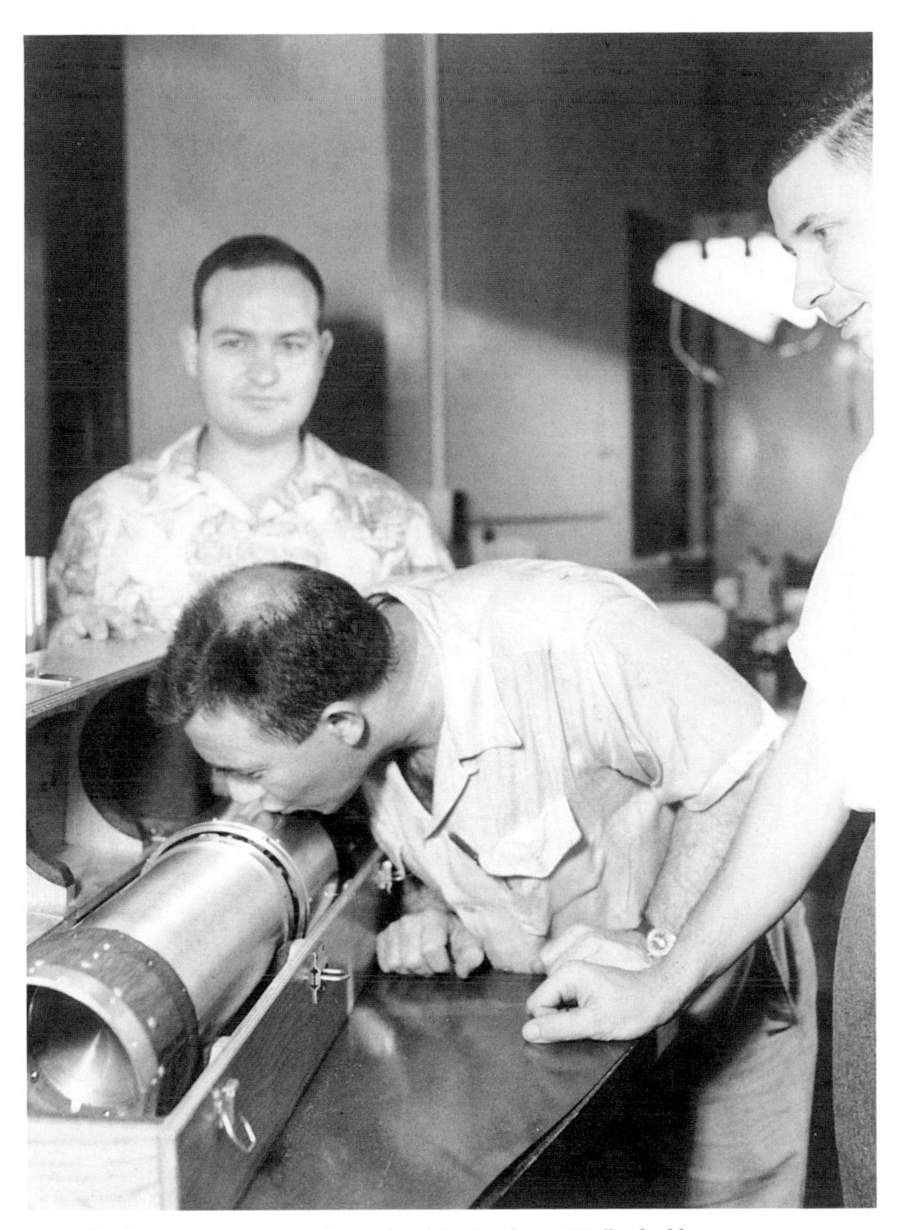

Van Allen kisses Iowa's scientific payload for Explorer IV, *flanked by graduate students Carl McIlwain (left) and George Ludwig (right). Courtesy of James Van Allen.*

Van Allen assesses data from Explorer IV *that poured in from Singapore, Lima, and other locations. Photo by George W. Black, Iowa City, courtesy of James Van Allen.*

Injun 3 *launched on December 12, 1962. This was one in a series of satellites built by the University of Iowa, the first university to undertake successful spacecraft design and construction starting with* Injun 1, *launched on June 29, 1961.*
Courtesy of the Department of Physics and Astronomy, University of Iowa.

University of Iowa student Don Gurnett's VLF (very low frequency) radio receiver on Injun 3 *captured a chorus of sounds that could be traced to radio emissions from the Van Allen radiation belts. The belts literally sang a song composed in outer space. Courtesy of the Department of Physics and Astronomy, University of Iowa.*

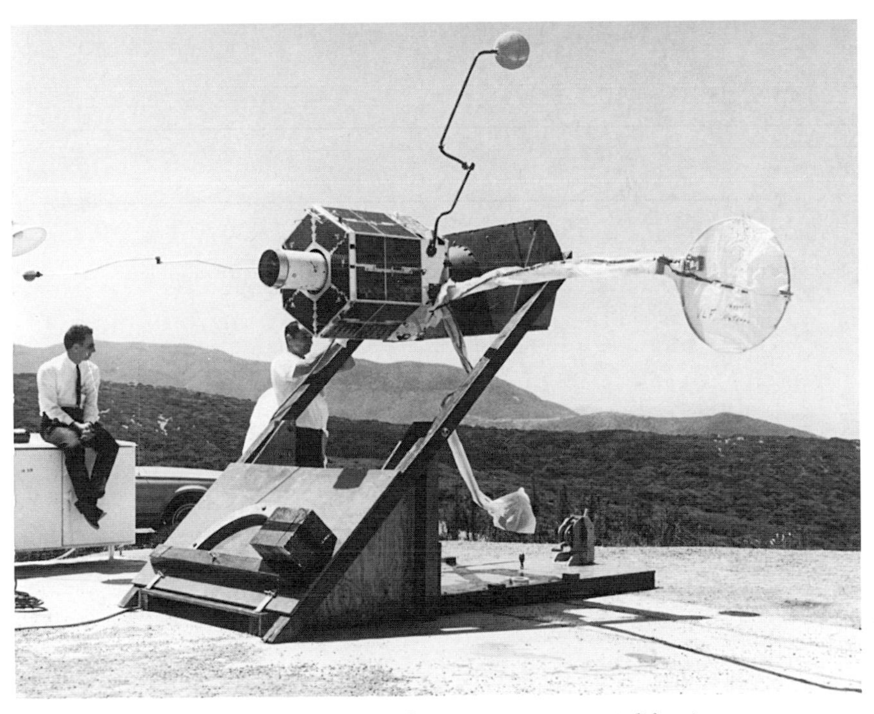

Outdoor testing of Injun 5, *Pacific Missile Range in Lompoc, California, 1968. Courtesy of the Department of Physics and Astronomy, University of Iowa.*

Drafting technicians, left to right: John Birkbeck, Jeanna Wonderlich, Sala Snipfer, Everett Williams, and Joyce Crisinger with Iowa's Hawkeye *satellite, 1974. Courtesy of the Department of Physics and Astronomy, University of Iowa.*

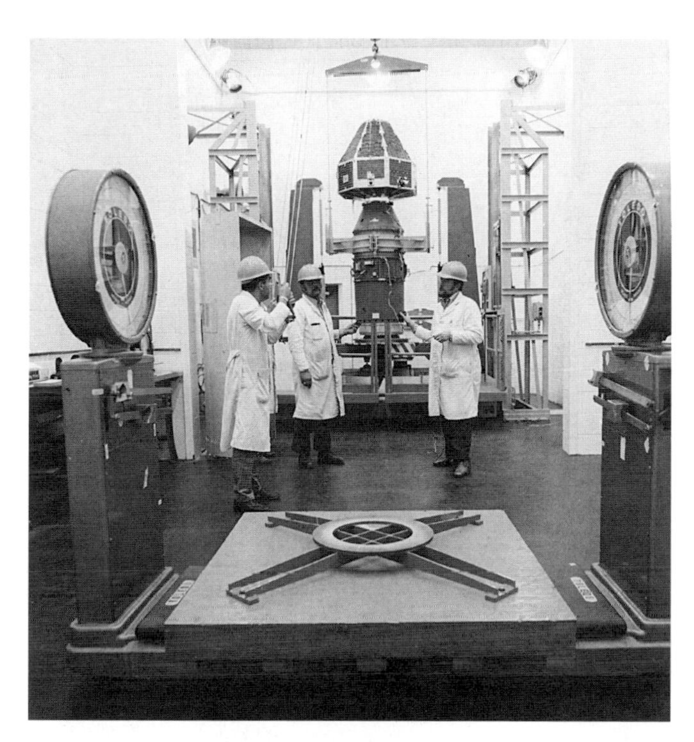

Both photos: Hawkeye *satellite undergoes testing, 1974. Courtesy of the Department of Physics and Astronomy, University of Iowa.*

Hawkeye *satellite testing activities. Left to right, standing: Roger Randall, Mike Nowack, and Doug Byal; left to right, front: Elwood Kruse and Lowell Swartz, 1974. Courtesy of the Department of Physics and Astronomy, University of Iowa.*

Sketch of the Pioneer 10 *space probe, which journeyed nearly 8 billion miles, transmitting data from the last functioning experiment—Van Allen's charged particle detectors—through its thirtieth anniversary on March 3, 2002. The spacecraft fell silent a year later but continues to travel toward the red star Aldebaran, the eye of the Taurus constellation. Courtesy of James Van Allen.*

Van Allen's motorized model of the radiation belts of Jupiter, a big hit at press conferences held at NASA's Ames Research Center when Pioneer 10 *flew by the planet in December 1973. The University of Iowa shop hastily made the model and shipped it to Van Allen in California with a set of wires that he fashioned into the belts. Courtesy of the Department of Physics and Astronomy, University of Iowa.*

Van Allen received the National Medal of Science from President Ronald Reagan at a White House ceremony on June 25, 1987. Official White House photograph from James Van Allen's collection.

Abigail and James Van Allen in Stockholm, where Van Allen received the Crafoord Prize from the Royal Swedish Academy of Sciences in 1989. Courtesy of Abigail Van Allen.

His majesty King Carl XVI Gustaf awards Van Allen the Crafoord Prize, 1989. Courtesy of Abigail Van Allen.

Van Allen's family celebrated the honor of the Crafoord Prize in Stockholm, 1989. Front row: Andrew and Elizabeth Cairns. Second row, left to right: Sarah Van Allen Trimble, His Majesty King Carl XVI Gustaf, Her Majesty Queen Sylvia, Abigail H. Van Allen, Hilary Schaffner, Cynthia Van Allen Schaffner, and Margot Van Allen Cairns. Back row, left to right: Gerald Trimble, Andrew Cairns, Robert Schaffner, Peter Van Allen, James Van Allen, Thomas Van Allen, Lois Brady Van Allen, and Susan Carney Van Allen. Courtesy of Abigail Van Allen.

Van Allen at the National Science Foundation's National Radio Astronomy Observatory at North Liberty, Iowa, originally operated by the University of Iowa under Van Allen's auspices. Photo by and courtesy of Tom Jorgensen, University of Iowa.

Van Allen at the Van Allen Day dinner, October 9, 2004. Photo by the author.

Van Allen swaps anecdotes with past and present colleagues and friends of the University of Iowa space program who gathered from around the world on October 9, 2004, for Van Allen Day. The scientific colloquium, public lecture, and dinner celebrated his ninetieth birthday. Front row, left to right: Stamatios (Tom) Krimigis (Van Allen's former graduate student, currently at the Applied Physics Laboratory of Johns Hopkins University), James Van Allen, Abigail Van Allen, and Carl McIlwain (Van Allen's former graduate student, professor emeritus at the University of California, San Diego). Peter Van Allen stands behind his parents. Photo by and courtesy of Arthur A. Caudy.

James Van Allen in his office in Van Allen Hall, 2006.
Photo by and courtesy of Tom Jorgensen, University of Iowa.

Discovery, 1962, *painting by Robert Tabor. Left to right: Carl McIlwain, James Van Allen, George Ludwig, and Ernest Ray pore over charts of radiation data from* Explorers I *and* III, *their first indication that radiation belts encircled the earth. Original in color. Courtesy of the University of Iowa Museum of Art.*

With Iowa's IGY program firmly underway and the satellite instrument under Ludwig's care, the Van Allen's planned the annual family vacation. "We traded in our 1950 Ford station wagon and bought a new 1956 Ford Country Sedan station wagon" for $3,124 in preparation for the trip, Van Allen noted in his journal on August 8. "On 10 August, 1956, Abigail, Cynthia, Margot, Sarah, Thomas and I set out for our long-awaited vacation at our camp on Peconic Bay in Long Island. Driving our new car with luggage on top [and inside]. Went by way of Mount Pleasant for lunch with Mother Van Allen and with George and family."

Abbie's brother Charles Halsey was building a new room and deck and expanding the two-room cabin with much-needed space for the growing family. But even on vacation, Van Allen couldn't stay put. He gave an all-day series of lectures on the scientific uses of rockets and satellites at MIT on August 15 in Cambridge, Massachusetts, met Abbie and Cynthia in New York on the seventeenth, flew to Washington for a meeting of the Technical Panel on the Earth Satellite Program and then headed to the IGY planning session in Barcelona on September 8 where he delivered a report on the scientific uses of Earth satellites. In Barcelona, the Russians made continuing claims that they would soon have a satellite in orbit but, despite the call for public disclosure as part of IGY, they gave few details and the Americans continued to discount their remarks.

No one from von Braun's team attended the Barcelona conference as they prepared for an all-important test launch at Cape Canaveral. The group was now closer than the Russians to launching a satellite. Von Braun, Stuhlinger, and the Huntsville team made last-minute checks as the Jupiter C stood 68.5 feet high against a gantry at the cape on September 20, 1956. At blastoff, the rocket rose into the air flawlessly. Stuhlinger tried out his new apex calculator that determined when the rocket had crested and his detonator to fire the upper stages, consisting of clusters of Sergeant rockets from JPL. The second- and third-stage clusters broke away in turn, soaring to an altitude of 682 miles, traveling 3,400, and propelling an inert fourth-stage payload that could have been America's first satellite. The rocket had reached the thrust and altitude needed to inject a live fourth-stage satellite into orbit. But the nose cone carried only dead weight this time—military orders. Washington followed up on the orders with observers who visited regularly to make sure the Huntsville team didn't "accidentally" launch a satellite that the Defense Department had officially halted.

As the Jupiter C achieved its hollow victory, Pad 18A nearby stood empty. It was assigned to the Vanguard.

• • •

Indian summer lingered in Iowa City. In the glow of a warm crimson sunset on October 26, 1956, Van Allen and Tyndall closed an era at the university with a private memorial service for George Stewart. They walked slowly around MacLean Hall, distributing Stewart's ashes at the foundation of his physics building to fulfill his last wishes. After trips to Long island, Barcelona, and Washington, D.C., Van Allen continued digging out from under a mountain of page proofs, plans for experiments, and departmental obligations. Then the call came from Stuhlinger with the report that the Jupiter C could launch a satellite.

Van Allen promised to advocate adopting the army vehicle as the primary IGY satellite project. "I talked with Pickering several times in late 1956 and early 1957 and we agreed that this is what needed to be done." Van Allen said. At the December TPESP meeting, the Working Group on Internal Instrumentation (WWII) finally assigned the order for launching the four top priority experiments. The NRL instruments to measure solar intensity and the space environment got the first launch slot with the Iowa experiment scheduled second. NRL's experiment to measure magnetic fields got the third seat. An experiment to measure the natural radiation balance of the earth placed fourth. The University of Wisconsin and the U.S. Weather Bureau jointly backed this experiment.

Now all America needed was the vehicle to carry the instruments. The TPESP headed to the Glenn L. Martin Company on December 4, 1956, and saw the three stages of the Vanguard in various degrees of development. They saw models and mock-ups of the motors being built elsewhere. While the complete Jupiter C had already made a successful launch, the Vanguard remained a jigsaw puzzle of pieces in various stages of completion. The first test of the Vanguard scheduled a few days later involved the Viking booster, the only stage that was operational, but Glenn Martin and the NRL still promised to deliver on all the promised rocket launches during the IGY.

As delays in the Vanguard program became evident, Stewart steadfastly recommended switching to the Orbiter or developing it as a backup. Independently, Van Allen advised the IGY Technical Panel on the Earth Satellite Program about the need to reconsider projects, as did Pickering.

During an interview in April 1957, Navy Secretary Thomas Gates estimated that it might be twelve to eighteen months before the first launch. Van Allen made his move. He sent a Western Union telegram to every member of the TPESP on April 18. "Public announcement yesterday—Navy Secretary Gates. 18 months before the first satellite launching. Consider this method of informing working participants an outrageous violation of good faith. Strongly recommend reorientation [of] IGY scientific program to use Jupiter for initial work." Odishaw, as IGY executive director, attempted damage control in a wire he sent out to the panel the next day. "Re Van Allen wire to panel. No such public announcement made. Reporter asked schedule. Reply year or year and a half. Reply appears unpremeditated. Am looking into. Urge privacy Van Allen communication be respected. Suggest refer inquiries this office. Topic can be discussed coming meeting."

Whether the comment was made in an official announcement or as the response to a question, the result was the same. The TPESP met in May and Porter raised the issue of Gates's remarks himself to Paul Smith of the NRL. Smith said all the firings would be completed by the end of 1958 but admitted that most of them would occur toward the end of the year. Pickering once again reminded the committee there was another rocket ready to go: the Jupiter C, now available to put a thirty-pound satellite into orbit due to a "kick" to the thrust. He said the Vanguard instrument packages might be adapted to the army rocket—and kept mum about the fact that one instrument—Van Allen's—was already being adapted to do the job. Van Allen left the meeting pessimistic about any prospects for the change. "The response I got to my telegram was that no one wanted to buck the system," Van Allen recalled.

So Van Allen and Ludwig lived double lives. They made formal reports on Iowa's Vanguard instrument at TPESP and rocket panel meetings and worked behind the scenes on the Jupiter C payload with the Huntsville team. In February, the ABMA rocket team went over a detailed description of the Iowa instrument that Ludwig had sent. On April 19, Stuhlinger, Josef Boehm, and a few others visited Iowa to go over technical specs on the payload and share details of the satellite. Ludwig planned adjustments in the weight and dimensions of the cosmic ray experiment and sent Stuhlinger the final dimensions on May 3, shortly before a second Jupiter C reentry test. Stuhlinger arranged a secret clearance for Ludwig to visit Huntsville on July 9–12. He had just returned from the July 1, 1957, opening sessions of the International Geophysical Year in Washington, D.C. With the IGY officially underway, the race into space with a satellite quickly climaxed.

10 Countdown to *Explorer I*

USS *Glacier*—Sunday, September 22, 1957. Van Allen made a final trip to shore at the Boston Navy Yard that Sunday to call home one more time before he headed for the Antarctic on Iowa's IGY's rockoon expedition. As soon as he got back on ship, Iowa graduate student Larry Cahill arrived and settled in. Cahill couldn't believe he was assigned a cabin above the waterline on the same upper deck as Van Allen. "The first time the fog horns blew, I realized why I was so lucky. It was installed right above my door," he said.

The USS *Glacier* pulled up the gangplank shortly after 10 A.M. Most of the passengers—a melting pot of American and foreign scientists and reporters—planned to "winter over" at the Little America research station in the Antarctic where they would be cut off from departure during a three-month stay. The navy sponsored the international research station and, as a sponsor of Van Allen's rockoon research for IGY, assigned his team to the *Glacier* as well. The Iowa scientist planned to sail as far as New Zealand with their cosmic ray experiments and be home by Thanksgiving.

• • •

Once en route, Van Allen quickly turned his attention to the ammunition barge that arrived with fifty-two Loki rockets and sixty-three igniters to add to the cache already on board. Captain John Cadwalader, one of Van Allen's wartime shipmates on the USS *Washington,* arrived with the barge to pay a visit. He and Van Allen reminisced briefly about the old days; then Cadwalader laid it on the line. The rockoons couldn't delay the ship's arrival in Little America with much-needed supplies. Van Allen started his launches immediately.

The ship reached the equator early in October, a crucial point for Van Allen's IGY mission, since he had never gathered cosmic ray readings there before. Here the *Glacier*, designed to break through ice in the most unforgiving climate on Earth, crisscrossed tropical waters for the launches. The northern tips of the Galapagos Islands, where Charles Darwin had pondered the mysteries of evolution, jutted above the surface of the water.

"While Dr. Van Allen was probing scientifically upward, I was probing scientifically downward," said ship's oceanographer William Littlewood, who took the rockoon stops as an opportunity to collect tropical water samples from different depths in the region.

A brisk wind picked up as Van Allen and Cahill prepared for the first equatorial launch on Friday, October 4. The ship sailed full throttle in the wrong direction to correct for the wind at first, in effect, creating a gale on deck. With this oversight corrected, Iowa's Skyhook balloon lifted off at 1:39 P.M., about 700 miles off the coast of Columbia. The rocket fired flawlessly an hour later. Afterward, Van Allen went down to the lab, stooping low beneath the air conditioner mounted in the doorway to relieve the unbearable heat and humidity in the hold. "Launching again rather hairy but finally got it off," he noted, and described an antenna concept to improve reception. In the time it took to write the two-page assessment of rockoon flight Number 76, the world changed forever. Cahill burst in to tell him the Russians had successfully launched a satellite into orbit. He and Van Allen raced to the radio shack to confirm the report.

"Radio reports said [*Sputnik*] was transmitting at 20 megacycles/second so Larry Cahill and I went up to the radio shack and tuned in at 20 mc/sec [and, alternately, at 40]," Van Allen wrote in his log. "My first reaction was: Could it possibly be true that this was the satellite's transmission? [Not a spurious effect of some kind]?"

Lieutenant Stephen Wilson, Van Allen, Cahill, and the ship's captain Commander B. F. Lauff, quickly found *Sputnik*'s signature beep—beep—beep. They pooled their equipment to identify the source. Van Allen went down to his lab and got one of the brush tape recorders used to record counts from his detectors. He soldered together a circuit linking the recorder to the ship's radio receiver so that the radio signal would drive the arm across the paper tape. The arm rose and fell, charting graphs that corresponded to the transmission signal being reported by the Russians. But that didn't mean that the transmissions came from a satellite in orbit. To check, Lauff suggested they measure the Doppler shift, the rise and fall of the frequency of *Sputnik*'s beep as it approached and passed overhead, the same shift that accounts for the change in tone of a train whistle as it approaches and passes. *Sputnik*'s Doppler shift measured from the *Glacier* matched up perfectly with the altitude the Russians reported for *Sputnik*. Van Allen wired his verification to IGY headquarters and cut the brush tape he had made into 11 3/4″ segments that he glued into his field log along with his account of the satellite.

Van Allen's field logs contain hundreds of entries dating back to the 1940s. A set of thick brown-backed ledgers filled an entire shelf in his office. On only one occasion in all those pages did a bristling commentary break through the dry, detached scientific descriptions of procedures, experiments, and results. That occasion was the launch of *Sputnik* on October 4, 1957. Few Americans were as isolated as Van Allen from the public outcry that erupted over *Sputnik* and few people had more at stake. Cut off from policy decisions that directly affected him, he wrote a seventeen-page appraisal of the Soviet triumph.

"Yesterday night—the 4th—was very exciting for me (as well as for the civilized world in general)," he began. He gave a lengthy technical description of the satellite—and calculated that the launch vehicle had to be nine times heavier than the Vanguard. Then he summarized steps taken on the *Glacier* to track *Sputnik*'s orbit and added his own thoughts, excerpted here.

"Brilliant achievement!"

"Tremendous propaganda coup for USSR—also coming during CSAGI Rocket and Satellite Conference in Washington."

"Confirms my disgust with the Stewart Committee decision to favor NRL over the Redstone proposal of [August] 1955!!"

"May lead to intensified U.S. effort—our 5-year plan."

"Causes me to be very sorry to miss the inevitable reconsideration and perhaps marked changes of the U.S. program. May be a genuine loss of opportunity for us at [Iowa] to assume a larger role in the future. By T.W. X. news today, the CSAGI proposed the setting up of an international committee not to exceed six men for the coordination of rocket and satellite programs!"

"?? Where do we stand now on Vanguard?"

"The decision about two months ago to 'regroup' the Vanguard program and [Newell's deputy John] Townsend's hint in Cambridge of Security Council action = advance knowledge in U.S. of status of the Russian program."

"The pompous character of the White House announcements!"

"Our highbrow choice of 108 mc/sec and the weak signal, out of reach to most amateurs and the average population."

"The astute choice of USSR frequencies [ham radio frequencies], which literally millions of people can hear directly."

"Russians have a very great scientific lead on us."

• • •

International Geophysical Year scientists sipped vodka chilled to a frost and feasted on caviar at a cocktail party in the grand ballroom of the Soviet embassy in Washington, D.C., on the evening of October 4. They had just finished a five-day round of meetings for the first CSAGI conference devoted to rockets and satellites. Ludwig delivered a report on the Iowa satellite experiment and hoped to pin down more details at the cocktail party about the Russian satellite launch. Soviet scientists said it was "imminent" and announced the radio frequencies of it. He didn't have to wait long for answers. Shortly after 6 P.M., an embassy official summoned *New York Times* science reporter Walter Sullivan to the phone during the party.

"Radio Moscow has just announced that the Russians have placed a satellite in orbit 900 kilometers [about 560 miles] above the earth," Sullivan's Washington bureau chief told him. Informed by Sullivan, Berkner climbed on a chair, clapped his hands to call the revelers to order, and informed the embassy staff as well as everyone else in the room of the *Sputnik* victory. He warmly congratulated the Russian scientists as their colleagues from all nations cheered them. Then the room went silent as someone turned on a shortwave radio and everyone listened to *Sputnik*'s beep.

Reporters poured into the embassy, a fortress of secrecy that usually closed them out. Now, drinks and news about the satellite flowed freely through the night.

Sputnik literally meant "traveling companion" or "companion to the earth," as the Russians poetically described their creation. The name drove home the point that they had launched into orbit a sister satellite to the moon. The satellite measured only 22.8 inches in diameter. Yet the metal sphere, polished to a high gloss, could be seen with the naked eye crossing the heavens like a shooting star for an hour or two before sunrise and after sunset. Pickering and Porter left the cocktail party and went to the IGY office at the National Academy of Sciences to issue a press release telling New Yorkers where they could look to spot the satellite at dawn.

"We didn't realize we were in a race with the Russians. We may have had some intelligence to that effect, but the public certainly didn't share it. We in the press didn't share it. The flight of *Sputnik,* that little beeping satellite circling the globe, really set us back on our heels. This country took a psychological beating," recalled Walter Cronkite, who was covering the newly created "space beat" for CBS News.

Americans swallowed a bitter pill. They, like their leaders, complacently assumed that the U.S. would be first in space. Now the Russians had taken the prize. The regular beeps emitted by *Sputnik,* caught and rebroadcast by radio receivers across the globe, raised the specter of every imagined terror of the Cold War.

• • •

Von Braun left the CSAGI conference early. He, General Medaris, General James M. Gavin, and Army Secretary Wilbur Brucker spent October 4 urging incoming Defense Secretary Neil McElroy to back the Jupiter C for the satellite program as they showed him around the Redstone Arsenal. Once more, they spelled out the delays in the Vanguard effort. It was too late. As they relaxed over cocktails, Base Public Relations Officer Gordon Harris rushed in with the news about *Sputnik*.

Von Braun turned to McElroy. "We could have been in orbit a year ago," he said. "We knew they were going to do it! Vanguard will never make it. We have the hardware on the shelf. For God's sake turn us loose and let us do something. We can put up a satellite in 60 days, Mr. McElroy, just give us the green light and 60 days." Medaris countered with ninety days but McElroy, who had not been confirmed yet, had no immediate power to back their proposal.

• • •

In Gettysburg, Pennsylvania, President Eisenhower received the news about *Sputnik* and took it complacently. The president's statement that "one small ball in space" didn't disturb him did little to calm intermingled media frenzy and public fears of the rest of the country. "He was therefore stunned and surprised by the intense concern expressed by most of the American people and much of the world press," York wrote. "In the minds of many, this seemed to prove not only that things were bad, but also that the current leadership could not understand that simple fact."

• • •

At the LBJ Ranch near Johnson City, Texas, Senator Lyndon Baines Johnson heard the news about *Sputnik* while hosting a Texas barbeque and immediately grasped the political volcano that was about to erupt. He phoned Sol Horwitz and demanded a Department of Defense report on *Sputnik* and the American satellite program by October 8. Johnson contacted other senators of the Armed Services Committee and called for a bipartisan investigation of *Sputnik* with public hearings before the Preparedness Investigating Subcommittee, which he chaired.

• • •

Just north of Harare, Rhodesia (now Zimbabwe), Alan Rogers swiftly pedaled home from high school on his bicycle. The *Sputnik* orbit had been announced over the school's public address system and Alan found his father John Rogers, a ham radio enthusiast, already picking up the satellite's signal. "There had been some early articles in the ham radio magazines speculating on the frequencies for the satellites," recalled Alan, now an astronomer with the Haystack Observatory at MIT in Cambridge, Massachusetts. Whipple and the IGY had enlisted the help of ham radio volunteers to fill in gaps in the global fence of tracking stations and John Rogers was among these volunteers. The volunteers had to install an extended array of antennas to capture the satellite signals. "Dad was a bomber pilot in the British Air Force. After the war he settled in Rhodesia and ran a general store with the tracking station on the side," said Alan.

• • •

In the house on Ferson Avenue in Iowa City, Abbie Van Allen shared the feeling of near panic that swept across the country as radio stations endlessly relayed the now-familiar "beep-beep-beep" to their listeners. "The people in the United States were absolutely shocked by what the Russians had done and terrified by this thing—a satellite—orbiting around continuously right over our heads," she said. *Sputnik* only added to the anxiety of that summer and fall when she couldn't reach her husband for weeks at a time during one lengthy rockoon expedition to the Arctic and then the one to the Antarctic.

• • •

In Barcelona, Soviet satellite scientist Dr. Leonid Sedov basked in the victory while attending the International Astronautical Congress. "The American loves his car, his refrigerator, his house. He does not, as the Russians do, love his country," he told reporters. But he also expressed surprise. "One thing we [in the Soviet Union] could never understand is why you chose such a complicated, difficult, and really marginal design for your satellite vehicle," Sedov told an American colleague at the congress. "Why, for heaven's sake, didn't you take one of your powerful [military] engines that you had been flight-testing for a considerable time?" The overflight issue hadn't bothered the Russians. They bundled *Sputnik* into a launcher for an intercontinental ballistic missile.

• • •

The approximate *Sputnik* launch date wasn't difficult to predict. It coincided within a few weeks with the one hundredth anniversary of the birth of the beloved Russian rocket pioneer Konstantin Tsiolkovsky, who demonstrated mathematically in 1903 that a rocket could launch a satellite into orbit. The Russians had been promising for months that such a launch was imminent. On Monday, September 30, Soviet scientist Anatoly Blagonravov announced at the opening session of the CSAGI meeting that the Russians were ready to place their satellite in orbit and, that same day, Moscow radio gave transmission frequencies and invited ham radio operators to tune in.

The banner headlines on October 5 screamed the news of *Sputnik*. "Satellite Flashes Past D.C. Six Times—Russians May Have Ultimate Weapon," warned the *Washington Post*. "Soviets Fire Earth Satellite Into Space; It is Circling the Globe at 18,000 M.P.H.; Sphere Tracked in Four Crossings Over U.S.; Visible with Simple Binoculars," read the four-deck banner of the *New York Times*. Interviewed on television, physicist Edward Teller, father of the H-bomb, told Americans that they had lost "a battle more important and greater than Pearl Harbor." The furor over *Sputnik* increased by the day. "The famed beeps have varied in length, frequency and spacing. These variations concealed in code the observations of the satellite's instruments," reported *Time* magazine.

"There were all kinds of wild stories—the Russians were spying on us with that little satellite, taking pictures of our defense establishment and the possibility that they could launch bombs with these satellites. It scared the devil out of us," Cronkite recalled.

The rumors even crashed the stock market for 1957. The public blamed Eisenhower for allowing a "missile gap" at the height of the Cold War, and voters who had paid scant attention to missile development could now quote the amount of thrust used to launch *Sputnik* into orbit. America didn't have a rocket with that much thrust. As for a satellite, *Sputnik* packed eight times the weight of the Vanguard satellite. Extra batteries accounted for much of *Sputnik*'s 185 pounds—batteries needed to power the low-frequency transmissions that allowed radio amateurs the world over to easily locate it.

• • •

Cut off from the furor, Van Allen refocused his attention on the rockoons and the cosmic ray readings he was making across two hemispheres. Then, on October 7, he fell ill with a 103-degree fever and pain shooting through his right leg. A leg injury on board ship had caused an infection and the ship's doctor gave him massive injections of penicillin to deal with severe inflammation and a high fever. It took until October 15 for him to recover. To make up for lost time, he and Cahill initiated a marathon series of eight rockoon launches starting at 8 P.M. October 16. They launched flights at roughly midnight, 2 A.M., 8 P.M., and 11:45 P.M. on October 17. They were ready for the next launch by 9:30 the next morning and prepared for more launches at 7:43 P.M. and again near midnight. They launched four more flights on October 19 and 20, giving the Iowa team a comprehensive series of data recordings from near and at the equator. Then the *Glacier* headed due south.

Ludwig, back in Iowa City after the IGY meeting in Washington and an eleven-day stint at NRL to work on the Vanguard instrumentation, fielded an immediate call from Eberhardt Rechtin at JPL. He alerted Ludwig to a rapid-fire turn of fate for the Jupiter C and announced immediate plans to travel to Iowa City.

Frank McDonald, Kinsey Anderson, and Ernest Ray joined Ludwig and Rechtin in the physics building for the first few minutes of the October 23 meeting to plan "Deal I." It was the first time Ludwig heard the code name "Deal" applied to the *Explorer I* satellite, a name offered by JPL's project manager and poker ace, Jack Froelich. "When a big pot is won, the winner sits around and cracks bad jokes and the loser cries 'deal!'" Froelich quipped.

That same day, NRL made a launch test with the Vanguard. But while the Russians already had a satellite in space, the Vanguard test used dummy second and third stages with a live Viking launch. Rechtin met with Defense Secretary McElroy on Friday, October 25, to offer JPL's services to launch a satellite with the Jupiter C carrying a University of Iowa experiment, all of which were ready to fly. McElroy gave the greenlight to use the Jupiter C for a backup satellite program on October 28 and Medaris ordered rocket Number 29 out of "long-term storage." In photographs of the *Explorer I* launch, the letters *UE* show on the side of the rocket. *U* and *E* are the second and ninth letters, respectively, in Huntsville, camouflaging the Number 29 "with a system that owed more to the Captain Midnight secret decoder ring than to modern cryptology," Paul Dickson noted in his book on *Sputnik*.

With McElroy's decision, the U.S. National Committee of the IGY quickly transferred top priority for the Iowa satellite instrument to the Jupiter C. Homer Newell, science program coordinator for the Vanguard, firmly backed up the decision within the NRL where tension rose over the change. Insiders, such as Porter and Newell, now realized for the first time that Van Allen had hedged his bets, building an instrument ready to fly in either the Vanguard or the Jupiter C. Putting rivalries aside, they helped remove the bureaucratic roadblocks that might have delayed the transfer. Only one piece was missing—Iowa's authority to switch Van Allen's experiment from the Vanguard project to JPL. Rechtin called Ludwig. With the clock ticking on a ninety-day deadline and the Iowa instrument the only one available for the launch, Rechtin pressed a graduate student for a decision on a national policy plan.

"He asked if I had authority to switch our project from Vanguard," Ludwig dutifully reported in his notebook. "I said no." Only Van Allen had authority.

Suddenly, the search was on for Van Allen, skirting along the edge of the Antarctic ice pack on the USS *Glacier.*

Pickering tried to contact his friend through U.S. Navy channels and failed. "It wasn't important enough for them—it just stayed at the bottom of the stack," Pickering said. He reverted to a simple Western Union telegram that Van Allen received on October 30. "Would you approve transfer of your experiment to us with two copies. Please advise immediately," Pickering wrote, in vague, security-laden prose meant to protect the changing status of the Jupiter C. Van Allen had no idea about the change of status and he didn't know that the Iowa satellite instrument had been approved as a payload. He wired back a noncommittal message. "Unable interpret your word transfer due ignorance recent development. Our apparatus for original vehicle nearly finished."

He returned to the business at hand. "November 1, 1957. We have been charging along through loose pancake floe ice of thickness two to six feet at a rate of 10–14 knots with seldom a hold up. The *Glacier* is a fabulous performer in the ice!" Van Allen noted in his log. High winds delayed a rockoon launch but Van Allen worked one in later in the day. After the launch, like a man living in two universes, he picked up a second wire about the Jupiter C from Pickering. "Present planning suggests most of existing equipment but transferring responsibility to us instead of carrying two programs," Pickering wrote. He asked Van Allen to phone him as soon as he arrived in New Zealand.

Van Allen read and reread the message. "Pickering now has me completely confused!" he wrote in his log. "It was . . . difficult to believe, though conceivable, that the Vanguard program had been cancelled and that primary responsibility for the IGY satellite program had been transferred to JPL-Redstone. Pickering's message might readily be so interpreted." If such was the case, Pickering's cryptic wire raised his fears that JPL at Caltech meant to simply take over the scientific research program instead of partner with Iowa. "It occurred to me that WGII [Working Group of Internal Instrumentation] had reduced Iowa's priority and given a higher priority to Caltech's cosmic ray experiment, though I was reluctant to believe this in view of our relatively much more advanced status," he noted." "I wish that I had some straight dope from Ludwig! May get a letter in New Zealand. Will then call Pickering" (Field Log, November, 1957).

Ludwig stalled for time to hear from Van Allen as Pickering and Rechtin pressed Ludwig to leave for an immediate and extended stay in Pasadena. "I was now torn between waiting for specific instructions from Van Allen and losing the chance to use Iowa's experiment," Ludwig said. In addition, university policy forbade payment of his paycheck for work performed in another state. Pickering offered a simple solution—come work for JPL for the time being.

Then, on November 3, the Russians launched *Sputnik 2* with the dog, Laika, as a passenger. National outrage forced Eisenhower to respond. He went on television on November 7 to calm fears over *Sputnik* and promise Americans that the United States would soon redeem national honor in space. He demonstrated American space technology by showing a Jupiter C nose cone that could be substituted with a satellite launched into orbit. The nose cone had been recovered from a reentry test of the Jupiter C but Ike didn't mention the real import of this nose cone: it proved the United States could have beat the Russians into space by more than a year.

With *Sputnik 2* in orbit, the Eisenhower administration formally sanctioned the army satellite project that McElroy had already unleashed. Medaris and von Braun got official permission to prepare the Jupiter C for launch. Medaris had pushed audaciously forward on McElroy's authority alone and preparations snapped into action. More than two years after the Stewart Committee rejected the Jupiter C in favor of the Vanguard, the administration turned to the Jupiter C to rescue American prestige. Now JPL and Huntsville had to make good on von Braun's ninety-day promise to get a satellite in space.

Von Braun assumed his team would finally build the satellite he had planned most of his life. Pickering came to Huntsville to hammer out details of the division of labor on November 9 and he quietly cornered Medaris in his office shortly before the meeting. He argued that JPL should design the satellite to fit both the upper stages of the rocket and the Iowa satellite instrument, which he said was being transferred to the lab. Pickering's almost casual approach to a major mission redirection matched historian Cargill Hall's description of him as "spare, reserved and, in a quiet way, implacable."

Medaris announced at the meeting that JPL would build the satellite. "Von Braun swallowed hard but did not comment. Boehm and his coworkers swallowed even harder," Stuhlinger recalled. Von Braun knew that Pickering had helped save the day for the Jupiter C, devising the reentry test program that allowed low-profile development on the launch rocket and the upper stages during the two years when Project Orbiter had no official sanction. Behind the scenes, both Pickering and Stewart had continued to lobby for backup status of the Jupiter C for IGY, and von Braun never forgot it. As for the payload, Pickering announced that the Iowa experiment would fit the Jupiter C/JPL satellite, maintaining the civilian character of the satellite. The military brass from Washington in attendance at the meeting nodded approval and von Braun feigned surprise.

"You don't say," he noted sagely of Van Allen's participation. Now everything was in place except for Van Allen's permission. Pickering sent off another wire, which Van Allen received after arriving at Port Lyttleton, New Zealand, on November 10. Still wary of the terms of the instrument transfer, he cabled Ludwig. "Question: Is Pickering plan for our experiment agreeable with you?"

By the time the cable arrived, America's first satellite experiment—Prototype Number 4—was en route to Pasadena in the trunk of Ludwig's 1956 Mercury. Ludwig, his wife Rosalie, and their two daughters hastily packed a few family belongings along with the experiment, tape-recorder parts, and other electrical and mechanical components. George Ludwig headed west on November 13, his thirtieth birthday.

Ray responded to Van Allen's cable with the news of Ludwig's departure. "George quite happy (with) Pickering plans. Hope you say yes." Van Allen did. Then, he and Cahill traveled to Christchurch and consulted with New Zealand scientists on the rockoons. Satellites never came up in the field-note observations about warm welcomes and enthusiastic discussions. The New Zealand scientists wanted to meet Van Allen and applaud the man who wrote a how-to guide on low-budget rocketry in "The Inexpensive Attainment of High Altitudes with Balloon-launched Rockets." The Christchurch visit provided a peaceful interlude before Van Allen returned to Port Lyttleton, packed up the makeshift laboratory on the *Glacier,* and hopped the first in a series of flights back home.

The Antarctic readings confirmed and vastly extended cosmic ray data with fifty-four successful rockoon launches (rockoon flights 56 through 109), nearly doubling the number of launches from the previous five years. In addition, Cahill found new evidence about the natural electrical currents rushing through the lower ionosphere in the area of the equator. The earth's magnetic field acted as a natural electrical generator as charged particles from the sun moved through it and created this electrojet.

But that summer and fall had been a long haul for Abbie Van Allen. Jim boarded the USS *Plymouth Rock* on July 7 for the Arctic rockoon expedition, headed home on August 30 and then left again September 18, shortly after Tom's second birthday. Cynthia was ten, Margot eight, and Sarah five. Abbie deftly juggled the care of the four children, the home, and volunteer commitments. Her mother, Helen Halsey, now remarried after Jesse Halsey's death, helped her during a short visit. As her mother prepared to depart, Abbie's sense of isolation overwhelmed her. She stood in the hallway and cried. Helen sat down with her daughter and told her the story of her own missionary past, living in Labrador in the small port town of St. Anthony.

"She told me that when the harbor in St. Johns opened to the south, my father would leave by dog team and go down there and get a schooner and go to Boston for supplies. It would take a long time and once he left she had no communication with him. She was pregnant with her second child and had a two-year-old to care for. That was on his first trip. The trips happened year after year and she had no idea whether he was eaten by a bear or adrift on an iceberg. The ice in the harbor at St. Anthony broke by the time he was returning and she'd walk out on the cliff to see if his ship was coming in. Well I snapped out of it after that story. I thought I had no reason to be a cry baby. This was all nothing compared to what my mother went through," recalled Abbie Van Allen.

• • •

Van Allen arrived in Iowa City on Thursday, November 21, bombarded suddenly by a world caught up in *Sputnik* mania and demands for vindication. He finally talked with Ludwig directly in a lengthy telephone conversation. His concern about Iowa retaining a lead role in the instrumentation escalated when he realized JPL had hired Ludwig for the instrumentation project rather than working with him as a liaison of the university.

JPL's leap to take over the satellite project seemed to justify the suspicions he had had about retaining full partnership in his own experiment. Van Allen finally settled the matter in a gentleman's agreement with Pickering. "I talked to Pickering about it off and on. It was civil. I was fond of Bill and the feeling was mutual. The bond between Pickering and myself and the fact that we [Iowa] had the IGY sponsorship resolved this. IGY sponsorship was a fundamental point. They were making the supporting spacecraft but it was an IGY-Iowa experiment," Van Allen said. "He was part of the IGY so he recognized that."

Yet Ludwig's position as a junior member of the JPL staff rather than a fully partnering research associate with the university created tension. When conflicts arose over decisions regarding the instrument, Ludwig called upon Van Allen to intervene. "If I had been here, I would never have allowed Ludwig to go work for JPL," Van Allen said. The work went forward, though, as the spotlight turned to the U.S. Senate.

• • •

Lyndon Johnson, as majority leader of the Senate, quickly parlayed *Sputnik* into political currency for the Democrats, based on his own instincts and those of aide George Reedy. Reedy spelled out the political potential of *Sputnik* in a lengthy memo on October 17 and advised Johnson to "plunge heavily into this one." He stressed that *Sputnik* opened a new era in history: Rome cemented power with its roads, Britain with its navy, and America with its airplanes, Reedy pointed out. Now, whoever ruled space ruled the future, he predicted.

Sputnik also offered a much-needed diversion for the Democrats and Johnson that fall. Pushing with all the might of his power, Johnson bludgeoned the Senate into passing a sweeping civil rights bill on September 9, and TV cameras caught a country torn by racial hatred in the aftermath. As unpopular as the bill was among his own southern block of senators, Johnson believed in civil rights and needed the law passed to transcend the southern image that might hamper his aspirations to the presidency. The lengthy series of Preparedness Investigating Subcommittee hearings allowed him to take direct and continuing leadership in a national security crisis. *Sputnik* was a godsend.

Johnsons's hearings convened on November 25. Edward Teller, father of the H-bomb, and Vannevar Bush, who had coordinated wartime science research including the Manhattan Project, testified that the United States had fallen behind the Russians in military preparation of missiles and in science and technology in general. Witness after witness testified that infighting between the various branches of the military undermined satellite and ballistic missile development. One cartoon from the period showed U.S. military officers glancing up at a Soviet rocket and expressing their relief: "Whew, at first I thought it was sent up by one of the other services." Witness after witness also testified that Eisenhower's civilian mind-set for the program got in the way. Launching a civilian satellite was considered an essential accommodation to the Cold War but civilian status bumped the program from top priority in terms of both national policy and funding.

The hearings broke for Thanksgiving break and for the much anticipated Vanguard launch at Cape Canaveral on December 6. The press covered this and other early missions like unwanted guests. "The military was still in charge at that point. We weren't even permitted out on the cape. They didn't even tell us when they were making a launch," recalled Walter Cronkite. "We went onto the breakwater just outside the canal that separates the cape from the mainland. Those are just great granite rocks [there] and it was very unpleasant to try and find a comfortable place to sit or to put a tripod with a camera. We got no cooperation from the military." The press arrived in full force for the Vanguard launch. The rocket lifted into the air for two seconds and exploded into flames. The spherical satellite rolled a few hundred feet away and beeped pitifully from the ground. "Kaputnik!" read one headline above a photo of the burning ruins of the Vanguard. But there was a solution, Medaris told the nation as Johnson's hearings resumed on December 13. He explained about economizing in his general funds budget so he could keep the Jupiter C in development, now the best hope for carrying a U.S. satellite into orbit.

Then von Braun testified. The 2,087 pages of transcripts from Johnson's hearings offer very dry reading most of the time, but the exchange between Johnson and von Braun still crackles with wit and venom. The two charismatic opportunists sized each other up and threw each other lines like two Method actors who knew how to capture an audience.

Von Braun described the interruption in the Army Ballistic Missile program that stymied the best hope for a satellite launch in 1956.

VON BRAUN: "The termites got into the system."

JOHNSON: "The what?"

VON BRAUN: "The termites."

JOHNSON: "Termites!"

VON BRAUN: "As General Medaris mentioned before, the money [for the Jupiter C] was allocated—but somebody withheld it, we got only part of it."

JOHNSON: "When did the termites come?"

VON BRAUN: "I would say we had clear sailing for about a year."

JOHNSON: "When did the termites come?"

VON BRAUN: "The difficulty began when the roles and missions assignment for the IRBM [intermediate-range ballistic missile] was given to the air force [in November 1956], and more and more people doubted whether the Jupiter would really go into production and so people were withholding final approval."

EDWIN WEISL [chief special council]: "Did you have difficulty getting money, the money that you needed?"

VON BRAUN: "Yes, sir."

WEISL: "Did you have difficulty in getting the money on time?"

VON BRAUN: "Yes."

JOHNSON: "Get that date! That is what I want. About what time, what month, and what year that the termites got there."

VON BRAUN: "I would say it began about November 1956."

Then von Braun looked past present defeats and spelled out victories for the future. He proposed a space agency and threw down the gauntlet for America to grab a bigger first than *Sputnik* by sending a man to the moon.

Johnson, the political genius, recognized in von Braun the perfect ally to turn the *Sputnik* debacle and a vision of space travel into election currency.

Von Braun, the patrician pragmatist, recognized in Johnson the power he needed to realize his dreams for a space program and ensure that nothing interrupted the Jupiter C this time around. Six weeks after McElroy sanctioned the Jupiter C program, funding remained at a trickle.

"The mills in the Pentagon are grinding slowly and we hope that the goodwill is followed up with a check sooner or later," von Braun concluded at the hearing with a light tone that set his audience laughing.

"It is nothing short of disgraceful that you would get an order to go full steam ahead and then Medaris would have to pull a sleight of hand stunt in the dark of the moon and go over to some other agency to get money," Johnson thundered in reply. "And I hope that as a result of the hearings and the testimony that was brought out here, that the government of the United States will supply the money." Shortly after the hearing, the cash began to flow.

• • •

The Deal I experiment neared completion in early December at JPL. Ludwig consulted almost daily with his colleagues in Iowa by telephone, Sentinella continued to build custom parts JPL needed, and Van Allen locked in funding and provided technical coaching as Deal I went into final spin tests. After all the retooling, it remained fundamentally a single Geiger-Mueller tube to survey cosmic rays, though everything was miniaturized, from the counter to the transistors. And the apparatus needed a long life span. Rockets gave instruments only a few minutes of useful data collection, but satellites could give them months and even years.

The Geiger tubes were critical. Looking through the window of the wood-burning stove in Mount Pleasant as a boy, Van Allen could have found a clue to his future in space. "The window was covered with mica. It's a natural material and can withstand heat, cold, the damp and the dry. It's really fantastic stuff—it has great mechanical strength," good for a trip in a rocket. Van Allen scouted for durable Geiger tube detectors for *Explorer I* and the mica-coated design in the specs from Anton Electronics in Brooklyn, New York, impressed him. Nicholas Anton developed the detectors for nuclear physicists and labs analyzing nuclear reactions. But the detectors had one big advantage over others for a trip in a rocket to count cosmic rays: Anton's proprietary design sealed the mica window to the tube of the detector with a fine beading of glass around the rim. The tiniest leak renders a Geiger tube useless since the tubes are filled with gas. A cosmic ray passing through it ionizes a gas molecule and generates an electric pulse that can be counted. Van Allen tracked down Anton's lab to a storefront that looked like a dingy electronics repair shop.

"I made Geiger tubes myself when I was in graduate school but nothing of the quality that Anton achieved. It was pretty amazing what came out of this little hole in the wall, grubby place," he said.

The Iowa satellite instrument used an Anton Geiger tube that could operate over a wide range of temperatures without deteriorating. All the electronics were wired to a new generation of mercury batteries and the entire assembly drew less than one watt of power to operate the instrumentation and transmit signals. The scientific package incorporated additional instrumentation, including JPL's temperature sensor and the Cambridge Research Center's micrometeor detector. It all fit snugly in the 6 x 21-inch sleeve of *Explorer I* attached to the fourth-stage rocket. The whole satellite weighed 30 pounds.

Van Allen received word in mid-December that the first Jupiter C firing would take place in early February. He worked feverishly to begin to analyze the rockoon data and helped finalize arrangements for Iowa's other IGY programs. Carl McIlwain was heading to Fort Churchill in Canada to shoot detectors directly into the aurora with Nike-Cajun rockets to obtain more definitive measurements of the primary particles that make up the aurora.

On January 9, Van Allen filled in his colleagues on the Working Group for Internal Instrumentation on the swift changes made to ready the Iowa experiment for JPL's satellite. On January 10, Medaris, testified once again at Johnson's hearings and emphasized the scientific nature of the Jupiter C's satellite program. It would detect cosmic rays with an experiment developed by James Van Allen at the University of Iowa, he said. Van Allen's phone began to ring off the hook as the press sought details about the physicist from Iowa who was hunting down cosmic rays from intergalactic space.

The massive Redstone rocket, capable of burning 50 gallons of liquid propellant per second, was already at Cape Canaveral by then. Installation of the upper stages began in mid-January with the Redstone anchored in place on the launching pad. A cluster of eleven Sergeant rockets provided the second stage with a cluster of three for the third stage and a single Sergeant for the boost of the fourth-stage satellite into orbit. The long slender spindle of the satellite protruded from the top but the Iowa experiment still wasn't in it. Ludwig encountered last-minute calibration flaws and on January 26—three days before the latest launch date—he took the equipment to his home near the lab to work on it away from other electrical noise. "Received Nuclear Corp. Model 161 Counter from Van Allen, which seemed to solve all the problems," he wrote in his notes the next day. He headed to Cape Canaveral on January 28 where he joined the rocket firing team of scientists from JPL and Huntsville.

The January 29 launch got scrapped due to high winds at the cape. Everyone readied for the launch again on January 30 but the firing was scratched again at 7:55 P.M. due to the wind. The blustery afternoon of January 31 felt equally hopeless but the winds subsided by evening and the unexpected decision to fly sent everyone from JPL and ABMA running for position.

Fueling started at 8:30 P.M. with a "jazzed up" liquid fuel called Hydyne, much more powerful than the ordinary rocket fuel, von Braun wrote in the *Des Moines Sunday Register.* At 9:45, there was a hold. Someone had seen something dripping from the rocket. Huntsville's propellant expert Albert Aeiler crawled into the rocket tail to determine the cause and found that it was just a harmless spill and wiped it up. With the scaffolding around the Jupiter C removed, the rocket stood luminous in the floodlights against an ebony sky. Then Kurt Debus, rocket launch center director at the cape, gave the word to start the automatic firing sequence. The Jupiter C launched at 10:48:16 and rose with a stream of orange flame at its tail.

The Huntsville team cheered from the ABMA hangar but braced for the next critical step of the flight. Stuhlinger stood before a makeshift "spatial altitude control system" that helped him calculate velocity and determine the precise moment for the split-second radio command to fire the second stage of the rocket. Even a second of error would jettison the flight. He hit the button right on target and the second stage of the rocket accelerated. Seconds later the third stage ignited, and then the fourth, quickly accelerating to nearly 18,000 miles per hour—the speed needed to inject the satellite into orbit. Medaris watched the launch from a bunker and then went to greet the press, invited on this rare occasion to the theater at the Patrick Air Force Base adjacent to the cape.

Only Ludwig, who had played so pivotal a role in the drama, had nowhere to go that night. He wasn't part of either the JPL or ABMA launch efforts. Putting rivalries aside, his friends at NRL invited him to the Vanguard's Hangar S where he watched the launch and followed the counting rate of the Iowa instrument as the rocket climbed. Radio tracking stations in NRL's Minitrack fence and JPL's Microlock network started to pick up the *Explorer I*'s signals almost at once.

"Within two minutes, Antigua [in the West Indies] reported that the satellite had passed overhead and soon similar reports were coming in from other down-range stations. Someone in the Pentagon group suggested announcing but von Braun and Pickering opposed the idea until they could be absolutely sure" that *Explorer I* had achieved orbit, noted Ludwig.

• • •

Van Allen, von Braun, and Pickering who had partnered so long for this moment, didn't observe the launch. They gathered in the War Room of the Pentagon with Army Secretary Wilbur M. Brucker and General Lyman Lemnitzer, army vice chief of staff. There were no televisions or even a loudspeaker in this inner sanctum of the building. But there was a phone. A call from the cape reported the launch looked successful and the satellite had been injected into an orbital path moving eastward around the earth. The group in the Pentagon bantered in high spirits at they waited for the next call to confirm the satellite had gone into orbit.

"Boys, this is just like waiting for the precincts to come in," Brucker wisecracked.

But a mood of tension gripped everyone as time passed.

"It was a really anxious period and a silence settled over the whole group. We drank coffee and chewed our nails and wondered what had happened because the expectation was that the satellite would go into orbit and should come around the earth [and reach California] in about 91 minutes, but 91 minutes passed and we got no reception, no reception," Van Allen recalled.

"We are out of coffee and cigarettes. What now?" Pickering wrote in his log as time passed.

A series of calls from Medaris broke the grim silence. Pickering had no answers for the general and he finally called JPL assistant director Frank Goddard and small-talked while everyone stared at him. More than 105 minutes passed, then 106, then 107 as Goddard awaited word from one of the telemetry stations or even a ham radio operator. The San Gabriel Radio Club, gathered in an old brick building on Broadway Street with their antennas on the roof and their receivers calibrated, picked up the *Explorer*'s signal first on the West Coast, 108 minutes after liftoff. Seconds later, the signal beamed to the Earthquake Valley station in California. "We've got the bird," Goddard said.

"Pickering announced the good news and everyone was jubilant and everyone slapped one another on the back. Then we had to leave for a press conference at the National Academy of Sciences," Van Allen said. It was now close to 2 A.M. on February 1 and a military chauffeur drove Van Allen, von Braun, and Pickering through the deserted Washington streets in a steady rain. He dropped them off at the back door of the academy. "We wondered if anyone would be there," Van Allen said. "We walked in and there were cameras and reporters—the room was full—and we stood there for two hours answering questions."

The press needed a good picture and the three men spontaneously picked up a prototype of the tubular *Explorer I* satellite and hoisted it over their heads in a victory pose. The photo became an instant icon of the space age. But Van Allen barely had time to savor the moment when the lengthy gaps he noticed in his scientific data suggested that the cosmic ray counter on *Explorer I* had failed—or he had discovered something in space that no one on Earth knew existed.

11 Celebrity Scientist and the Birth of NASA

The White House—February 4, 1958. Abbie Van Allen's life changed forever a few hours after *Explorer I* went into orbit. She awoke that Saturday morning with a long list of chores, errands to run, four kids clamoring for her attention, and a phone that never stopped ringing. Between phone calls, the doorbell rang and a Western Union delivery man handed her a telegram—from the White House.

"The President and Mrs. Eisenhower hope you can come to dinner at the White House on Tuesday, February 4, at eight o'clock. Stop. White tie. Stop. Please wire reply. Mary Jane McCaffrey, Social Secretary."

The phone rang again. Abbie Van Allen recognized the deep, resonant voice of University of Iowa President Virgil Hancher. "I thought he was calling about the invitation but he was calling in reference to the launch the night before. I just assumed that if we were invited to the White House, the Hanchers would somehow be going too. But he didn't know anything about the invitation."

President Eisenhower sent the Van Allens, the von Brauns, the Pickerings, and General and Mrs. Medaris an impromptu summons to a state dinner that had been scheduled months earlier.

"I haven't heard from Jim. He probably won't want to go," Abbie Van Allen told Hancher.

"It's a command performance. You have to go," he said.

"I don't know if we can afford it," she said.

"I'll take care of that," Hancher assured her and immediately arranged for the University of Iowa Foundation to fund the trip. With the dinner now a certainty, Abbie turned to her closet. She had plenty of suits and dinner dresses but nothing in a full-length formal and nowhere to shop for one. The nice dress shops in Iowa City politely closed their doors on weekends.

The phone rang again and Abbie poured out her dilemma to Irene Tyndall, wife of Van Allen's mentor and master's thesis advisor, Professor Edward Tyndall. The indomitable Mrs. Tyndall, a social fixture in town, got out the word that Abbie Van Allen needed a dress for dinner at the White House. She arrived at Abbie's doorstep with an armload of dresses and a pair of long white kid gloves in short order. Abbie chose Mrs. Tyndall's ice blue satin gown with a scoop neck trimmed in mink, three-quarter length sleeves, a fitted bodice, and gracefully flaring skirt. A seamstress took charge of quickly cutting it down to a size 8.

• • •

Abbie and Jim Van Allen boarded the plane from Cedar Rapids to Chicago for the connection to Dulles Airport on Tuesday, leaving the children in the care of Katharine Chapman, wife of Sydney Chapman. The Chapmans had adopted the Van Allens as a second family when Sydney taught classes as a visiting professor at the University of Iowa in the mid-1950s. Katharine arrived for a long-awaited visit just as the Van Allens left for the whirlwind trip to the White House. They arrived the day of the dinner at the Hay Adams Hotel where the von Brauns also had a room.

They arrived at the White House for a reception in the stately East Room. Jim Van Allen drifted off to greet Medaris and Pickering, and Abbie made appropriate replies to the two ladies sitting with her on a blue velvet Chippendale sofa. She tried to free most of her attention to absorb the glories of the room.

The president showed up at exactly eight o'clock wearing a black tie after a fruitless search for his only white tie. An aide had loaned it to von Braun. In any case, Mamie Eisenhower captured most of the attention. "Mrs. Eisenhower wore a red tulle dress which had a tiered bouffant skirt and strapless bodice," Abbie Van Allen wrote. After introductions to the President and Mrs. Eisenhower in the receiving line, she promenaded down a long hallway to the dining room on the arm of her prearranged dinner partner, General Randolph Pate, commandant of the Marine Corps. A marine band played marches and Abbie took her seat at a horseshoe-shaped table ablaze in candlelight. Medaris sat on her other side and Ike and Mamie sat together at the head of the table with a menagerie of gold figurines stretched out before them for a centerpiece.

But Abbie and her borrowed dress grabbed more headlines than the dinner in papers from New York to Omaha. "Science Gone Folksy," the *Omaha World Herald* mused. "Two sidelights of the launching of the American *Explorer* seemed especially pleasing. On the next day, Sunday, Dr. James Van Allen who designed the satellite gadgets that record scientific data in outer space, drove to a farm home near Iowa City to discuss the triumph with his assistant, the farmer's son," reported the *Herald.* "Two, when Dr. and Mrs. Van Allen left Iowa City for a White House party held last night, he planned to rent a dress suit in Washington and she packed a formal dress borrowed from a neighbor. We surmise such folks will do more to popularize science in schools than a whole bale of federal cash."

The local press hurried en masse to Ferson Avenue for pictures of Abbie Van Allen in the borrowed dress and a full account of the dinner to celebrate a victory that the president, at first, hoped to downplay.

Press Secretary James Hagerty broke into Ike's late-night bridge game to tell him *Explorer I* was orbiting smoothly overhead. "Let's not make too big a hullabaloo over this," Eisenhower said, still annoyed with the army's mavericks in Huntsville. Ike's son John, in his memoir, *Strictly Personal,* noted "a feeling in the White House that the army represented an upstart, second only to the Soviets."

Medaris advocated supremacy in space with religious zeal, calling the Soviets the "anti-Christ" in one interview. But mostly Medaris thought it was the army rather than God that should lead the crusade. "The army exploited the fact that it had Wernher von Braun on its team, and sent him out on a barnstorming mission to present the army's ideas and hopes to anyone who would listen and then write his congressman about it," wrote physicist and presidential science advisor Herbert York. Medaris took up the mission himself on *Meet the Press* on February 9, soon after *Explorer I* went into orbit and CBS followed up with a two-part television program called *Shooting for the Moon* that predicted how von Braun, the ABMA, and Redstone contractor Chrysler Corporation would blaze the trail for a journey to the moon.

The public relations blitz didn't prevent the navy and air force from pitching their own supreme role in space with arguments that ranged from the strategic to the comic. The navy claimed space as an extension of the sea because it would be traversed by ships—spaceships. The air force claimed space as an extension of the air and "proposed renaming itself the U.S. Aerospace Force," York recalled.

From the standpoint of practical military politics, the air force had already cemented its role in the development of intercontinental ballistic missiles with the Atlas as the key rocket. York became a key military advisor on missiles and space as chief scientist for the newly formed Advanced Research Projects Agency (ARPA). But the President's Science Advisory Committee, chaired by MIT's James Killian, promoted a civilian space program and Killian had an ulterior motive to use civilian science as a cover. The committee wanted to safely fence off the science satellites from covert military programs, including the U-2 spy plane, and from military upstarts, such as John Medaris.

On the day Van Allen arrived in Washington for the state dinner, Killian and York (also a member of the president's advisory committee) met with Eisenhower to deliver an update on the "missile gap." The term blitzed "political discourse with great force and frequency" after *Sputnik*, York noted.

At their meeting, Killian and York sketched out a blueprint for space that would create the National Aeronautics and Space Administration (NASA). Killian tapped his inner circle of scientists and commissioned Harvard physicist Edward Purcell to cobble together the structure for NASA in a mere two weeks. The Senate addressed the same task by forming the Special Committee on Science and Astronautics. LBJ quickly parlayed his visibility from the Preparedness Investigating Subcommittee hearings into the chairmanship of the new Senate committee.

Space had become a hot political commodity. Until *Sputnik* beeped in celebration of Soviet ingenuity, politicians largely left space in the hands of fiction writers, the military, and scientists. Now the formation of a civilian space agency became the focus for collisions between the military, the President, Congress, and the National Academy of Sciences. Van Allen symbolized the leadership role civilian science could take. His success "demonstrated clearly and unequivically that a university physics department could design and build instruments that would operate in space and produce significant scientific results," wrote NASA scientist John Naugle in his book *First Among Equals*.

The rocket panel, renamed the Rocket and Satellite Research Panel for its final term of service, described an ambitious civilian space agency in two seminal reports. The proposed agency relied on a cross section of scientific leadership that had been at the core of the success of the rocket panel. In addition, the November report, *A National Mission to Explore Outer Space,* called for the "eventual habitation of outer space" and the December report, *National Space Establishment,* set the goal of a "manned expedition to the moon by one or two men by 1968."

Van Allen, Pickering, and Whipple lobbied for their proposal in Washington early in December before their second report was even completed. Van Allen served on Killian's presidential science committee and rocket panel members played key roles in the space advisory network mushrooming through Washington. They quickly realized that Killian favored the metamorphosis of the National Advisory Committee for Aeronautics into a space agency. NACA director, Dr. Hugh Dryden, soon climbed to the top of Washington's A-list to lead a new space agency, though even he didn't feel that his woefully understaffed and under-financed civil service organization could undertake the job at hand.

The government formed NACA in 1915 to promote civilian aeronautics research for the fledgling airplane industry and for the military. NACA coordinated research through several civil service laboratories that also served as a conduit for military collaboration. The agency directed the research agenda through the direction of a volunteer seventeen-member advisory committee, a staff of 170 at NACA's Washington headquarters, and the committee's legendary full-time clerk John F. Victory, whose service spanned the forty-three-year history of the committee.

By 1958, Dryden had created several space technology groups based on the programs of the IGY. Van Allen chaired the space research objectives group, giving NACA a selection process for prioritizing programs and experiments for space, based on the functioning examples of the Rocket and Satellite Research Panel and IGY's Technical Panel on the Earth Satellite Program [TPESP].

Van Allen and other panel members continued to lobby for a comprehensive agency that could incorporate NACA and independent scientific institutions such as universities rather than hanging all the pieces of the space establishment on NACA. Simple math supported their position. NACA labs had a combined budget of $355 million and a staff of nearly eight thousand people for fiscal year 1958 while scientific and educational grants to independent institutions came out of the small NACA headquarters budget—about $3 million a year. Van Allen brought his proposal directly to Lyndon Johnson in a meeting with Johnson's committee staff in February. Van Allen realized Johnson "felt the administration should take the initiative to make a space proposal to Congress and that the Killian committee is the group to frame the Adminstration's point of view."

Van Allen read the intent loud and clear. Even with strong civilian participation, the administration and Johnson never expected to really give up military control of the space program and pushed the controllable NACA model. Fear that the Soviet Union would take the missile gap to the far corners of the solar system anchored the agenda for both the civilian and military buildup in space. Even scientists agreed with this point of view. "We do not wish to see the satellites of Mars or Jupiter become the satellites of the Soviet Union or the possession of any other country," wrote the governing council of the American Federation of Scientists in a letter to Johnson. Johnson introduced the administration's space program as the National Aeronautics and Space Act of 1958, S.B. 3609. The bill paid public lip service to civilian control but gave only an advisory rather than a governing role to a NASA board in terms of space policy. In giving the real power to one head, appointed by President Eisenhower, the bill bowed to the position of Air Force General Bernard Schriever and others that "the military side of space technology, like pitching in baseball, was 75 to 90 percent of the game." It essentially condoned two space programs—one civilian and scientific and one military.

Van Allen recognized firsthand that the military intended to take an active role in space exploration regardless of the civilian programs in place. The Eisenhower administration quickly added a series of moon probes to the IGY missions and three of them, under air force supervision, bypassed Van Allen's own working group to select instruments, though he participated in the missions. Members of the rocket panel lobbied for more civilian control and Van Allen worked with the Iowa congressional delegation to draft a NASA bill with more teeth for such control—and more likelihood of university participation. Efforts by the panel and others to challenge the Senate bill helped shape a more powerful and independent NASA for House Bill 11881 promoted by Massachusetts Democrat and House Majority Leader John W. McCormack. Both Johnson and McCormack invited Van Allen to testify. With his latest instruments on board *Explorer III*, successfully launched on March 26, 1958, Van Allen came to the hearings as the foremost space scientist in the country.

The House of Representatives held hearings first, starting them on tax day, April 15, with a packed audience and TV cameras awaiting the first witness: Wernher von Braun. Von Braun came to Congress with a complicated web of IGY and military space projects under his wing for 1958 and no funds at all for any space exploration in 1959. He had a ready rocket design to launch an astronaut into orbit around the earth and the initial design for the rocket needed to get to the moon. But the army had no authority and no money to undertake either of these efforts. Von Braun found himself in much the same position as when the country turned to the Jupiter C in desperation after the launch of *Sputnik* six months earlier.

Von Braun reiterated the successes of the IGY *Explorers* and explained how *Explorer III* delivered a continuous stream of data to earth with an ingenious tape recorder that George Ludwig had developed. Hit the playback command to "interrogate" *Explorer III* and "we get approximately 100 times more information" than with *Explorer I*, he said. "The cosmic ray people involved in the *Explorer III* experiment have made some discoveries that were quite unexpected," he noted, making the first vague reference to the discovery of the inner radiation belt that Van Allen formally announced at the National Academy of Sciences two weeks later.

Von Braun played out his role as space visionary but also urged Congress to focus in on realistic budgets for realistic projects that could be rapidly achieved, such as launching an astronaut into space. "Some people have taken the position that it would be nothing but a stunt. But we've been told the same when, back in 1954, we first suggested a minimum satellite with the Redstone missile," he said, predicting that "one of the next surprises the Russians are going to pull on us is placing a man in orbit." The Redstone rocket could save the day again, taking a man higher into space and sooner than the X-15 space plane assigned to the task. Uncertain who his next sponsor would be, von Braun tacitly supported a military and civilian partnership for a space agency.

Unlike von Braun, Van Allen was in a position to sharply challenge the military agenda and stake out a separate role for independent science. "When one speaks of placing scientific apparatus in orbit about the moon, I feel that this is much more clearly a purely scientific undertaking," he testified. "Military importance is very difficult if not impossible to demonstrate." He offered immediate and practical civilian applications of space as well, predicting major new commercial industries with his comments.

"One example is in the study of weather on a worldwide scale and in the reliable predictions of weather. The civilian applications of such predictions are vastly more important than the military ones. As a second example, there is the clear prospect of high speed communications on a worldwide scale by the use of satellites," he said. From the most basic scientific standpoint, he identified space as a "vast area of ignorance" with unknown physical phenomena that must be explored through the new space agency.

Van Allen reinforced what others said about lack of funding for space and complacency over Russian technology. He placed his support squarely behind a civilian agency with civilian power over decision and at least a $500 million budget. The cost of ignoring such an opportunity would be far higher indeed, he predicted.

"It is quite easy to visualize, if we don't get cracking on this field, that my young son, for example will want to go to Russia to study physics," Van Allen said, responding to a comment that America had mistaken material might for intellectual superiority.

By the time Johnson's hearings on NASA convened on May 6, Van Allen carried to the Senate hearing the added clout of his announcement of the discovery of the earth's radiation belts, vast zones of charged subatomic particles that are the subject of the next chapter. The discovery delivered a public relations coup for the politicians promoting a civilian space agency. But on May 15, the very day Van Allen testified before the Senate, the launch of the 2,926-pound

Sputnik 3—a satellite a hundred times heavier than *Explorer I*—gave America another inferiority complex and inflamed the hearings. Pickering testified that day and explained to a spellbound and appalled audience that it required a rocket with 600,000 pounds of thrust at takeoff to launch *Sputnik 3*. The rockets used to launch *Explorers I* and *III* had 6,000 pounds of thrust at launch, he said.

Still, Van Allen's discovery demonstrated America's research lead over the Soviets no matter what size satellites they launched and it proved the need for a national program to explore space, Pickering emphasized. "Dr. Van Allen has given us some completely new information about the radiation present in outer space. The information was completely unexpected. Radiation levels a thousand times or more greater than the cosmic ray background have been measured," he said. "This is a rather dramatic example of a quite simple scientific experiment which was our first step out into space." Then Pickering laid it on the line and appealed to America's entrepreneurial roots. He called a civilian space agency a "national necessity" and a "forerunner to the commercial exploitation of space or the military exploitation of space." Van Allen testified next and Johnson buttonholed him on key policy points.

"Do you believe that any of the present, permanent research authority of the Department of Defense with respect to the space weapons systems should be transferred to this new Agency?" Johnson asked.

"Yes, sir; I think a good bit of it might well be and should be," Van Allen said.

"Which part?" Johnson probed.

"I think a good bit of what is being done within the Department of Defense in this field at the present time is on a stopgap basis and is mainly and dominantly of a civil character. But it is being done by the Department of Defense as a matter of expediency because there is no other agency in the country at the present time able to undertake it," Van Allen answered. He advised an accelerated time frame of transferring programs to the new agency.

The House approved its bill June 2, giving the new agency the full administrative status. The Senate approved Senate Bill 3609, however, with the creation of a National Aeronautics and Space Council to oversee national policy in addition to a space agency. On July 7, Eisenhower brought Johnson to the White House to hammer out a compromise. Johnson deftly suggested that Ike himself chair the council as a way to oversee the stiff civilian mandate in the House bill but Ike saw the council as a publicity stunt that would focus too much attention—and too much of the president's popularity—on space missions. In the end, Johnson got a behind-the-scenes council and McCormack got his more independent space administration. Eisenhower signed the bill creating the National Aeronautics and Space Administration on July 29, 1958, and it took over the National Advisory Committee for Aeronautics.

Scientists, working through the National Academy of Sciences, assumed that they would continue their role in developing a space science agenda through NASA instead of the IGY. Academy president Detlev W. Bronk generated the Space Science Board to ensure the direct role of scientists in space policy. He named Berkner to head it and Berkner, Porter, Van Allen, Dryden, and other key figures from the IGY's technical panels moved swiftly on a mandate to involve talented scientists from across the country in experiments that could be launched through NASA's programs. Berkner's Fourth of July telegram in 1958 galvanized a response from two hundred scientists, widening the sphere of space science nearly tenfold compared to the number of participants in space science experiments at the start of the IGY.

"It came as a bolt out of the blue," recalled Dr. Kinsey Anderson, who was still a research associate at the University of Iowa when he received the call to propose an independent research project. "I have a vivid memory of it—someone actually asking me to do science in space," Anderson said.

As projects poured in, Berkner asked Porter to prioritize the ones requiring satellites and probes and asked Van Allen to prioritize those involving sounding rockets, as the rocket panel had done. By the end of July, the Space Science Board prepared and adopted a complete space science program. Dryden's preliminary organizational charts for NASA treated the board as a conduit into NASA but Dryden swiftly lost support as his critical attitudes about sending an American to the moon telegraphed through Washington. Ike's science advisors, Johnson, and both houses of Congress had settled on this undertaking as a must for the space race. The administration squeezed Dryden out of the running to lead NASA and made T. Keith Glennan, Killian's hand-picked choice, as the head instead.

Glennan moved into the Dolly Madison home near the White House to begin planning his mission and quickly adopted the Vanguard crew at NRL to form the corps of a nonmilitary science team at NASA's first space center, near Beltsville, Maryland, soon renamed the Goddard Space Flight Center. Newell became NASA's assistant director of the Office of Space Science on October 20, 1958 and Van Allen quickly drew him into the Space Science Board. Thus started what Newell described as a "love-hate" relationship between NASA and the academy of sciences. Glennan's organizational chart called for an advisory role from the Space Science Board and other agencies, including the military.

“With NASA in the driver’s seat, but the scientific community serving as navigator, so to speak, a tugging and hauling developed with a mixture of tension and cooperation,” Newell later wrote in *Beyond the Atmosphere.*

Newell succinctly summed it up. “Decisions concerning the space science program could not be made on purely scientific grounds,” he stated, given the needs for funding, facilities, and lobbying in Washington. The formation of NASA built in a “tugging and hauling” on three fronts. First, NASA now stood in competition with the country’s space scientists and the revered National Academy of Sciences. Second, NASA soon found itself split between human and robotic space exploration in a clash for political and financial clout. Finally, NASA’s ambiguous relationship with the Department of Defense meant that almost any space program could be leveraged away under the umbrella of national security.

The multiple roles York and Killian played as science and military advisors demonstrated the built-in conflicts between NASA and the Defense Department from the start. York, serving as a key architect of the civilian space program with Killian, also worked closely with McElroy and Quarles to draft a military blueprint for space. York claimed the grand goal of a human spaceflight to the moon for both NASA and the Advanced Research Projects Agency, with its focus on missile defense. The air force already had a third human spaceflight program of its own.

Moreover, the Defense Department became the training ground for astronauts as NASA responded to the politically expedient goal of planning to send a crew of Americans to the moon. One of NASA’s first reports flowed from a committee that included Pickering, Van Allen, and von Braun and placed scientific research, advancement of space technology, and development of human spaceflight with a high-thrust rocket as NASA’s top goals. Glennan lobbied the president for $100 million to build the rocket, now called the Saturn.

Von Braun stayed with the army, still holding the keys to the development of the one million ton thrust Saturn “moon” rocket. But Glennan wanted von Braun’s team, just as he wanted and took over JPL for NASA late in 1958. Medaris stonewalled NASA to keep von Braun, but Glennan knew that von Braun’s first loyalty was to space. After a visit to Huntsville, Glennan reported back to York that von Braun was a man who was talking army but thinking moon. Glennan and York pried loose von Braun and his inner circle away from the army and his team that now had 4,800 members remained with the space program as the ABMA became NASA’s Marshall Space Flight Center. Glennan lobbied for the money to build the new moon rocket. The downpour of cash for human spaceflight to the moon compared to the funding for other programs quickly sparked tension.

The human spaceflight priority promised to eventually eat up most of the resources for space research. Behind the scenes, Van Allen began to challenge human spaceflight, though he didn't go public with the rift as yet. Instead, with contacts on all fronts, he pushed for new programs such as probes to other planets—to places where humans couldn't travel but could touch from afar. With the probes, Van Allen transcended both sides of the spaceflight debate and he identified the plans with clear goals and tight budgets that found a friendly reception among the space scientists and NASA. "Among other things that I learned in the navy," Van Allen noted, "was how to make a sound decision when the basis for a decision was diffuse, inadequate, and bewildering."

• • •

Space and satellites invaded everything from home décor to popular slang after the first satellites went into orbit. People in the know were "in orbit." Brainy folks were burning their "jets." Hot cars had "rockets." The cache of space placed Van Allen on everybody's short list of people to call—not just reporters, but scientists, the military, and aides throughout the government.

He continued to monitor the *Explorer I* tapes and divided up his "space" faculty and dedicated graduate students among projects that stretched them to the limits. "I'd been working for years with rockets and rockoons doing a latitude survey of cosmic rays and now we had the satellite data to totally supplant years of work. We were waiting for enough accumulation of data so we could do a worldwide survey. Hanging in the air were all the instruments we had to prepare for other missions. We were trying to assimilate data and prepare for the next mission," Van Allen said. "Then reporters were calling asking us, 'What are you finding out? How is the instrument working?'"

Van Allen's pipeline of space projects hit high gear as NASA took shape. He rolled out experiments for several subsequent *Explorers* and the first *Pioneer* moon probe mission. He discovered the radiation belts (the subject of the next chapter) and documented the creation of artificial radiation belts from a series of U.S. nuclear blasts. The basement of the overcrowded physics building housed a pipeline of instrument construction, data reduction, and experiments involving his master's and PhD students.

"Other physics departments might view the master's degree as a consolation prize, but I viewed it as significant training to do research," Van Allen said. Upstairs, he navigated classes, staff, budgets, and negotiated an unexpected split between the departments of mathematics and astronomy, ending a decades-old partnership. "Math department has decided to seek a new head if I still agree (as I did several years ago) that the astronomy department might be incorporated in the physic department," creating the Department of Physics and Astronomy, he wrote in his journal. Van Allen and Iowa's lone astronomer, Dr. Hugh Johnson, backed the new department with the idea of building up astronomy as a strong field to partner with space science.

Van Allen's celebrity status after *Explorer I* added to the load. Requests for speeches, lectures, and invitations to lunch with the likes of Robert Frost poured in. His research won him election to the National Academy of Sciences in 1959 and honorary PhDs from the University of Michigan, Northwestern University, and a dozen other institutions. Streams of reporters found their way to the physics building, attracted to the space program at the heartland of the country's corn belt. They arrived as Van Allen hand-finessed electronic circuits with his graduate students, shirt sleeves rolled up, and paper tapes spilling new secrets about the solar system across the cement floor. The gifted teacher won out over the essentially shy man during the interviews. And the stories brought the exploration of outer space back home to where a likeable physicist made science seem simple. "If you can't explain it to your grandmother, you don't understand it yourself," Van Allen said time and again.

He didn't seek out the press but didn't ward them off either. "We were like the heroes who raised the flag over Iwo Jima, sort of temporary heroes, rescuing the honor of the United States in this great Cold War with Russia by having a successful satellite," he said.

Walter Cronkite arrived for the first time in March of 1958. "I remember I was very much impressed with Dr. Van Allen. He was a very modest man—but very personable. I felt that we had become friends on that first meeting and indeed we did occasionally talk by phone," Cronkite said.

Time magazine covered the discovery of the inner radiation belt in May 1958 and steered Van Allen into a cover story in 1959. The article followed him from his 6:45 wake up call, as three-year-old Tommy jumped on his bed, to his arrival at the physics building where even the janitor addressed him by his nickname "Van." After descriptions of his office and a quick tour de force through his lecture on transformers, the article turned to the frenetic pace of instrument fabrication, testing, and data analysis in the now famous basement. The writer described Van Allen's "battered pipe," his wartime bravery, and his "deceptive skill in threading his way through Washington's bureaucratic jungles."

Even a gifted writer couldn't improve on Van Allen's closing quote that summed up the urge to go into space: "The satellite is a natural extension of rockets, which are natural extensions of planes and balloons, which are natural extensions of man's climbing trees and mountains in order to get up higher and thus have a better view."

Unpretentious himself, Van Allen had little patience for pretentious inquiries about his work, however. "One that lingers in my mind is the following telephone conversation. 'This is Jim Lear, science editor of the *Saturday Review of Literature* calling from New York.' Heavy emphasis on 'calling from New York,' then a long pause waiting for me to recover from the thrill of hearing from such an important person, in New York, no less. Actually I did know who he was and had often characterized him as the anti-science editor of the *Saturday Review*. He continued: 'I read of your recent report of the discovery of the radiation belts of the Earth and thought that I would do a piece on the subject. What I found remarkable was that such important work was being done at a midwestern state university.' Heavy emphasis on the 'midwestern state university.' Well, I don't think I responded with any profanity but I did manage to convey a suggestion as to what he could do with his piece and hung up," Van Allen wrote years later. Lear complained to University of Iowa President Virgil Hancher, who called Van Allen for an explanation and closed the conversation by congratulating him for his response. Van Allen's folksy comments, modest surroundings, and forthright manner made him the perfect candidate for the scientist as a celebrity who offered a trusted voice on several fronts. He calmed fears regarding the use of microwave ovens and punctured the false sense of security promised by the Cold War civil defense that anyone could survive a nuclear attack in a $30 homemade bomb shelter.

Iowa physics professor Max Dresden blasted the bomb shelter myth. Van Allen lured the Northwestern University physics department chair to the University of Iowa with the promise of a free reign to pursue theoretical physics. Theory aside, Dresden challenged the comic naivete of the Kennedy administration's plans for backyard and basement fallout shelters. The administration offered how-to plans and optimistic words penned by Nobel Laureate Willard Libby, a chemist and a leader of the Atomic Energy Commission. Libby's articles ran nationwide and the *Iowa City Press-Citizen* picked up more than a dozen of them.

"For $30, I built a fall-out shelter in my backyard. It gives my family 100 times more chance of surviving nuclear fallout than if I had done nothing," Libby wrote in one article.

Dresden lashed back at the absurdity of such advice in a letter to the press, cautioning the public about the "extremely dangerous" impression that families could survive nuclear war for $30. Eight University of Iowa faculty members, including Van Allen, signed the letter. It pilloried Libby and the Kennedy administration for presenting the gruesome survival of a nuclear holocaust as a "two-week vacation in a model home."

The *Iowa City Press-Citizen* ran the dramatic confrontational letter on the front page rather than the editorial page. The national press spread the news, exchanging Dresden's role in the juicy debate for a national face-off between two celebrity scientists: Van Allen and Libby. While other academics had previously voiced similar concerns over the fallout shelter fantasy, the fallout from the Iowa letter helped shift national policy to a civil defense strategy of community shelters in places such as Chicago's subway tunnels and the cavernous basements at the University of Iowa.

But space exploration offered a vision beyond the Cold War, a promise to touch the moon and distant planets for the first time. Van Allen's discovery of the radiation belts with America's first satellites dramatized the mysteries to be found when he announced the discovery in the spring of 1958.

12 Discovery of the Radiation Belts

Cape Canaveral—March 26, 1958. The Jupiter C stood iridescent in the dawn sky, shrouded by plumes of liquid oxygen a few hours before launch time. George Ludwig, Jack Froelich of JPL, and other JPL satellite engineers watched it with subdued anxiety when it finally soared skyward in a bolt of scarlet flame. *Explorer II,* launched three weeks earlier, had failed to fire into orbit. But eight minutes into *Explorer III*s launch, the ground stations began to catch a high-pitched hum. The JPL team burst into jubilant cheers but Ludwig focused most of his attention on the Iowa Geiger counter. For him and Van Allen, *Explorer III* held the promise of vindication from a baffling mystery and suspicions of a possible instrument malfunction on *Explorer I*.

• • •

The Geiger tube on *Explorer I* counted cosmic rays like a faulty adding machine. Ground stations received data in a chorus of rising and falling tones that correlated to levels of cosmic ray counts. Then, the instrument seemingly blanked out and the counts dropped to zero before returning to reasonable levels. Sparse data sets of a minute's duration transmitted from uncertain altitudes only intensified the gaps in the puzzle.

"I've got bad news for you. My people tell me that your counter stopped working," Pickering called to say as the pattern emerged.

"I thought he was wrong because we had all these pieces where it was and wasn't working," Van Allen recalled, though he couldn't explain the anomaly either.

"When I got back from my Fort Churchill expedition, they were busy looking at the data and scratching their heads," McIlwain noted. "I pointed out that another possibility was that the flux might sometimes be very high, driving the Geiger tube into such hard saturation that it didn't count at all."

Getting a satellite in space within four months of the first *Sputnik* required nothing short of an engineering miracle but Van Allen candidly described *Explorer I* as a "shakedown" operation that succeeded on "fool's luck." Yet even its fragmentary data produced miles of reel-to-reel audiotapes that delivered cosmic ray counts in the form of a staccato symphony of rising and falling tones. The tapes arrived via regular mail days after they were recorded and the backlog piled ever higher in the basement of the physics building where tapes played nonstop on Crown Royal console tape players. Pens on long arms pulsed and slithered across paper tapes to graph the audio tones. A squadron of students measured the peaks and valleys of the graphs and converted them to numerical counts recorded in notebooks. Van Allen, Ray, Kasper, and McIlwain scanned the paper tapes for a transition from the rising counts to the sudden plunges. If a transition existed, it was lost in one of the gaping holes in the data.

But *Explorer III* had Ludwig's data recorder to pin down the roller-coaster ride of the cosmic ray counts. The recorder collected continuous data. A ground station transmitted a single radio command and triggered the recorder to dump all the data from a complete orbit of Earth in a matter of six seconds. Van Allen pinned his hopes on the recorder as *Explorer I* delivered messages that marked a major technical failure or a mystifying new property of space.

The tape recorder had already captured the public's imagination. Walter Sullivan of the *New York Times* visited Iowa in October 1957 as Ludwig completed the prototype of the Deal 1 Geiger tube, a "masterpiece of miniature electronics," Sullivan wrote. "Most impressive to the layman was the tape recorder which was to give the satellite its memory. It was the size of a small alarm clock, designed so that the (40.5-inch) magnetic tape would jump forward once a second for as long as two hours, winding a spring as its wheels revolved."

A satellite with a memory was just the thing Van Allen needed to figure out how a journey through space could paralyze a Geiger counter and then restore it to normal activity. Ludwig started working on the recorder in mid-1956. In addition to lining up miniaturized watch parts, he switched from Mylar recording tape that could stretch in high temperatures to a rugged "0.001-inch thick phosphor bronze recording tape with an electroplated nickel-cobalt recording surface." The tape operated on a time-release take-up spool, moving in short stops like the hands on a mechanical clock. It recorded 200 seconds of data per inch of tape until the playback command from the ground released a spring that pulled the tape across a playback head. Then the recorder deliv-

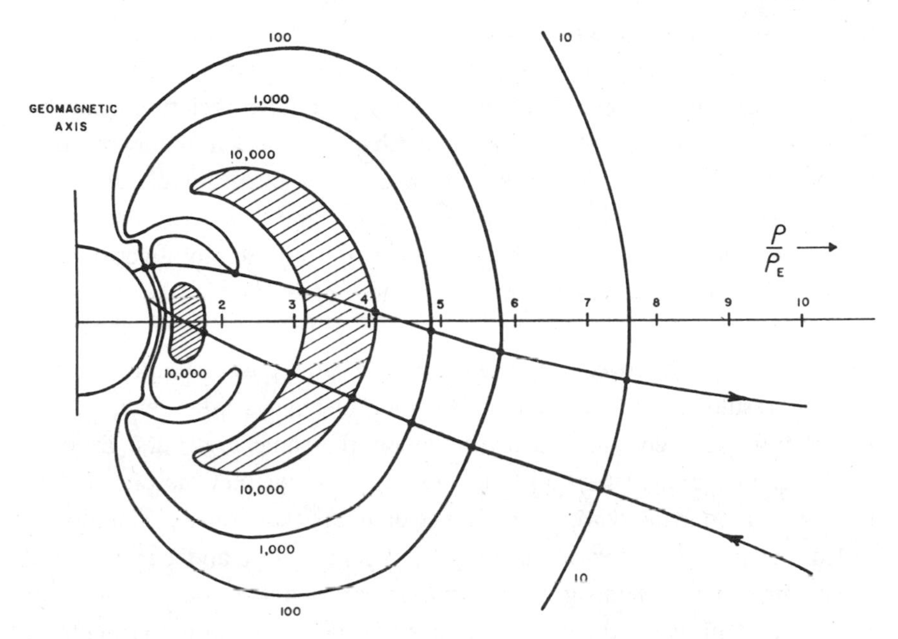

Early schematic of the radiation belts (crosshatched areas). The lines show the outbound and inbound paths of Pioneer 3. *Courtesy of James Van Allen.*

ered a global epic of data before the recorder erased it and wound the tape back in position to record again. The whole device weighed less than 8 ounces and consumed less than 1/35,000 of a single watt of power—power conservation that permitted the satellite to operate for a full forty-four days. Ludwig and the instrument makers in the University of Iowa shop had spent some 2,000 hours working on the recorder for the Vanguard satellite.

Because of design differences in the *Explorer 1* (Deal 1) compared to Vanguard and due to the frantic effort to make *Explorer 1* flight-ready, the recorder wasn't included as part of the final payload. "The primary objective for the first Deal 1 was to orbit a satellite as quickly as possible, while achieving somewhat limited scientific objectives," Ludwig recalled. The Deal 2 satellite included his full instrument package, including the recorder, on the failed March 5 launch of *Explorer II*. The disappointment when the satellite didn't achieve orbit added to the national gloom of a second Vanguard failure in February when a control system malfunction after 57 seconds of flight drove the vehicle into a dive. The Russians remained one *Sputnik* ahead. America quickly learned to accept the mood swings of the space program with the successful launch of a Vanguard on March 17. It reached record altitudes of more than 2,500 miles and ushered in solar-powered satellites that didn't need batteries. Van Allen believed a successful *Explorer III* flight, complete with the recorder, held the potential for America's next breakthrough in space.

Cold War competition didn't keep Van Allen from appreciating what the Russians had done. He took a break on March 21 when a clear night promised a chance to visually spot *Sputnik 2* over Iowa City. "A beautiful majestic sight as it moved across the sky—silent smooth motion," Van Allen wrote in his journal the next morning. "Took Cynthia, Margot, Sarah, and Tommy together with Ed Nelson and Harriet and Teddy to the west edge of the married student housing area to watch."

The night watch proved prophetic. *Sputnik 2* spotted a surge in cosmic ray readings resulting from the radiation belts but the Soviets didn't realize at the time that they had encountered anything more than an energy burst. *Explorer III*, launched five days later, sealed the discovery for Van Allen, despite the fact that Ludwig found faulty wiring in the instrument package during countdown.

"My reaction to that was to heat up the soldering iron and start rewiring the wiring channel," Ludwig recalled in a University of Iowa documentary on Van Allen. "This is not the way things are normally done in preparing for a rocket flight or a spaceflight, so a number of JPL people relocated to a trailer to find out what they should do about this madman who was rewiring the satellite during countdown."

NRL scientists once again pitched in to optimize data capture and two days later, on March 28, NRL's San Diego station downloaded the first complete global recording of cosmic rays found by the Geiger tube on *Explorer III*. Ludwig briefed Van Allen Sunday, March 30, en route to Pasadena, and Cronkite arrived at the physics building the next day for an interview.

On Tuesday, Van Allen traveled to Washington, D.C., to confer with NRL scientist Jack Siry, who was still struggling to calculate *Explorer III* orbit altitudes over different spots on the globe. On Wednesday, Van Allen learned that the first *Explorer III* data had arrived in D.C. At the end of the day, Van Allen took a taxi to the Vanguard data reduction center on Pennsylvania Avenue and picked up paper tapes of the global cosmic ray readings collected from *Explorer III*.

He stopped at a drugstore to buy graph paper and a ruler before returning to his room at the Dupont Plaza Hotel with the paper tapes that showed the peaks and dives in radioactivity as the cosmic ray counts rose and fell for the 102 minutes it took *Explorer III* to orbit the globe. He used a slide rule and Siry's fresh estimates of the orbit to calculate counts as a function of the latitude and altitude of the satellite. He carefully plotted fifteen minutes of rapid rises in the counts from predictable levels to the maximum the instruments could process. There were forty-five minutes of zero readings, then readings that rapidly decreased, as though the instrument had a switch to turn it on and off.

"At 3 A.M., I packed my work sheet and graph and turned in for the night with the conviction that our instruments on both *Explorers I* and *III* were working properly but that we were encountering a mysterious physical effect."

A microfilm copy of the San Diego data arrived in the mail at the University of Iowa while Van Allen was gone. Kasper, Ray, and McIlwain grabbed the reel and put it on a microfilm reader. "We were looking for a clear transition, the switching points from rising counts to zero. That's what we couldn't make out with the *Explorer* data and now, there it was," McIlwain said.

"So we knew at once that there was something of very high intensity out there. I immediately took the spare payload and put it in front of an X-ray machine," McIlwain recalled. "The results showed that fluxes that would ideally produce more than 35,000 counts per second, instead drove the rate to zero." Essentially, McIlwain's trial showed the levels of radiation that would saturate the instrument and choke off readings.

Van Allen returned to Iowa with his graphs and calculations in his briefcase. "I was wildly busy, just exhausted. I don't even have a journal account of what happened that day," he said. When he got to the office, he noticed a note from Ernest Ray left on his chair.

"Space is radioactive," it said. Ray's conscious exaggeration summarily announced that he and McIlwain were onto something too.

Van Allen laid out his graphs and McIlwain laid out his X-ray results. The three men looked over the papers cautiously at first. Everything pointed to an abrupt boundary crossing into a zone in space where cosmic rays suddenly rose to unimagined levels. "Then it clicked right away. That was the moment when the lightbulb went on, the eureka moment," Van Allen said.

Ludwig returned from "foreign duty" at JPL on April 11, Van Allen noted with obvious relief in his journal. Rosalie Ludwig, newborn George Junior, and the couple's two daughters took an airplane back to Iowa and George drove home in the Mercury.

Data analysis from the satellites topped a feverish agenda when he returned to the physics building. Van Allen put aside the pressing fabrication of multiple instruments and spread cascades of paper tapes from *Explorers I* and *III* across a table as he, Ludwig, Ray, and McIlwain hammered out a description of a radiation zone in space—the first discovery of the space race.

Speaking for all four of them, Van Allen announced the discovery at a joint meeting of the National Academy of Sciences and the American Physical Society on May 1, 1958, in Washington, D.C. Everyone anticipated cosmic ray counts would rise steadily at the altitudes satellites could reach and this proved this to be the case. "The counting rate was more or less sensible-looking at the start and then blanked as it came along in here and this transition occurred," Van Allen said. Then he described the abrupt climb of the satellite across an invisible boundary about 600 miles above and into a field where radiation intensities vaulted nearly 1,000 times beyond the area just below it. Here, the earth's magnetic field trapped a dense blizzard of charged subatomic particles into a region between about 600 and 6,000 miles above the earth. This band above the earth extended outward for thousands of miles from the region of the equator and particles spiraled back and forth within it, traversing half the globe in a matter of minutes.

"I'd like to make sure I understand," one incredulous scientist asked after the presentation. "The idea is that the only mechanism you can conceive of by which the stuff is kept at high altitudes and kept from getting down to lower ones is magnetic confinement?"

"Yes, that's the essence of it," Van Allen agreed. With the counts changing so drastically just below the band, "we just think we just have to have the magnetic field holding them up."

Richard Porter then ushered Van Allen to another room for a press conference where Van Allen explained the intricacies of "geomagnetically trapped corpuscular radiation." Reporters grasped the idea: a dramatic band of radiation girdled the earth with particles capable of piercing sheets of metal as though they were silk. Still, they needed a simple, visual term to explain the concept.

"You mean they circle the earth like a belt?" asked science reporter William Hines of the *Washington Evening Star.*

"Yes. That's right," Van Allen said.

At a scientific conference in Europe soon after, NRL physicist Robert Jastrow used the term *Van Allen radiation belt* for the first time and the name quickly defined a permanent new landmark in the heavens. The Russians sent up the first satellite but America made the first scientific discovery there, the most momentous discovery of the International Geophysical Year, according to Walter Sullivan. The discovery opened up a new mapping of the solar system through new fields such as magnetospheric physics, devoted to the magnetic fields of the planets and sun, and plasma physics, devoted to the study of the solar wind of hot gases radiating outward from the sun.

It also proved a 1907 prediction made by Norwegian physicist Carl Størmer that a charged particle caught in a dipolar magnetic field, such as the earth's, could become trapped in a crescent that would arc out into space from one pole to another. The particles would spiral through this zone, Størmer demon-

Early rendering of the radiation belts. Courtesy of James Van Allen.

strated in a classic illustration with which every one of Van Allen's physics students became particularly familiar.

"The work of Størmer, at his desk at the University of Christiania, was an awesome demonstration of the capabilities of man's intellect for he was calculating the behavior particles far too small to see in a region far to remote to visit," wrote Walter Sullivan. Throughout his career, Van Allen credited Størmer with the "foundation of much of my own research." But Van Allen didn't have the luxury of spending much time at his desk to even fully review all of the data from *Explorers I* and *III*.

With the May 1 announcement behind them, Van Allen and his team rushed to complete and test the instrumentation for *Explorer IV* and the secret Argus bomb blasts (the subject of the next chapter). Though Argus was top secret, the actual IGY *Explorer IV* launch was not and gave adequate cover to the work in the cramped space instrument laboratory webbed with electrical wiring. The *Explorer* tapes lined an entire room by now, neatly labeled by tracking station: San Diego, Pasadena, Havana, Singapore, Lima, Peru, Antofagasta, Chile, and other points around the globe. The hallway to these enclaves served as an additional laboratory and the unused elevator shaft doubled as a repair shop. The official machine shops fabricated parts nonstop on the second floor and patrons of the hardware stores near campus became accustomed to harried graduate students rushing in to buy wire or miniscule screws for instruments.

Van Allen's graduate students pushed themselves, taking the idea of a hands-on education to the limits of endurance. "George Ludwig and I let our poor bodies work as long as they would let us but we both found that sixteen hours is the limit, anything past that and you're doing negative work and it'll take longer the next day to clean up the mess you've made," McIlwain said. Both lived near campus and snatched odd hours to get home to their families. McIlwain, who married as an undergraduate, had a newly adopted little girl. But the excitement of new discoveries carried them through the long hours and family sacrifices.

Explorer IV, launched July 26, orbited the earth in a northward to southward direction that took cosmic ray counts at high latitudes. It had an array of detectors that could directly count the blizzard of particles in the radiation belt and could help identify the particles. McIlwain quickly determined that the radiation belt announced May 1 banded the earth from a latitude of 50 degrees north to 50 degrees south, a distance that stretched from central Canada to Patagonia. No sooner had Van Allen and his team firmly pinned down the contours of one belt when they found telltale evidence of another, larger and higher radiation belt arching over the inner belt and downward toward the earth in a cutaway shape of a crescent moon. A slot of space with "normal" cosmic ray counts separated the two belts of trapped particles.

Ernest Ray traveled to the Fifth General Assembly of the IGY from July 30–August 9, 1958. *Sputnik 3,* a 2,926-pound behemoth launched on May 15, set off a new round of American paranoia about the missile gap and suspicions that the Russians must be getting close to sending a man into orbit. But one of *Sputnik 3*'s main missions was to identify specific subatomic particles in space with a variety of cosmic ray detectors that helped pinpoint protons as the particles in the lower belt. The Russians had made no announcements about any unusual cosmic ray findings with *Sputnik 2.* Now Sergei Vernov, Van Allen's Soviet counterpart, talked at the conference about bursts of variable cosmic ray readings in space. He made no attempt to interpret them. Ernest Ray summarized for the IGY audience the announcement of the radiation belt.

Ray didn't mention the outer belt. Not even a satellite could measure a cross section of an area encircling the globe at an estimated 10,000 miles above the earth. The detectors on *Explorer IV* could only measure particles in the horns of the outer belt as they arched inward toward the polar regions. Van Allen cautiously hedged his bets about announcing the outer belt until he could be sure of what he would be describing. Satellites had eclipsed more than ten years of rocket and rockoon expeditions. Now, Van Allen needed still another tool—a probe that could travel thousands of miles upward into space and cut directly through both radiation belts. He got his wish. The air force ushered in a prelude to interplanetary missions in the fall of 1958 with a lunar probe, *Pioneer 1*. It launched on October 23, 1958, carrying Iowa detectors, cutting through the radiation belts but failing to reach the moon. *Pioneer 2,* launched November 8, carried an array of JPL instruments but the probe barely reached the altitudes of the orbiting satellites before it fell back. *Pioneer 3* belonged to Iowa again and Van Allen kept his instrument package simple—a Geiger counter and a backup Geiger counter. On December 6, the probe soared about 63,000 miles into space and fell back to earth. It was another failure—except for Van Allen. His instruments etched the upper and lower boundaries of the radiation belts as the rising probe transmitted particle counts. The lower belt maintained a relatively uniform boundary between 600 and 6,000 miles above the earth. At its most intense swathe above the equator, the upper belt covered more than 5,000 miles, from altitudes of about 10,000 to 15,000 miles above the earth. It stretched from the Arctic to the Anarctic Circles. Van Allen announced his latest finding on December 27, 1958, at a meeting of the American Astronautical Society in Washington, D.C.

With a composite picture pieced together from thousands of readings in space, Van Allen presented a cutaway diagram of the two radiation belts and estimated the maximum counts recorded by his instruments at 25,600 per second, with each count recording more than one particle. "This was equivalent to 160,000 particles striking the area of a postage stamp every second," reported Sullivan. Van Allen described how particles spiral up and down in a corkscrew pattern inside the belts, traversing thousands of miles from one hemisphere to the next in a second. The inner belt is filled mostly with protons while the outer belt traps electrons, he reported. As particles lose energy, they precipitate along the magnetic field lines of the earth. The solar wind, the plasma of hot ionized gases radiating outward from the sun, constantly replenished the belts with fresh particles, "a sort of leaky bucket constantly refilled from the sun and draining away into the atmosphere," Van Allen wrote in *Scientific American.*

He had one problem with his leaky bucket theory—the tremendous energies of many particles in the belts far exceeded anything measured in the solar wind. He later found that galactic cosmic rays fueled the earth-encircling inner radiation zone with protons.

The press relied on a variety of descriptions to grapple with the concept of radiation belts. "Find space doughnuts," heralded the whimsical headline in the *Science News Letter*. Van Allen's composite picture of the radiation belts, pieced together from thousands of cosmic ray counts across the globe, raised fears as well as national pride. "Radiation Belt Dims Hope of Space Travel," read the headline reporting the discovery in the *Chicago Tribume.*

"I think people thought of it as something ominous. The whole idea of the Van Allen belts I think presented itself as one of the major handicaps to ever putting a man into space. It was believed that a man could not survive in space. This looked like a major block. It kind of took its place alongside the myth that we couldn't exceed the speed of sound," said Cronkite.

Van Allen warned about the risks the belts posed to astronauts and a University of Iowa press release explained that the radiation levels would be enough to kill a human being (orbiting through them in a satellite, for instance) in forty-five hours. Knowledge of the belts helped NASA identify safe orbits for astronauts and northern and southern "escape" routes into space where the belts tapered off.

• • •

By the summer of 1959, the Soviet press began to cover the radiation belt story and lay claim to part of the discovery. The debate came to Van Allen's attention for the first time in March 1959 when *Pravda* published an article suggesting that Soviet scientists were the first to announce the readings that located the outer radiation belt. Robert Toth, a reporter for the *New York Herald Tribune,* contacted Van Allen about an article translated from *Pravda* and written by Vernov and a colleague. "In particular, the second page of the translation says the Soviets reported finding [*sic*] of the two zones of radiation to the IGY assembly last summer," Toth wrote Van Allen. "I wonder if you could give me some comment on this matter and the article in general for a possible news story. I would appreciate hearing from you as soon as possible."

Van Allen responded the next day. "The first public announcement of the discovery of the existence of the trapped radiation was my paper before the joint session of the American Physical Society and the National Academy of Sciences on May 1, 1958. Up to that time, there had been, to my knowledge, no announcement of Soviet radiation observations with *Sputniks 1* and 2, which gave any indication of the existence of the high intensity of the trapped radiation. *Sputnik 3* was launched on May 15, two weeks later," Van Allen wrote.

He pointed out that Ray had attended the Fifth General Assembly of the IGY from July 30–August 9, 1958, and heard papers by Vernov and A. Chudakov, the first exchange of Russian and American scientists regarding the radiation belts. They offered no report on the outer belt and no graphic rendering of the belts as shown in Van Allen's reports.

"In retrospect, I should say that both the Soviets and we independently did have the basis—as of late August—for speculating on the structure of the [outer] radiation zone," Van Allen wrote. "But neither group had the perspicacity to do so at the time." Van Allen waited until he had the direct and confirming observations of *Explorer IV* and *Pioneer 3* before announcing the outer radiation belt on December 27, 1958, at the American Astronomical Society in Washington D.C. "I conclude that the statement of Vernov and Chudakov that their Figure 2 was reported in August 1958 to the IGY conference contains a considerable admixture of hindsight. Nonetheless, I wish to express my considerable admiration for their work. And, as a scientist, I am delighted with the confirmation of our observations which the Soviets are obtaining."

Van Allen's comments on hindsight covers sentiment on the radiation belts outside of Russia. "I'm sure you've heard the story that Vernov discovered the Van Allen belts," said Michael Neufeld of the National Air and Space Museum.

But the Russians continue to revisit the issue, evident in a December 2000 article by Ivan Zavidonov of the Russian Academy of Sciences. Zavidonov stated he relied on previously unavailable Soviet records and reported that Vernov and his colleagues published a paper describing two separate zones of radiation in October, 1958, in the Russian publication *Iskusstvennye Sputniki Zemli*. The Soviets censored the issue at the time. Suspicions by both the Soviets and the Americans that nuclear ballistics testing by the "enemy" artificially created the radiation belts may have initiated the censorship.

However, in the paper, Vernov differentiated between the zones of radiation by latitude rather than by altitude, identifying a polar and equatorial zone separated by a gap, as though he was describing vertical shafts of radiation rather than belts that encircle the earth. Nothing in the paper matches the signature profile of two concentric radiation belts that Van Allen reported in December 1958. Vernov focused the paper on the so-called soft radiation in the northern region of the aurora, as his title suggests: "A Study of the Soft Component of Cosmic Rays Outside the Atmosphere."

But in 1960, Vernov and Chudakov published another paper claiming discovery of the outer belts with *Sputniks* 2 and 3. Zavidonov's article makes it clear that, as far as the Russians are concerned, Van Allen discovered the inner radiation belt and Vernov discovered the outer belt, for which the Soviet Union gave him the 1960 Lenin Prize. Zavidonov dismisses any confusion over the matter to Cold War politics and "misrepresentation of the events in the mass media."

The fact remains that Vernov's early reports on satellite research discuss cosmic radiation levels rather than radiation zones. The Soviets didn't have the data to map the zones until later, as Van Allen points out in the letter to Toth, the earliest explanation of the matter. "The Soviet Cosmic Rocket 'Mechta' was not launched until about a week later—namely on January 2, 1959. It was impossible for the Soviets to have a sound observational basis [for the diagram shown in] the March 6 *Pravda* article prior to the flight of Mechta," Van Allen wrote to Toth.

Van Allen gave frequent credit to the combination of readings from Russian and American space missions that reinforced his own mapping of the inner and outer radiation belts in December 1958. "In retrospect, these firings represent a more or less interlaced series of events which occurred during an important period of nip-and-tuck developments by the Russians and ourselves. All this has permitted an unprecedented dovetailing of data," Van Allen told the audience of the Institute of General Semantics where he was invited to speak in 1959. "While the Soviets and the U.S.A. might not agree in regard to the disposition of Berlin, it is certain that we can reach ready agreement in the realm of physics."

The radiation belts continue to allure scientists. "We had a great magnetic storm after the launch of *Explorer I*—still a pretty unique event in the history of space. It was a great solar flare and great increase in the solar wind—increase in hot gas from the sun. And we saw a lot of effects on *Explorer I* data. We published a paper revisiting this two or three years ago. It's an active subject at the present time," Van Allen said.

The *Explorer* program, managed by NASA's Goddard Space Flight Center, lives on and continues to make momentous discoveries. The *Cosmic Background Explorer* (COBE), launched in 1989, discovered a ripple of microwave background radiation that could be traced back some 15 billion years to the big bang. Explorer missions continue to study how solar storms reconfigure the earth's magnetic field, radiation belts, and the aurora.

Original magnetic audio tapes from the first *Explorer* rested for decades in the cavern that once housed the Van de Graaff accelerator in the basement of the University of Iowa's physics building on the Pentacrest. Computer technician Bob Brechwald, retired from the physics department, set out to transfer the data to CD-ROMS. He located an old Crown Royal reel-to-reel tape recorder to play the tapes and a tech-savvy student who needed a summer job. *Explorer*'s audio ballad of the first discovery in space is preserved.

13 Space Shield for the Cold War

Moscow, Cosmic Ray Conference—July 1959. To the astonishment of Sergei Vernov, Leonid Sedov, and other Russian physicists present at the academy meeting, guest scientist James Van Allen spelled out detailed findings of the top secret Project Argus, the prototype project that tested the creation of a space shield with artificial radiation belts.

• • •

Van Allen didn't have to travel all the way to a meeting of physicists in Moscow to see the faces of the skeptics as he began to discuss the radiation belts. Plenty of people on his side of the Cold War fence viewed the belts he had discovered the year before as telltale evidence of nuclear testing rather than as a natural phenomenon. The Americans blamed the Russians. The Russians blamed the Americans. Now Van Allen could settle the political confrontation. The Argus nuclear blasts gave him proof that no one was to blame for the natural belts arching through thousands of miles of outer space. Nuclear weapons that could cause immense destruction produced rather puny, short-lived artificial belts by comparison. Van Allen knew because he had measured the artificial belts Project Argus generated with instruments carried on *Explorer IV* in 1958. The battery of instruments did triple duty: studying the inner radiation belt, identifying the outer radiation belt, and surveying the impact of five separate nuclear bomb blasts for the Pentagon.

The narrow bands of the artificial belts decayed in a matter of weeks while cosmic rays and solar plasma replenished the natural belts. The artificial belts also showed the fingerprints of radiation characterized by a nuclear blast. The message was clear. While accountability remained a controversy for a general nuclear test ban treaty, nuclear tests in space would leave a signature that a satellite loaded with cosmic ray detectors could monitor.

Van Allen's four detectors on *Explorer IV* studied the radiation belts directly for the first time. The single Geiger tube detectors on the previous *Explorers* revealed the belts with huge gaps in the data because readings from the instruments, saturated by radiation, blanked to zero with each entry into the belts. But *Explorer IV*'s scientific program measuring the radiation intensity in these zones gave cover to a hidden agenda—the survey of the impact of several nuclear bomb explosions. The swift declassification of key aspects of the bomb blasts—apparent around the globe and reported in the press—allowed Van Allen to discuss them in Moscow. But Van Allen focused on the science rather than the bold Argus test meant to generate an anti-ICBM "Astrodome" around the earth with a shell of artificial radiation belts. The U.S. Air Force had exploded the low-yield nuclear missiles above the atmosphere and electrons from the blasts did indeed create the artificial belts intended for the shield. Sustaining these ephemeral shells would require numerous and continuing bomb blasts, however, so the project was abandoned.

The *New York Times, Time* magazine, and other publications touched on the military significance of Argus in articles in March 1959. They introduced Nicholas Christofilos, a brilliant and visonary Greek elevator mechanic who rose to the highest circles of nuclear physics at Berkeley's Livermore Laboratory. Trained as an engineer, the shock of *Sputnik* sparked Christofilos's original idea for defending democracy with a space shield, a concept dusted off and refined thirty years later with the Reagan administration's "Star Wars" defense model.

"Nick was more strongly moved by [*Sputnik*] than anyone else I knew," wrote York in his memoir *Making Weapons, Talking Peace*. Christofilos took the experimental results of confining particles in an earthbound nuclear reactor and applied the concept to outer space. From the mushroom clouds of numerous blasts, electrons would radiate around the globe like a huge nuclear umbrella following the earth's magnetic field, Christofilos reasoned. This could weave a space shield that could burn up spy satellites and prematurely trigger enemy bombs so that they would disintegrate before ever reaching their targets. York was the one who compared the idea to a global Astrodome.

"Only a few weeks after *Sputnik*, he [Nick] came in to tell me all about it. In essence, he proposed to explode a large number of nuclear weapons, thousands per year, in the lower regions of the earth's magnetosphere, just above the upper reaches of the atmosphere," York wrote. "He expected that this region would extend over the entire planet except for a small region at the poles. Nick had, in effect, invented a version of the naturally invented Van Allen belt before it was discovered."

Christofilos, born in Boston in 1916, moved back with his parents to their native Greece when he was seven. He later earned degrees in engineering from the National Technical University and went to work in Athens as an elevator mechanic. The Germans confiscated the elevator company and its employees as a repair facility during World War II and Christofilos, restless with his new work, studied German and German physics journals that focused on particle accelerators. After the war, he started sending American scientists a stream of innovative ideas about accelerators until the Brookhaven National Laboratory hired him in 1954. No sooner had he settled in at Brookhaven when he lost interest in accelerators and turned to a new onslaught of plans to build a fusion reactor. His hypothetical designs won him a job at the Livermore Laboratory at Berkeley, now the Lawrence Livermore National Library.

Then came *Sputnik* and Christofilos shifted gears again. He worked away on calculations for the space shield, a tight protective mesh of electrons, generated by nuclear blasts that would continue to be fired to sustain such a shield. "Nick always thought in terms of big numbers and what other people would regard as outrageous ideas," York said.

Christofilos was aware of the dangers of his plan if too much radiation resulted from a blast. "A-bomb explosions must be carefully designed to avoid creation of hazardous radiation," Christofilos wrote. "Fortunately, [for an experiment] a much smaller yield, namely in the kiloton range, is sufficient to yield detectable quantities without creating any radiation hazard at all."

York, still at the Lawrence Laboratory with Christofilos at the time of the *Sputnik* launch, came to Washington at McElroy's request in the crisis response to the Russian satellite. York joined the Pentagon's newly formed Advanced Research Projects Agency (ARPA), as chief scientist. Now he had the power to authorize Christofilos's proposed experiment—Project Argus. Van Allen visited JPL headquarters in Pasadena in March 1958, just as Pickering first heard about the secret plan. Pickering had already suggested Van Allen as the man with the detectors to monitor the Argus blasts.

By early April, Van Allen knew he had discovered the natural radiation belts that Christofilos hoped to create artificially. His discovery seemed to confirm Christofilos's premise. Van Allen got an informal go-ahead for the Argus project from York on May 1, the same day he delivered his report at the National Academy of Sciences on his discovery. On May 10, he and Ludwig traveled to JPL to meet with Richter, Boehm from the ABMA in Huntsville, and representatives of NRL.

"That was a very important meeting, all undocumented except for these notes," Van Allen said, referring to his eleven-page handwritten summary of the meeting on Argus. In the course of the freewheeling meeting, Van Allen matter-of-factly confirmed that Iowa would build all the payload instruments in the basement of the physics buildings. "Agreed: [Iowa] will coordinate payload assembly!" he wrote exultantly in his notes. While he had fought against military control of scientific space research, he had no qualms about partnering with the military for a defense mission. He knew what he was looking for this time around and developed instruments accordingly. "The really revolutionary thing was the new system of four detectors, designed to cope with the [radiation] intensity that we had previously found to be present," Van Allen said. "In [planning] *Explorers I, II* and *III* we had no idea the radiation belts existed, they were not forecast. They were a total surprise and our instruments were essentially overwhelmed by the radiation. Still we got good results and, by the time we designed *Explorer IV*, all the detectors were designed with the fresh knowledge we had and we were very eager to go out there with more discriminating detectors" that could give more details about the intensity and energy levels of the charged particles, Van Allen said. The detectors were also much smaller, miniaturized from the size of a D battery to the size of a triple A.

Boehm and ABMA took responsibility for the satellite for the mission, using the previous *Explorers* as models but slightly redesigning them to allow for more instruments, more battery power and a total weight of 37 pounds instead of about 30. Jet Propulsion Laboratory and the military boosted rocket performance and a launch plan to maximize the range of the orbit from fifty degrees north to 50 degrees south of the equator, a range that would cover the earth from central Canada to Peru. Previous orbits reached to only 33 degrees north and south, barely blanketing the southern half of the United States. National Research Laboratory and JPL added new ground stations in Nigeria, Singapore, and other locations to receive the transmissions.

"It was all decided on a handshake so to speak. We didn't have to ask anybody about our decisions. We understood each other well enough to just go out and do it," Van Allen said. But the meeting assigned Iowa an unbelievably tight schedule of deadlines. "June 6—First complete prototype! . . . July 1, two flight units. July 15, two flight units," Van Allen wrote in his journal. Flight testing at ABMA and the cape needed to be completed right after that. It was seventy-seven days to launch when Van Allen and Ludwig left Pasadena that evening in May, a schedule that even Van Allen could see required "lots of miracle work."

The Iowa team took up the task immediately, not waiting for the formal contract approval of $123,000 to build all the instrument packages. This included

$13,060 for 2,500 hours of engineering labor and $5,224 for 3,200 hours of manufacturing labor. The laborers were the familiar team of McIlwain, Ludwig, and Ray with backup from Sentinella's crew and spot help from any wayward physics students who happened to step into the basement. "Even undergraduates could count on finding part-time employment," said Bruce Randall, Van Allen's longtime research associate whose carpentry skills got him his first job in the physics department as a freshman in the early 1960s. Van Allen also hired a corps of women whose dexterity at sewing seemed to provide the perfect training for wiring together tiny transistors and electrical components the size of grains of wheat.

The Iowa scientists designed, built, tested, and field-tested the promised four packages of the suite of four instruments for *Explorer IV* in those seventy-seven days, working furiously under a sign that read "This job is so secret even I don't know what I'm doing." Ludwig, Boehm, and Stuhlinger commuted back and forth between Huntsville and Iowa City as debate raged over transmitters, batteries, and other details to make the instruments responsive to the intense radiation of the radiation belts and the weaker artificial belts likely to occur. Pentagon brass showed up—and went home incredulous. "Visitors to the University of Iowa during the spring and summer of 1958 were astonished to find that a crucial part of this massive undertaking had been entrusted to two graduate students and two part-time professors working in a small, crowded basement laboratory of the [1910] physics building," Van Allen noted. Despite the secrecy, the Iowa team could build everything in the open since both the satellite and the instrumentation officially served as programs to further study the radiation belts. Porter and Kaplan quickly arranged IGY sponsorship. Only the double-duty aspect of the mission to study the nuclear blasts was secret and only McIlwain and Van Allen had the full security clearance to be informed of all the details.

Explorer IV carried the time instruments designed to study radiation in the belts directly rather than deducing their presence from a data blackout. "We had no idea what was up there. What we knew about radiation belts then was that Geiger tubes would saturate upon entering them," McIlwain wrote.

The new instruments could sort out the cosmic ray environments of space inside and outside the belts. Van Allen ordered Geiger tubes designed to withstand the possible onslaught of 40,000 electrons or protons per square centimeter second in the intense radiation zones of the belts. The secret to the sensitivity was in the scaling. One Geiger tube would release one tone for every sixty-four particles counted, perfect to detect the relatively low intensity of cosmic rays outside the belts. In addition, a lead shield on this tube would reveal the energy levels of particles—whether they had enough energy to literally pierce through lead.

"A smaller Geiger tube was scaled to release a tone for every 2,048 particles counted, perfect to monitor the massive flux of radiation in the belts. The system offered a much more discriminating count than the 'simple-minded' detectors on *Explorers I* and *III*," Van Allen said.

Anton Electronics in Brooklyn built the miniaturized Geiger tubes at a total cost of $316. In addition, RCA built scintillators with photomultipliers that produced an electronic pulse each time a single electron or proton passed through its crystals, but only high-energy electrons and protons would trigger a pulse. But another scintillator was filtered to measure the total energy input rather than charged particles, a good way to measure changes in energy as a result of the nuclear blast. When two parts of the tiny photomultiplier tubes failed in vibration tests, RCA immediately redesigned them and put them into production for a new standard for satellite and rocket experiments. Iowa commandeered the service with a DX-A2 top-priority contractor status guaranteed to cut through red tape, work shortages, back orders, or delivery delays with any defense contractor. Hughes Aircraft, Raytheon, and Texas Instruments provided parts along with dozens of big-name contractors and family-run Iowa City hardware stores.

McIlwain concentrated on the sensitivity of the instruments and Ludwig took charge of overall assembly. Ludwig strapped whole setups of instruments on the vibration table at the southeast corner of the basement, tests that often required additional hours of work to secure any part capable of springing apart. "Continued hard work in the laboratory on potting," Ludwig wrote. Potting referred to the palm-sized discs of electronic components "potted" in pink plastic foam that hardened to hold them in place. He also checked channel wiring, power, and weight tabulation. "Got one-day delay in delivery of prototype to June 7," he wrote.

That morning, Ludwig, Van Allen, and McIlwain loaded the completed cylindrical instrument package Number 1 into a shipping case. Before they closed the lid, Van Allen gave the metal cylinder a kiss for good luck. McIlwain and Ludwig escorted it via army aircraft to Huntsville. They returned with it to Iowa to make repairs on June 14, backtracked to Hunstville for new rounds of testing June 19, and returned to Iowa City on June 21. With the prototype now operating to everyone's satisfaction, intense work on the flight units began on June 23.

The Soviet *Luna* launch to the moon failed on June 25, followed by a Vanguard failure on June 26. In addition, contact was lost with *Explorer III*, but there was no time for the Iowa team to give much thought to the mishap. Everyone stayed focused on completion of the instruments for inspection rounds starting early in July. One package of instruments went to NRL for the telemetry testing and three flight units went to Huntsville. Temperature testing, calibration testing, and vibration testing of the units proceeded in unending succession until July 16 when two of the three instruments went to the cape via military aircraft. Carl and Mary McIlwain flew down for the launch. Ludwig drove down with Rosalie in the Mercury. After the strained weeks of unremitting work, they gave way to almost giddy levity during the brief retreat of the ride. "We sang variations on the Purple People Eater song en route," Ludwig recalled.

McIlwain and Ludwig completed their final inspections of the instruments on July 18. By July 20, flight units one and two stood ready to launch, ready until an NRL engineer showed a tendency of flight unit two to "squiddle," his term for instability in maintaining radio frequency for data transmittal. The potentially critical problem required a complete set of new tests to pinpoint the problem at channel 4, the transmission channel from the filtered scintillator designed to measure total energy flux. With the problem corrected, flight plans proceeded on schedule.

Ludwig and McIlwain spent the final days before the launch in assessing ground-station readiness and testing the flight unit spare still on hand. "Once, curious about a Redstone rocket on a neighboring launch pad, we climbed the gantry and found a dummy test capsule for manned flight. There, high above the ground inside the capsule, we tried to imagine what it would be like to have the rocket beneath us ignite and carry us into space," McIlwain recalled.

Van Allen traveled to IGY headquarters in Washington for the *Explorer IV* launch while McIlwain and Ludwig remained at the cape. They worked in tandem with JPL and army crews who had upgraded the Jupiter C to allow for the wider orbital latitudes that promised Van Allen a cinematic "view" across the radiation belts.

Explorer IV launched successfully into orbit on July 26. Shortly after, McIlwain and Ludwig went to a press conference at the cape with von Braun. Von Braun captured most of the media attention, McIlwain recalled, even though he deferred questions regarding the radiation belts to his two young colleagues. Over lunch in a café, von Braun made light of his star treatment. "You are the important ones," he said. "I'm just the trucker."

A downpour of *Explorer IV* tapes quickly began to arrive in Iowa adding to the load of reducing the *Explorer III* data. With the Pentagon pushing for top-priority analysis of the *Explorer IV* data, a major part of the *Explorer III* tapes simply gathered dust and no one mourned the fact that the satellite had stopped sending information about the shadowy region in space where the Geiger tubes always stalled. Now *Explorer IV* gave the shadow substance. Lots of guesswork had been removed from the system, particularly the guesswork about altitude. In a matter of six months, NRL had turned orbit determination into a precise science and closed many of the gaps in real-time data acquisition that backed up the recorder.

The expanded ground-tracking system promised to deluge the physics building basement with tapes but Van Allen found the perfect person to manage the load. "Mrs. Annabelle Hudmon started work yesterday in charge of data reduction," Van Allen wrote in his journal on July 31. The twenty-two-year-old Annabelle Welsh Hudmon, a native of Shenandoah, Iowa, had proven analytical skills and even a security clearance from her recent job at the Naval Ordnance Laboratory in Corona, California. "I was working on a computer for the navy pilots to tell them when to fire the Sidewinder missiles," she said. She knew about Van Allen from all the *Explorer I* publicity and, when her husband, Stanton, took his medical residency at the University of Iowa, the enterprising Annabelle decided to join the space program. Ordnance laboratory director F. S. Atchison, himself a physicist with an Iowa PhD, recommended her to Van Allen for any job in computational analysis. He noted that she planned to be in Iowa for four years while her husband pursued medical studies.

Van Allen interviewed her, hired her on the spot, and told her to pull together the staff she needed for a satellite data reduction center. Then he promptly left for an out-of-town meeting. Hudmon stretched tables across the lobby of the already cramped basement, strung lights above the area, and gave scores of undergraduates and graduate students an impromptu math quiz to prove their skills for her team. McIlwain mentored her efforts and helped with the assembly line she put into place. She arrived with the janitors every morning at seven and student helpers trickled in and out for the rest of the day. "We'd start out playing these tapes and there were all these graph plotters running to record what the tapes said on graph paper. I had someone actually putting the reel-to-reel tapes that came in from say, Australia, in the [plotter] and then it would graph," she recalled. Hudmon and the students measured the graphs and put their measurements in standard MIT lab notebooks, writing in pencil. "That [data] would go in an envelope with the date and time and a description of what the instrument was counting."

"When I paid another visit, the rolls of tape recorded from the satellites by the far-flung network of stations filled shelves that reached to the tall ceiling of their archive," wrote Sullivan, who described the data plotted nonstop on nine consoles. "Nine needles—quivering, pulsing or wandering—inscribed nine lines on a rapidly moving roll of paper [tape]. The pulsing needles marked the passing seconds."

The data drifted higher and higher, the deadlines for other missions pressed down on the lab, and reporters continually invaded the physics building for the latest word from space. Yet every morning he was in town, Van Allen calmly came down to the basement smoking his pipe and asked Hudmon to bring in the latest data. "He loved that pipe," she said. "So even though they were in the midst of all these different projects and all the hiatus with the radiation belts and the explorer missions, this was still maintained as a very calm intellectual environment." She brought in the latest envelopes of data. Van Allen, Ludwig, and McIlwain huddled over them, piecing together the picture of the natural versus the artificial radiation belts from Hudmon's reduction of 200,000 data points from *Explorer IV* and 500,000 points from the 1959 mission *Explorer 7*, according to McIlwain. He summarized the daunting task in a letter of recommendation he wrote when Hudmon left in 1961. "There can be no doubt that she will prove to be a valuable asset to any organization she chooses to join," he wrote. Van Allen estimated that it took 12,000 hours to read and tabulate one year of data from the *Explorer IV* tapes at an average cost $1.33 per hour for wages.

The first readings from *Explorer IV* clearly identified the outer radiation belt, the slot between the inner and outer belts, and then observed the effects of two 10-megaton nuclear missile tests prior to the Project Argus blasts. The missiles were fired at about 48 miles and 27 miles above uninhabited Johnston Island in the Central Pacific. The first blast, code-named Teak, unleashed an apocalyptic fireball that covered nearly 20 miles of the sky within a fraction of a second on August 1. The plan to create artificial radiation belts with the upcoming Argus bursts remained secret. But the Teak blast, clearly photographed from more than 700 miles northeast in Hawaii, knocked out radio communication from Sydney, Australia, to Vancouver, British Columbia, in an electromagnetic tidal wave. It couldn't remain a secret for long. The Associated Press sent out the first dispatch about the blast in space. The second blast, code-named Orange, seemed to blot out the sky over the Central Pacific on August 12, as it exploded some 20 miles lower than Teak.

The *Explorer IV* readings from the Teak blast confirmed instrument reliability to observe the impact of the three Argus nuclear blasts. Van Allen left for the drive to Long Island with his family on August 4. He was still on vacation when *Explorer 5* launched on August 24 as a backup to *Explorer IV*. He monitored the launch from the Mackey Radio Station on Long Island and promptly heard the bad news that the final stage of the Jupiter C rocket failed to ignite. A whole suite of instruments, working to perfection in early transmissions, burned up in the earth's atmosphere as the satellite plummeted back to Earth. The entire Argus mission now depended on *Explorer IV*. The air force launched Argus 1, 2 and 3—each about a 1.5 kiloton bomb—between August 27 and September 6. The Defense Department decided on the shipboard launch from the USS *Norton Sound* for reasons of secrecy and safety. Van Allen specified the launching locations in the South Atlantic to maximize the potential formation of artificial belts in the slot between the natural belts.

Despite their yield and the havoc they produced, Teak and Orange created marginal belts that lasted only a few days, due to the relatively low altitudes of the blasts, Van Allen noted. The Argus blasts detonated approximately 300 miles above the Earth, sparking a fireworks of auroral lights at the North and South poles. And, as Christofilos had predicted, electrons from the blast quickly radiated around the globe and created three thin, artificial radiation belts above the inner natural belt. The *Explorer IV* detectors indicated that the earth's magnetic shield forged only about 3 percent of electrons from the Argus blasts into these belts while most of the rest decayed into a shower of particles as they hit the atmosphere. Christofilos's bold conception for an experiment and Van Allen's stunning instrumentation documenting the impact of all five blasts helped remove any doubts about the absolute need to ban above ground nuclear testing and proved that breaches to a ban could be documented.

In the end, the artificial belts lasted only about a month and the Argus test proved that the earth's magnetic field wasn't powerful enough to hold in place a shield of a strength that would be necessary to damage an incoming ICBM. But Argus was an audacious experiment by almost any other measure—it enveloped almost the entire planet, was carried out successfully after only four months of preparations, and made use of satellite techniques that, less than a year before, had been beyond human experience.

The Argus program reflected several aspects of Van Allen's character and background. As he participated in the Argus project, he fought on the side of those who pushed for a civilian space agency, the fledgling NASA formed in 1958. He was also a navy veteran who maintained navy contacts and he recognized the benefits as well as the conflicts of military and civilian space agendas in promoting space exploration. He could and did work both sides.

While Argus epitomized the Cold War, Van Allen, like many scientists, still looked to space as an arena for international cooperation. The International Union of Pure and Applied Physics made one of the first overtures to continue the international dialog on space after the IGY disbanded on December 31,

1958. It was the union that invited Van Allen to give a paper on the radiation belts at the Conference of the Cosmic Ray Committee in Moscow in July 1959. Van Allen was able to openly discuss the scientific hot potato of the Argus findings at the conference because the results were quickly declassified. American officials understood that ground stations in Russia would register the blasts just as stations in the United States detected Soviet blasts, Van Allen said. The National Science Foundation gave Van Allen a $1,200 grant to participate. A federal official, whom Van Allen believed to be from the CIA, paid a visit to enlist him for a special assignment. The federal agent asked him for a "trip report" covering questions in eleven areas of interest, including a full description of recent cosmic ray developments, names of institutions and leading scientists, leads on anyone who was secretive or evasive, and all the materials from the conference.

"On the basis of formal reports presented at this conference and on informal conversations with the delegates, evaluate the status of cosmic ray research in the Sino-Soviet bloc," the questionnaire stated. Special interest was expressed in Soviet estimates of the radiation risks in the "Van Allen bands." The agent also suggested Van Allen visit ten locations including the Institute of Physics of the Atmosphere in Moscow, the Air Force Engineering Academy in Moscow, Moscow State University, the Moscow Higher Technical School, and the observatory in Dalgoprudny.

Van Allen assumed that his Soviet colleagues were sizing him up based on similar requests from the KGB. But that didn't dim his fascination as he toured Soviet laboratories and met his counterpart Vernov, who pulled out the prototypes for *Sputniks* 2 and 3.

Sedov (the man who suggested that Americans were too busy worrying about cars and refrigerators to launch the first satellite) asked Van Allen to follow up on the cosmic ray conference presentation that brought him to Moscow. Sedov extended an impromptu invitation for a more detailed technical seminar on the radiation belts and Argus at the USSR Academy of Sciences. Still, the unscheduled stop for the session at a remote location of Moscow left Van Allen somewhat apprehensive for the first time on his trip. He invited conference delegates John A. Simpson of the University of Chicago and George W. Clark of MIT to accompany him.

"I figured that if all three of us disappeared, someone would certainly investigate," Van Allen said. At the academy, he spoke and showed slides to an expert audience. Y. I. Galperin, a physicist who became a close friend of McIlwain's, translated. The scientists asked pointed and searching questions about the natural origin of the pre-Argus radiation. "The Russians were just as suspicious as we were that the radiation belts had been injected into the geomagnetic field by clandestine bomb testing," Van Allen said. "But they could see for themselves from the slides how different the Argus belts were from the natural belts. That pretty much laid the question to rest."

The Russians returned the visit, traveling to the United States for an American Rocket Society meeting that November, and Van Allen asked to host them at the University of Iowa on November 23 and 24. "Suggest visit to our laboratories on Monday and Tuesday. Would be very grateful if Professor Sedov would give a general university lecture on Monday evening on space research in the Soviet Union," Van Allen wrote the Soviet embassy on November 18. The embassy okayed the visit and Van Allen hastily prepared for the arrival of Sedov, Anatoli A. Blagonravov (former lieutenant general of the Soviet Army), physicist Valerian I. Krasovskii and the interpreter V. G. Kostomarov.

Sedov had become the unofficial ambassador of the Soviet space program, widely respected as an astrophysicist and carrying the ambitious title of Soviet Chairman of the Interplanetary Communications Commission of the USSR Academy of Sciences Astronomical Council. The International Astronautical Federation had elected Sedov its president earlier in 1959. Sedov mastered the mix of humor, science, and Soviet propaganda. When the Russians withdrew a paper from a scientific meeting in London, Sedov told a Russian colleague the paper had errors made obvious by a British report. Pressed for a confirmation by Western reporters, Sedov sidestepped the errors issue and fired off a succession of other explanations. "He chuckled merrily at each new alibi," reported *Time* magazine.

Blagonravov and the other visitors were less well known in the West but, inside Russia, Blagonravov was the relentless space booster who had pushed for both missile development and space satellites from the start. Absent and nearly unknown in the West was Sergei Korolev, the scientist who had masterminded both the Russian ICBM and the *Sputnik* satellites. The Russians allowed Korolev a rare public appearance and the honor of announcing the imminent launching of the first *Sputnik* satellite. Then Korolev retired once again from public view, both the man and his work treated as state secrets.

But Korolev's work was the vaulting standard by which the Soviets measured what they saw in America. Korolev's *Sputnik 1* weighed in at 184 pounds out of necessity to accommodate large parts and heavy batteries. The miniaturized detectors and palm-sized discs of electronics "potted" in pink foam, all configured in the modest quarters of the physics building basement, impressed the Russians immeasurably. Van Allen displayed the streams of paper tapes rolling off the Iowa data analysis assembly line, streams of data not always available to the Russians because foreign ground stations didn't have access to the data transmission codes.

Sedov gave a general university lecture to a packed audience on Tuesday, November 24. His subjects of lunar flights and the sheer novelty of having a Soviet scientist on campus drew a standing-room-only crowd to the lecture. Another of the Russian scientists gave a physics department colloquium on the findings of *Sputnik 3*.

Abbie Van Allen invited the Russians to an American family dinner complete with four children—nearly five, since the birth of the Van Allen's youngest child was imminent. "Blagonravov entertained our young children by letting them listen to the ticking of his big pocket watch, a genial grandfatherly activity that I found to be an interesting footnote to his reputation as a tough-minded lieutenant general of the Soviet Army," James Van Allen recalled.

As he arrived the next morning for breakfast with the Soviets at a local restaurant, Blagonravov congenially translated the story about the Russian visit reported in the *Daily Iowan* student newspaper. It was a reminder that the space race, even amid the rivalries of the Cold War, offered a relatively safe haven where scientists reached toward the boundaries of the universe rather than toward the chasm of extinction.

14 Space as a Cottage Industry

Fairfax, Iowa—April 1962. Only the crickets broke the stillness of the night as University of Iowa electrical engineering senior Don Gurnett headed across his father's farm in Fairfax. He carried a handmade radio receiver with a loop antenna constructed of fifty turns of wire. Gurnett came to the fields to try to detect whistlers, the sounds of natural, very low frequency (VLF) radio waves produced by bursts of lightning.

Safely beyond the electrical power line interference of Iowa City, he sat in the velvet darkness pierced only by the stars. But Gurnett didn't need a local lightning storm to catch some whistlers. Whistlers from distant storms—guided by the earth's magnetic field—dart back and forth between the hemispheres and then funnel back to earth producing a trail of whistling notes picked by VLF radio receivers. Radio pioneers had discovered whistlers in 1918 and a near mystical following of radio amateurs eventually headed to the open countryside to listen for them and catalogue other musical tones, static hisses, and sounds originating from space. Gurnett learned about whistlers from Roger Gallet, a scientist with the National Bureau of Standards who gave a talk at the University of Iowa and showed a basic design for a VLF receiver. Gurnett perfected his own design but carried it dejectedly back to Iowa City after the first night of his vigil in the fields. The receiver hadn't picked up a peep. He returned to the farm again the next night, listening hour after hour. Then, suddenly, the receiver delivered a series of faint, whistling tones, enough to convince him the gadget was working.

• • •

Gurnett had big plans for the receiver after that. He wanted to send it into space to study the cosmic events that shaped the natural radio sounds and he had just the spacecraft in mind. While studying engineering as a freshman in 1958, he approached Van Allen for a job in the physics department. Van Allen always had a spot for an enterprising engineering student, especially now when he had initiated a cottage industry of satellite construction. Gurnett got the job and eventually started working on Iowa's *Injun 1* satellite, the very first university-built satellite to go into orbit. He advanced to the position of project engineer for *Injuns* 2 and 3. *Injun 3* resembled two mirror-finish domes joined at the middle, with solar cells lining part of the lightweight magnesium skin. Gurnett knew the design inside and out. He knew just where he could mount the VLF radio receiver he wanted to fly on the spacecraft. He asked Van Allen for permission to add his gadget to the mission and got it as Van Allen recognized a whole new realm of potential research that nobody had touched.

But there was another potential disaster to consider—the possibility that the VLF receiver would blare with feedback from the satellite data transmitter just like an amplifier in a loudspeaker system sometimes breaks into a squeal. The backdrop of low frequency interference from electrical power lines in Iowa City made it impossible to test for the problem there. So Gurnett and the *Injun* team loaded up the satellite and made the test behind the barn on the farm in Fairfax. The test went well in this unlikely location. Electricity came late to Fairfax, and Gurnett, now troubleshooting a satellite, had completed homework assignments by the light of oil lamps a little more than a decade before. *Injun 3* studied solar radiation and the aurora with sixteen Iowa instruments that included twelve tried and true charged particle detectors, three photometers, and the VLF receiver.

One thing was certain as soon as the spacecraft went into orbit. Whistlers trapped in the earth's magnetic field sang loud and clear without any of the long periods of silence that had frustrated Gurnett's tests with the receiver on the farm. "Instead, we had this tremendous chorus of strange signals and all sorts of radio phenomena that frankly had never been heard before," Gurnett said. British researchers coined the term *dawn chorus* to describe a steady stream of natural radio tones from space because the concert of sounds reminded them of the chorus of chirps sung by birds when they awake at dawn. But no one could account for all of them. Now Gurnett, Van Allen, and the Iowa team quickly realized that the "chorus" of VLF radio sounds they were hearing from space could be traced to the radio emissions from electrons in the Van Allen radiation belts. The sounds dramatically changed in pitch and frequency with even mild solar storms, recording how the intensity of radiation in the belts changed in response to events on the sun. The radiation belts literally sang a song composed in outer space.

The earliest transmissions from *Injun 3* couldn't be heard because the satellite itself got lost in space after being launched from Vandenberg Air Force Base in California on December 12, 1962. "It went into orbit but not into the orbit we expected. We didn't know where it was. I thought it was a good test of the North American Air Defense Command (NORAD) to find it. They were supposed to be looking for Soviet missiles coming in over Canada and yet they didn't know where our spacecraft was. It was about three or four days before they found it," Van Allen said.

"The satellite was up there waiting for a command from the ground transmitter" before it would send data, Gurnett said. "We didn't know where to point the ground transmitter antenna" without a fix on the position of the satellite.

When the satellite finally was located, the sounds of space streamed in from Gurnett's VLF radio receiver. Then Van Allen turned skeptical. "Could the spacecraft be producing any of these noises?" he asked Gurnett, a graduate student in physics by now and already a veteran *Injun* engineer.

"I don't think so—that spacecraft is so little, so small compared to the magnetosphere out there," Gurnett replied.

"I don't think that's a good argument," Van Allen countered. "You know, a little microbe can make an elephant sick and die."

"It was a darn good question. One that I hadn't thought of, you see," Gurnett said, reflecting back forty years later on the lessons learned from Van Allen. It would be years and many missions hence before Gurnett could confirm with reasonable confidence that the spacecraft wasn't generating some of the sounds in its high-speed orbit.

Gurnett breathed a sigh of relief that the VLF data transmissions came in loud and clear. It had been a difficult problem to design a satellite data transmission system that combined both the audio and digital information. "There was a loss of signal strength—a penalty for combining the two signals," Gurnett said. "I was really worried about telling Professor Van Allen how, in doing this, I had thrown away half our power." To correct for the problem, he bought a one-watt transmitter, the highest-powered transmitter that had so far been used on a satellite. The gamble worked. Gurnett's radio receiver essentially provided headphones to listen to the sounds of space. It delivered a new tool to study space and plasma, the vaporized, subatomic soup of protons and electrons that radiate from the sun and the stars. Gurnett's findings helped pioneer the field of plasma physics. He continues this research as one of the foremost experts in the field, still teaches the subject at the University of Iowa and recently published *Introduction to Plasma Physics,* the latest textbook on the topic. The concerts of natural radio transmissions from across the solar system fill the collection of audiotapes in his office.

Plasma physics emerged after World War II as scientists, including Van Allen, worked to develop fusion reactors that would mimic the workings of the sun and generate virtually unlimited power in thermonuclear reactors. The VLF receivers took the study of plasma into space and gave space scientists an audio signature that correlated to other measures of plasma flow, solar storms, radiation belts, and the boundaries of magnetic fields. Gurnett's first spaceborne measurements of the dawn chorus and auroral radio sounds added to the huge coup of Iowa's already successful *Injun* satellite program.

• • •

With the campus's first computer system for data reduction installed with the help of the physics department and a university telemetry observatory in place in nearby Hills, Iowa, Van Allen started making satellites. The pastoral self-reliance idealized by Van Allen's generation, prized by his parents, promoted by his teachers, and reinforced by the war years made the idea of constructing a satellite at Iowa a natural, something in the blood. The halting, contradictory, and ultimately self-defeating policies that delayed *Explorer I* and Van Allen's own persistence made homegrown satellites an obvious progression for his mastery of techniques to design experiments. "It meant we could build satellites to fit the experiments rather than building experiments to fit the satellites," he said.

Van Allen began planning a homegrown satellite in 1958 and hoped to fly it directly through the auroral zones to extend the research done with rockoons and the *Explorers*. Brian O'Brien, a twenty-five-year-old Australian cosmic ray physicist with a PhD, had a similar idea. "I decided early in 1959 that I wanted to put a satellite above [each aurora] circa 1,000 km [about 600 miles] to look down at the light and up at the electrons and protons probably causing them," he said. Van Allen hired him as an assistant professor.

The turning point came because O'Brien fell in love twice in quick succession. He fell in love the first time with his wife Avril. The two amateur spelunkers met in a cave outside the University of Sydney where he earned his PhD at the age of twenty-three. Avril's family forbid the young couple to marry until she turned twenty-one so O'Brien went off to explore the Antarctic and fell in love again—with the aurora. Van Allen's work involving satellites, the aurora, and the radiation belts convinced O'Brien to join the faculty at Iowa. "Van sealed the best offer by giving us $1,000 for travel—we were very poor indeed." O'Brien married Avril three days after her twenty-first birthday, set off on a six-week voyage via London to Iowa, and felt right at home.

"We had a lovely, friendly reception in Iowa. Iowans were almost like Australians, except with funny accents, of course," O'Brien said.

As the satellite program got underway in 1959, the Iowa team was now focused on results from *Explorer 7*. Its data—from far more sensitive detectors than those previously used—showed that original estimates of electrons in the outer radiation zone that arced near the poles were too high. With the new measurements, Van Allen modified his "leaky bucket" model from 1958 that suggested the aurora at both poles might be caused by spillover or "leaks" of particles from the radiation belts. Now it appeared the belts had far tighter plumbing. "The precipitation of particles from the radiation belts was just inadequate to sustain the aurora—far too feeble—we had been off by a factor of a thousand" in terms of the leaks, Van Allen said. The new findings led to his "splash catcher" model. He concluded that "splashes" of energy from solar storms powered the fluctuating auroral lights and intensities in the radiation belts.

With numerous other missions in the works, the Iowa satellite program developed on the sidelines until it got a sudden boost from the navy. Van Allen's heartfelt support for a civilian space agency didn't stop him from networking with military contacts for grants. And the formation of NASA didn't stop the military from continuing space exploration on its own. An informal discussion with the Bureau of Naval Weapons (BuWeps) and APL offered him the chance to "hitchhike" on a navy satellite.

The navy required that the satellite fill thirteen inches and forty pounds of spare space in a nose cone of a rocket meant to launch two secret military satellites. The Office of Naval Research (ONR) offered in January 1961, to fund Van Allen's first satellite and return him to the aurora to resolve the source of the "soft radiation" it discovered with the rockoons ten years earlier. There was a catch, however. Van Allen had just two months to build the satellite, a backup, and all its instrumentation on a budget of approximately $100,000. He planned to name the spacecraft, *Hawkeye,* for Iowa's mascot and football team. But an ONR admiral called and asked Van Allen to change the name so that the satellite wouldn't be confused with the navy's new Hawk missile. "Let's call it *Injun,*" Van Allen suggested, drawing inspiration from the Cajun-sounding rocket and Mark Twain's Injun Joe.

O'Brien eagerly offered to help with the satellite and became Van Allen's assistant project scientist. The tall, lanky young physicist quickly enlisted a new group of students into the crash *Injun* construction program. The team included Gurnett, Curt Laughlin, Ray Trachta, Kent Hills, Lou Frank, and Bill

Whelpley—a group fondly called the "Injuneers" around the physics building. Trachta and Whelpley developed satellite plans that used lighter-weight magnesium instead of aluminum in the body. Frank and Laughlin built new twin-channel detectors to simultaneously look up at particles coming in toward the aurora and down at the aurora itself.

Gurnett developed the first digital data system used on a satellite. "*Injun 1* was the first spacecraft to use digital data transmission and the first to talk from the launch pad [via] computer to me in Iowa," O'Brien said. O'Brien worked on imaging the auroral effects and on temperature control to keep the "bird" cool.

O'Brien and the new team of whiz kids summed up the "recipe" for a state-of-the-art satellite, experiments, and detectors in their "Injun Cookbook." *Injun* also consolidated dramatic advances in detectors. The Geiger tubes did a good job of counting particles and mapping their direction. But it's difficult to tell the energy of a particle or even whether it's an electron or a proton with a Geiger tube, though filters can be used to screen for specified energy levels, the strategy used for the *Explorer IV* detectors. But solid state detectors offer an alternative, using semiconductors to generate an electric pulse from the charged particles that hit them. The pulse varies based on the charge and energy of the particle. *Injun* carried a full array of detectors and the tight deadline for crafting everything kept everyone working double shifts. Gurnett, like the rest, was punching eighty hours a week on his time cards.

The 40-pound *Injun 1* launched from the cape on June 29, 1961, the first university-built satellite to go into orbit. Sixteen vertical plates formed the sides of the cylindrical satellite that fit into a "cup" at the top of the navy's Transit 4A navigational satellite, developed to update the capabilities of the Polaris nuclear submarines. The *Injun,* in turn, supported the navy's *Solrad* satellite to monitor solar X-ray radiation. The *Injun* and the *Solrad* didn't separate in orbit, as planned, but both functioned with some reduced capability. With orbit achieved, the university switched on the transmitter with a radio command from Iowa City and the data started flowing. Jubilant with their success, the Iowa team was summarily confronted by North American Air Defense Command officers asking why radar was tracking two hundred orbiting objects instead of *Injun* and the final stage of the rocket.

"Here's what they discovered happened," Gurnett said. "We could have launched with the satellite transmitter on, which is what some people wanted to do so they could track the rocket during the launch. But we decided to launch with it off because we were afraid of a possible electrical discharge in the transmitter as the rocket went up through the atmosphere." The discharge could have permanently damaged the transmitter. The decision was a stroke of good luck because switching the transmitter on just before launch would have triggered the command-destruct system on the final stage of the rocket. The whole mission would have blown up on the launch pad during the countdown. But the spacecraft was safe once it separated from the final stage of the rocket, which did explode when the radio command activated the transmitter. That accounted for all the pieces NORAD was tracking. No one had tested for that sort of mishap.

And no one told Gurnett and fellow physics student Don Stilwell that they needed a government bill of lading to drive home a truck filled with Iowa's equipment. "We left the cape and just waved to the military police guards as we drove out of the gate," Gurnett said. "FBI agents were searching for us all the way back to Iowa City. They finally called Professor Van Allen and asked what was in the truck. They thought we had stolen some government equipment." Luckily, the local police who arrested them for speeding as they drove through a speed trap in Mississippi didn't know the pair was "wanted" by the FBI.

O'Brien and the student team quickly assembled *Injun 2* and the detectors for NASA's *Explorer 12*, a satellite powered by solar cells lining huge paddle-wheel arms. Gurnett went to the cape again for the 4:30 A.M. (EST) launch of the latest *Injun* on January 24, 1962. He thrilled as the launch of the Thor-Able Star rocket lit up the night sky but, a minute after launch, the rocket malfunctioned. "I was project engineer on *Injun 2*. That was my first role as project engineer and I watched that one go into the ocean," he said.

Due to the loss, *Injun 1* continued on alone to complete the most comprehensive survey ever made of the aurora pulsing and changing contours as solar storms flooded the area with soft radiation. But the *Injun* satellite had company while documenting the aftermath of the 1.4 megaton Starfish nuclear missile blast ignited in space about 260 miles above the earth on July 9, 1962. *Injun 1* tracked the formation of an artificial radiation belt resulting from Starfish that was much larger than the one that resulted from Argus and intense enough to markedly alter the inner radiation belt. AT&T's *Telstar* satellite monitored the blast with a detector on board as well. A combination of the communication system and Cold War politics soon locked Van Allen into a heated debate over the credibility of his *Injun 1* data and the dangers of the artificial belt from the Starfish blast.

Telstar, representing the first efforts at the commercialization of space, launched on July 10, 1962, from Cape Canaveral with the promise of relaying phone calls, news photographs, and television programs at a rate equivalent to one thousand words per second. The 34-inch tall satellite—with an anticipated life span of two hundred years—was damaged within hours by the radiation resulting from the Starfish blast. AT&T and Bell Telephone Laboratories, makers of the satellite, correctly blamed Starfish for the failure based on readings from a detector on board the satellite.

Telstar's detector, operating independently from the crippled telecommunications system, also became the basis for predictions that the artificial belt generated by Starfish would last longer and pose more of a risk than Van Allen had reported to the President's Science Advisory Council (PSAC) based on the *Injun* data. The PSAC now hastened to issue a warning for astronauts based on the *Telstar* findings. Van Allen swiftly challenged the *Telstar* claims, backed the *Injun 1* findings and questioned the timing of the PSAC's hasty scare. Van Allen pointed out that the PSAC's position ignored the basic facts that *Injun 1* and *Explorer 12* had established the pre-Starfish properties of the natural radiation belts. *Telstar* was launched after Starfish and had no data to use as a basis for comparing effects found after the blast. He estimated that most of the durable radiation would probably be undetectable by the summer of 1963, though vestiges of it were still detectable six years later. He emphasized that the peril to astronauts from the artificial belt was short-term.

The administration had indeed timed the PSAC scare to discourage the Russians from pursuing their own nuclear tests that fall. They paid no attention and exploded two bombs on schedule that fall, but the public furor over the perils of such testing on both sides led to the 1963 Treaty Banning Nuclear Weapon Tests in the atmosphere, outer space, and under water.

"The telling points up two things about Van Allen," noted the *Des Moines Sunday Register*. "He had no fear of taking on the government when he believed that the quick evaluation of the bomb's effects was for political purposes and he had no fear of being wrong." In Van Allen's case, no fear of being wrong, meant he had observations to lay on the table for others to scrutinize. Van Allen carried that philosophy, at the heart of scientific research, into simple home repairs, and into the national arena where he soon became one of the most vocal critics of human space missions.

• • •

McIlwain and Ludwig stayed out of the picture as new projects such as the *Injun* series ratcheted up. They had to get back to course work, complete their master's degrees, write their theses, and prepare for comps to earn their PhDs. Master's degrees, a consolation prize in many graduate programs, were the boot camp of Van Allen's program, training students to take on the independent responsibilities for research in a highly competitive field. Van Allen promised himself he wouldn't interrupt them with any new commitments and resolutely kept the promise, placing another momentous plan on hold. "I will not undertake pushing for the establishment of an Institute of Space Science," he wrote in his journal on August 31, 1958. "For the near future, McIlwain and Ludwig would be well nigh essential building blocks." He decided to wait a year or two before returning to the idea. But it was a turning point lost—he never revisited the issue as the department underwent a changing of the guard.

Van Allen hoped to keep both Ludwig and McIlwain on, grooming them for faculty. He offered each of them $10,000 for eleven-month contracts as post docs. "I well realize that these offers are considerably above the usual rate for fresh PhDs. But I consider both Ludwig and McIlwain to have extraordinary capabilities," Van Allen wrote in his journal. He feared that the university's program would slow down without them. McIlwain officially received his PhD in May, 1960, and Ludwig received his in July in electrical engineering rather than physics with a thesis based on the development of an advanced particle detector system housed in a complete satellite payload. Unfortunately, the launch vehicle that was to take the system into orbit failed on March 23, 1960. But the work laid the basis for the payloads for Iowa's own satellite designs.

After completing his doctorate degree, Ludwig informed Van Allen in April that he intended to join Frank McDonald and Meredith at the Goddard Space Flight Center. "Printing service is working on 300 copies of my thesis. They are to be finished August 25. Have been packing books, mostly at the farm, and cleaning up odds and ends," he wrote in his log notebooks on August 16, 1960. "Today I am finishing the packing of my office books and files. This is my last entry at SUI—the end of a wonderful era!!! What lies ahead?" He once again loaded up the Mercury and drove to Silver Spring, Maryland. His wife, Rosalie, arrived with the girls the next day and the family moved into a rental home that McDonald had found for them for $150 per month. "I left Van for the same reasons I ran away to the air force," Ludwig said. "I felt I had to strike out on my own."

The two men stayed close friends and colleagues, working together once again starting in October 1960 as NASA began planning for the *Mariner* missions to Venus and Mars. The impact of a major solar storm that fall only heightened the curiosity of what he might find on journeys to earth's nearest neighbors. "Abbie and I and the children saw an extraordinary aurora or air glow. The northern sky was a fine diffuse pink glow from the horizon—fading progressively in intensity" to the south, Van Allen noted in his journal. McIlwain agreed to stay for another year but left in 1961 to organize a space science program at the University of California at San Diego. In addition, Van Allen's old mentor Tyndall retired at the end of the 1960 term, the last of the physicists from Van Allen's own student days at the university. Sentinella retired in 1962 at the close of the school year. "A magnificent human being and splendid craftsman," Van Allen wrote of Iowa's master instrument maker.

Despite the impending losses of old friends from the scene, other longtime mentors renewed relationships. Tom Poulter, now directing the Stanford Research Institute in California, asked Van Allen to serve on the institute's advisory panel. Van Allen's old mentor Scott Forbush came to Iowa as a visiting professor for the 1960–1961 school year. Once shunted to the sidelines at DTM, Forbush now stood with Chapman, Compton, and Millikan among the pioneers who had helped lay the groundwork for observations of cosmic rays and the solar wind. From his start in observing cosmic rays in 1926 at DTM's observatory near the equator in Hyancayo, Peru, Forbush now happily helped interpret the findings of *Explorer 7*, nicknamed the "Heavy Explorer" with a payload of seventy pounds of experiments. The results correlated even mild solar storms with huge variations in the intensity of particles in the lower regions of the radiation belts. *Explorer 7* showed how common such events were by measuring particles over the polar caps.

Van Allen also welcomed John Rogers, who moved to Iowa from Rhodesia to help with telemetry and plans for an off-site observatory. To support the space research, Van Allen and McIlwain pushed for one of the first IBM computer systems at Iowa, one shared with Everet F. Lindquist, a leader in education research whose "brain derby" high school achievement competitions in Iowa evolved into standardized college entrance exams with the ACT test. "During the war, Lindquist had room after room in Seashore Hall of women grading these tests for the military. And he said to himself, testing will never survive if it's done this way," said former University of Iowa President Willard (Sandy) Boyd, who was provost at the time.

Lindquist figured out an optical scanning system to score tests, and he and Van Allen pooled grant money in 1961 to rent one of the era's high-powered computers, the IBM-7070. As they developed a permanent computer center with their grant money, they shared it with the whole campus, Boyd said. "Basically they both would run their computer programming at night so we would have it in the daytime for students and other faculty, so they were sort of the core of this. They were very loyal to the university. They made it possible for us to grow into the computer era" as the space program at Iowa expanded with Gurnett and others in a new generation of Iowa PhDs joining the faculty and research staff within the next few years.

Van Allen's family had grown as well. Peter Cornelius Van Allen, Abbie and Jim's youngest child, arrived December 12, 1959. Their nephew Jonathan was born to Maurice and Janet in Iowa City a few months earlier. Maurice, also called Van, headed neurology at the University of Iowa Hospital and taught in the medical school.

With five children ranging in age from newborn Peter to twelve-year-old Cynthia, Abbie turned her attention to building a dream home. Just after they celebrated their fifteenth anniversary on October 13, 1960, the Van Allens bought a large wooded lot on the hilltop ridge overlooking the Iowa River. Abbie planned a contemporary 3,500-square-foot home with a deck and floor-to-ceiling windows dramatically encasing the two sides of the home's centerpiece—a sprawling living room that overlooked the bluff. The panorama of treetops gave the room the romantic appeal of living in a tree house. The home cost $44,000 to build and rested exactly 1.9 miles from the physics building, Van Allen reported in his journal. Abbie named the street that circled up the hillside Woodland Mounds Road.

As many times as the Van Allens had moved, their family homesteads remained anchors in their lives, with the trips to Southampton every summer and drives to Mount Pleasant to see Alma once a month. "Oh yes, that was always great fun and my Uncle George and his family were always there. We had fun playing with the cousins and I have just vivid, vivid memories of Grandma waving good-bye from her back porch" as her family arrived, recalled Sarah Van Allen Cairns. "She cooked these unbelievable Midwestern dinners, you know, this unbelievable fried chicken and mashed potatoes. And her best thing—and I'm praying Margot still has this recipe—was a chocolate sour cream cake. I just loved being in her house."

Van Allen carried a full academic load in addition to research, administrative, and family responsibilities. His advanced classes included "Electricity and Magnetism" and "Solar and Terrestrial Physics." But in the early 1960s he began to teach the undergraduate "Introduction to Astronomy," his favorite class and a favorite among history, art, and economics students eager to fulfill their science requirement by studying with a celebrity scientist. He walked into class with lecture notes jotted on yellow legal paper and with a plastic inflatable globe. He used the globe to demonstrate the 23.5-degree tilt of the earth and show how the seasons occurred. He carried the same globe to talk about rockoons and the earth's auroral zones for his presentation when he received a lifetime achievement award from the American Polar Society in 2004, an award earned years earlier by Admiral Richard Byrd.

"He wanted to teach that course in an era when outstanding scholars did not teach intro classes," said Boyd, who remains active in the law school and not-for-profit programs. "He was sending a signal that every student should be entitled to the best there is at a university."

As with all classes he taught, Van Allen took a hands-on, experimental approach. Students recorded observations using department telescopes and recorded the phases of the moon. *New York Times* science reporter Jim Glanz recalled waking at 4 A.M. and climbing to the roof of his dorm to track the waning crescent moon. He described the experience while on campus to be honored as one of four University of Iowa Alumni Fellows of the College of Liberal Arts and Sciences for 2005.

Van Allen was up too, former students noted. No one could ever afford to claim they had seen the moon on a cloudy night, as the moon exercise became Iowa legend.

"The most famous part of this class was the moon plot exercise, which required students to go out every night for a month or two and plot the path of the moon across the sky. This led to many stories, such as reports of the police being called to investigate students who were standing in some dark parking lot holding their arms up toward the moon, as if carrying out some moon-worship ritual," said Gurnett. "They were actually holding a calibrated string to measure the angle between the moon and a nearby star. It soon got to be common knowledge around campus that anyone seen staring up at the moon in the middle of the night was just doing Van Allen's moon plot exercise."

In addition, Van Allen piled on the basic physics and equations involving the gravity and planetary motion. But he also gave students a chance to interpret the class in terms of their own intellectual passions. "You could write a short story or a poem for the class for extra credit. Van Allen wanted to make sure we thought about science in a broader way—that science can be both art and science," said Tom Boyd, son of Sandy Boyd, who took the class. "I got a C even though I was the son of the university's president. I was the one who proved that grading at the University of Iowa is truly anonymous."

Van Allen also grounded students in the discoveries of Newton, Kepler, and other great scientists of the past, bringing in antiquated instruments to reinforce his points. "You know he loved all those ancient pieces of equipment and also the discovery mode and he really passed on that good feeling about astronomy to the students," said Frank.

Another approach set Van Allen apart from most astronomy professors. He not only showed students the planets with telescopes. He touched them with a series of planetary probes that became the focus of more than forty years of research that included the entire seventeen-year period he taught astronomy until he retired in 1985.

Van Allen continued with the *Injun* missions throughout the 1960s, though the space program changed rapidly around him. "Once when we were out raking leaves, I saw a kid's kite with a spaceman on it fly into the air," Cynthia Van Allen recalled. "I said, 'do you think there will ever really be men in space, Daddy?' And he said of course there would be." Van Allen was one of the few parents who could answer such a question with firsthand knowledge. But President John F. Kennedy made it official in 1961 by setting national policy to send astronauts to the moon and bring them safely home again by the end of the decade.

The space scientists worked as a close fraternal group through the early years of NASA. But *Apollo* escalated the space program and focused on a dramatic national effort. As the field got bigger and the *Apollo* program took it to a grandiose scale, Van Allen fought ever harder for robotic spacecraft with economical instrumentation and a long life span. While NASA focused on the *Apollo* missions to send astronauts to the moon, Van Allen focused on the next logical step for robotic missions—journeys to the other planets. As the *Injun* satellites got underway, he fine-tuned the *Mariner* missions to Venus and Mars. The *Mariner* missions transformed Van Allen from a geophysicist concerned with cosmic rays and the magnetic field at the earth into a planetary scientist and astrophysicist with his sights set across the solar system.

15 The *Mariners*

Venus—December 14, 1962. The tiny *Mariner 2* spacecraft neared the end of a 109-day voyage to the brightest jewel then visible in the dawn sky. Centuries of human beings had greeted Venus as the morning star. Now JPL scientists sent a greeting to *Mariner* with radio signals that switched on the instruments for earth's first close-up exploration of another planet. Against great odds, *Mariner 2* hurled across 180 million miles of space with a small haven full of instruments prepared to unravel the secrets of Earth's cloud-shrouded neighbor. Newspaper articles speculated about finding clues "to the possibility of the existence of life on Venus."

The instruments flashed into action at about 10:55 A.M. Pacific Standard Time. Passing nearly 21,500 miles above the planet, they transmitted observations about the atmosphere and blistering temperatures at Venus. Twenty-second spurts of scientific data alternated with eighteen-second spurts of operating updates from JPL until the craft approached the dark side of the planet and temporarily lost touch with Earth at 11:37 A.M. In those previous, precious forty-two minutes, instruments measured the microwaves and infrared light streaming from the planet to determine the scorching heat at Venus and the composition of the gaseous clouds shrouding it. Van Allen's particle detectors looked for a magnetic field and radiation belts. Other investigators mapped the planet's rotation.

Despite the historic moment and Pickering's personal invitation to come to JPL for the fly-by, Van Allen had stayed home, crossed his fingers, and listened for word of *Injun 3*. The satellite still hadn't shed the final stage of the launch rockets and Van Allen's group agonized over the weak signals intercepted at the Iowa City tracking station. "Fuddled," Van Allen remarked in his journal, describing communications thus far.

To make matters worse, McIlwain called from the cape with disappointing reports about the Relay satellite, launched the night before with Iowa instruments to study the inner radiation belts. Battery problems on the satellite were already jeopardizing the detectors.

Van Allen stuck by the phone. Then Lou Frank stopped by for news of *Mariner 2*, eager for readings from the experiment he had helped build. Van Allen looked up. It was already dark outside, streetlights glistening in the cold, as *Mariner* sped beyond Venus and approached an orbit around the sun. Frank and Van Allen phoned Hugh Anderson at JPL for a rundown on the mission. "The SUI 213 Geiger counter showed—on quick inspection—no increase of counting rate," Van Allen wrote in his journal. There was no increase on JPL's Geiger tube either and no trace of a magnetic field on the University of California's magnetometer. The readings at Venus matched the background veil of outer space and meant that Venus simply didn't have a magnetic field to find. "We sailed by with our detectors as though the planet wasn't there." Van Allen said.

• • •

Van Allen began to speculate on the possibility of looking for radiation belts around Venus and Mars, our closest neighbors, almost as soon as the first satellites orbited the earth. "One of my driving aspirations right away was to push on with magnetospheric studies of the other planets and the solar system," Van Allen said.

With the first satellite missions, Van Allen helped define a new geography of space. It encompassed the radiation belts, the aurora and the region influenced by the earth's magnetic field, a region flattened as the solar wind slammed into it on the side facing the sun and flaring out on the side open to interstellar space. Physicist Thomas Gold coined the term *magnetosphere* to cover the new area Van Allen helped define to study the region of a planet's magnetic field.

The term defined one clear mandate for the probes to the other planets. Did they have magnetic fields and radiation belts? And what about the solar system? The idea of a heliosphere suggested a bubble encasing it and a boundary, or heliopause, where the solar wind lost momentum and slammed back like tidal waves from the onslaught of currents of cosmic rays. The concept made it unreasonable to consider Pluto, considered a planet until 2006, as the back door to the solar system. The only way to define the new boundaries would be to send out probes to explore—and to remap the solar system.

The *Mariner* missions to Venus and Mars, to the other planets and beyond the solar system offered a gold mine for scientific discoveries. Van Allen and other members of the Space Science Board, an advisory board to NASA, decided to congregate the space community for a comprehensive forum that promoted such missions in a wide-ranging agenda for space research. Van Allen agreed to chair the "Space Science Summer Study" as the board planned it early in 1962. "I was very keen on this and offered to be the host for the meeting and participate very heavily in it, and so, I made all the practical arrangements for the meeting in Iowa City," Van Allen said. He opened the doors of the University of Iowa that summer for the gathering of more than one hundred space scientists and government officials to the summer study that established wide-ranging research objectives in every arena of space exploration. Eighteen working groups met for eight weeks and Van Allen edited much of their 565-page blueprint that detailed research recommendations for rockets, satellites, probes to other planets, and astronaut missions. It called for the participation of scientists in spaceflight crews, international cooperation, the establishment of a mammoth Earth-orbiting space telescope, and the search for extraterrestrial life, including conditions that would support the origins of life. The seminal "summer study," funded by NASA, marked the first of many held at other locations and it gave Van Allen a forum to promote the study of cosmic rays, the solar wind, and the magnetic fields of other planets.

Aside from the scientific agenda, planning for the conference fell to Abbie Van Allen. Guest lists, housing checklists, and travel brochures cluttered her desk at home as she took on the logistics of lodging, feeding, entertaining, and cooling so many people. She found hosts for scientists, and sometimes their families, at houses and apartments all over Iowa City and located hotel rooms for those who would arrive for briefer stays. She booked Mississippi River cruises and even found air-conditioning units for some of the sessions. The new library, home base for the summer study, boasted central air conditioning. But other campus buildings where the scientists planned to meet didn't yet. The Van Allens considered this amenity essential for visitors unaccustomed to a blistering Iowa July when breezes off the river felt like a blast from a furnace.

On June 17, 1962, representatives of the U.S. space community converged in the library auditorium. Workmen prowled through the building to finish wiring and construction in time for the start of classes as the scientists poured into hastily furnished rooms. Van Allen welcomed them. Lloyd Berkner, Space Science Board chairman, addressing the tensions between human and robotic space mission advocates, offered advice on diplomacy in his opening remarks. "It is most important for the university and industry people to comprehend the many broad problems and decisions that must be faced by our government people. The two-way exchange of ideas may perhaps, in the long run, be one of the most enduring benefits to come from our efforts," Berkner said. "It is with this in mind that many of the key people from the government have agreed to spend time with us, and be part of our Study. Let us be careful at all times to listen to each other." He also promised that "your advice has been sought and—we are assured—will be carefully heeded."

For the eight weeks of the summer study, Iowa City served as an epicenter of the space program where scientists brainstormed, participated in their working groups, and drafted the comprehensive plan that detailed the role of space science in every aspect of planetary research, including a search for extraterrestrial life, fossil evidence, and compounds that support life in interplanetary space. Ground astronomy, long-distance probes, and samples taken from land vehicles should "be energetically pursued," the report recommended. "The search for extraterrestrial life has an obvious fascination for peoples of all nations. In a few short years, this topic—at one time merely science fiction—has been lifted from the category of escape reading and whimsical speculation and now stands as a serious objective. If life does indeed exist on another planet and we or the Russians find it, that discovery would have an enormous impact on people of every race and culture the world over, whether they are scientists or not," the report noted. "The more precise determination of the astronomical unit or the decision as to whether the moon has a liquid core may become the meat of philosophical arguments, but for 99 percent of the world's population such questions have little meaning." The study gave Mars the greatest chance of sustaining life and called for extensive robotic fieldwork on atmospheric and environmental conditions.

But the nuts and bolts of the report stressed more pressing needs, such as robotic space missions. The study emphasized the critical role of robotic space exploration as a tool for primary scientific exploration and as a critical part of the preparations for human spaceflights to the moon or other planets. Once the groundwork had been laid, the report specifically called for NASA to "train scientists as astronauts and vice versa" and to include scientists on every flight team. The chapter on "The Scientific Role of Man in Space Exploration" recommended that "one crew member of each *Apollo* mission should have scientific abilities and training." And scientists wanted that crew member assigned to the surface since "a maximum return is anticipated only if the scientist himself lands on the moon."

The summer session report contrasted with many of the national policy goals hastily cobbled together after the Space Science Board met in Washington, D.C., on February 11–12, 1961. Van Allen attended that meeting where NASA deputy director Hugh Dryden asked for a clear, national goal for space. Dryden requested a mandate "with enough dramatic appeal that the administration, the congress, the scientific world and the man-on-the-street will instantly recognize it as deserving their hearty support," according to the meeting minutes. The board obligingly stated that "the major national objective of the United States space program should be a manned expedition to the planets for the purpose of scientific investigation and exploration" and a search for evidence of life.

"The board strongly emphasized [during the meeting] that planning for scientific exploration of the Moon and planets must at once be developed on the premise that man will be included," Berkner wrote in a three-page report on "Man's Role in the National Space Program."

For a show of unity, all the members of the Space Science Board agreed to endorse the March 31 report sent to NASA's new head James Webb, even though Van Allen, Simpson, and proponents of robotic space science dominated the board. "Deeply felt doubts among many scientists about the scientific value of having man in space were not given voice in the report," notes Allan Needell of the National Air and Space Museum in his book *Science, Cold War and the American State.* "There was a growing feeling among scientists that manned space missions of the sort being planned by NASA and the military agencies were wasteful of resources."

But Berkner wanted the Space Science Board to maintain a strong role in all of NASA's planning and now he called upon old friendships to reach consensus on the "national goal" supported by a stellar list of America's space scientists. The report gave Webb and Vice President Lyndon Johnson the clout to convince Kennedy and Congress to endorse the *Apollo* moon landing as a national priority. Killian, Kennedy's bureau of budget director Howard Bell, and even the venerable Vannevar Bush expressed reservations about the cost of human spaceflights and the sacrifice of other programs. But timing is everything. The "Man in Space" report arrived just days before Yuri Gagarin orbited the earth on April 12 in the *Vostok 1*, a feat that eclipsed Alan Shepard's fifteen-minute suborbital flight on May 5 in the *Friendship 1*. Kennedy turned to Johnson demanding a victory in the space race. Johnson turned to NASA insiders including von Braun and political and business power brokers to back the moon mission at an estimated cost of $11 billion.

"I believe that this nation should commit itself to achieving the goal, before the decade is out, of landing a man on the moon and returning him safely to the Earth. No single space program in this period will be more impressive to mankind," Kennedy told a joint session of Congress on May 25. And no program would be so difficult or expensive, he admitted.

Van Allen hailed both Gagarin's and Shepard's achievements. This "will doubtless stand as one of the major milestones in the history of the human race," he wrote in his journal of Gagarin's flight. But he expressed his first public doubts on Kennedy's national priority in October in New York City where he accepted the American Rocket Society's first Research Award for the discovery of the radiation belts. Van Allen said at the reception that the "blunt" goals of the space program greatly outdistanced scientific competence and called for $100 million in funding at the university level to give such goals "a fine cutting edge of professional competence." The eight-week summer study defined the terms of achieving that competence with largely robotic missions.

While never contradicting or criticizing the national goal, the scientists working at the summer study made it clear that scientists believed human orbital flights such as *Mercury,* while a great achievement, would do little to get astronauts to the moon without the foundation of research from robotic missions.

"We are convinced that the current Earth-based and unmanned spacecraft programs will fall far short of providing information essential to the *Apollo* mission on the required schedule," concluded the summer study Working Group on Lunar and Planetary Research, which included Newell as a NASA "contributor" rather than a participant. As for the other planets, the report called for an extension of the *Mariner* program and reiterated that any future possibility of human voyages to the planets required the early and steady accumulation of data with robotic probes first. The plan diplomatically applauded the "national goal" for manned space flight but circled the wagons and established priorities that reflected the beliefs of many scientists on the Space Science Board. The message to NASA was loud and clear: don't undermine the basic space science programs while shooting for the moon and planets or you may not get there. Like a travel guide, basic science would give space explorers the safest route to their destinations, tips on what to bring, and pictures of what to expect when they arrived, Van Allen said. Robotic surveyors and probes could swoop in for photographs and drop to the ground floor of planetary surfaces to pick up a few souvenirs for chemical study and no one had to worry about how to get them home. He pushed to convene the summer session to drive home this point.

The collective connections of Van Allen, Berkner, and other members of the Space Science Board drew Homer Newell and some forty other NASA representatives to parts of the eight-week study session. The Department of Defense, the Atomic Energy Commission, the National Science Foundation, and the National Bureau of Standards sent representatives. Van Allen's circle of former graduate students and faculty working at NASA showed up en masse.

Cahill came from headquarters while Ludwig, Meredith, and McDonald all made trips from the Goddard Space Flight Center. Stuhlinger brought reports on the behemoth Saturn *Apollo* rocket under development at Huntsville.

Other old friends such as Lyman Spitzer and Fred Whipple arrived, too. Van Allen welcomed colleagues in cosmic ray research, including John Simpson of the Enrico Fermi Institute at the University of Chicago, and he introduced new faces. A young astronomer named Carl Sagan from the University of California at Berkeley made his debut at the summer session with a freshly minted PhD.

Van Allen chaired the study group for "Particles and Fields" with Cahill, Ludwig, and Meredith all participating in a sweeping agenda to explore magnetic fields of the other planets, in the solar system and, ultimately, in the Milky Way galaxy. Van Allen staked out research interests in all of these arenas and set out for the first planet shortly after the summer study ended on August 10. JPL's *Mariner 2* mission launched on August 27.

For *Mariners 1* and *2*, his team included Frank and other new graduate students. Given the distance and possible rigors of the journey, the team relied on the simple cosmic ray experiments developed for the *Explorers* using the miniaturized Anton Geiger tubes. The *Mariner* spacecraft itself incorporated a lot more uncertainties. It represented a remarkable metamorphosis of JPL's ill-starred *Ranger* series of missions sent to scout out the environment surrounding the moon.

JPL's *Ranger* spacecraft was designed to drop down toward the moon's surface with an RCA television camera recording the scene until it crash-landed on the surface. But in the push for a near term first over the Russians, NASA gave JPL the go-ahead to retool the *Ranger* for the historic *Mariner* missions to Venus and Mars. JPL planned on a larger, new design for the interplanetary trips with Convair Astronautics' new Centaur rocket for the second-stage boost after a liftoff in the air force Atlas. But development delays in the Centaur sent Pickering looking for an alternative. Venus and Earth would line up favorably late in 1962 and JPL wanted a shot at the planet then. Any delay would imperil the mission with overwhelming additional distances. Instead of making a new spacecraft, JPL decided to retool a *Ranger* to go to Venus and do it in eighteen months. The lighter, two-stage Atlas-Agena combo that launched the *Rangers* could serve the *Mariners*.

The decision seemed like a long shot. A quagmire of rocket and technical failures plagued the missions of the *Rangers*. The Agena rocket now commandeered for *Mariner* failed on the first two *Ranger* missions. An inverted symbol in a computer code and a sequence failure in the master clock demoralized the *Ranger* program on subsequent tries. The questions were obvious. Why would *Ranger* work any better with instruments carried to Venus, a 180-million-mile trip for a rendezvous 36 million miles from home? Would instruments respond to commands to begin operations? And what would the heat from such close proximity to the sun do to the entire instrument and communications network? No one knew. This was untracked terrain as JPL began to retool the *Ranger.*

In its *Mariner* mutation, the 447-pound space probe resembled a huge, robotic dragonfly with a 6-foot tubular body full of instruments attached to a hexagonal head carrying the control system, batteries, computer, and communications gear. Two winglike panels jutted out more than 8 feet from the base of the head with 9,800 solar cells, enough to generate 148–222 watts of power for the probe operations and instruments. With this configuration, the spacecraft needed lightweight, simple experiments—a signature Van Allen trait for the Iowa team's particle detectors and a hallmark of all the other instruments on *Mariner.* Van Allen's instrument with a single Anton Geiger tube would search for a high concentration of confined particles, a situation that defined a radiation belt.

Mariner 1 launched from Cape Canaveral on July 22, 1962. But the range safety officer destroyed it only minutes into the launch when the Atlas booster veered wildly beyond the planned flight path. A single hyphen, omitted in the computer coding of the guidance system, caused the problem. Van Allen heard the bad news in Iowa during a four-day break in the summer session. But *Mariner* 2 responded to every challenge. It was heading off course on a trajectory that would have meant more than a 230,000-mile planetary miss when JPL radioed a mid-course correction that worked perfectly. When one of the solar panels lost output for a week, resumed functioning, and then failed permanently, JPL engineers feared disaster. They had recently witnessed distress signals from *Ranger 5* and the spacecraft went mute, the solar panels short-circuited, and the batteries soon failed. But *Mariner*'s closer proximity to the sun saved the day because the solar panels recharged swiftly and the spacecraft got along on just one battery.

And *Mariner* 2 delivered results right from the start with instruments that combined the forces of MIT, Harvard, JPL, the University of California at Los Angeles, the Goddard Space Flight Center, and other institutions as well as the University of Iowa. Van Allen's detectors showed that the cosmic rays dropped in intensity closer to the sun, swept away by gusts of solar wind. His and other instruments offered a breathtaking scientific confirmation of the existence of the solar wind with four-months of detailed readings. At Venus itself, *Mariner* 2 confirmed what many astronomers and scientists had believed from long-range measurements. The scorching surface of the planet was hot enough to melt lead and clouds of carbon dioxide shrouded it in darkness.

Paul Coleman of the University of California at Los Angeles and Van Allen surprised the science community with the first real evidence that not all planets have a magnetic field or radiation belts. No magnetic field meant no belts, even though Venus was about the same size as Earth and had a molten core, Van Allen told reporters. *Mariner* confirmed the reason why.

Scientists believe that the earth's magnetic field is created deep within the molten core of the planet as it rotates. Elementary physics teaches that current flowing through a conductor creates a magnetic field and, conversely, moving a wire through a magnetic field creates a current, the principle that makes possible the illumination of every lightbulb. In the molten core of the earth, iron and nickel generate the electrical currents that produce the earth's magnetic field. But they generate current because the earth spins at more than 1,000 miles an hour, a swift dynamo of motion evident on Earth in the cycles of night and day. Without the planetary mass and pressure to produce a molten core and without the rotational speed to produce the dynamo effect, planets have a negligible magnetic field. Previous radar observations of Venus showed that it rotated only once over a period of 243 days, too slowly for the dynamo effect that sparked electric currents in the core of the earth, Van Allen explained.

Webb, flanked by U.S. senators, announced the *Mariner 2* victory of the first mission to another planet in Washington, D.C. Such success was worthy of an encore. On January 10, 1963, NASA publicly announced a *Mariner* trip to Mars, a plum sought after by the Goddard Space Flight Center. Goddard proposed a probe that could land on the surface of Mars but *Mariner 2*'s stunning fly-by convinced NASA to stick with JPL, reported JPL historian Clayton Koppes.

As *Apollo* tooled up, however, Van Allen still felt that space science objectives were losing ground. He groused at the Space Science Board over the favoritism shown to the new space flight centers and to industry over universities. In May 1964, he stayed in Washington, D.C., a week and "worked with George Derbyshire at the Space Science Board to draft a resolution favoring heavier emphasis on planetary experiments," Van Allen wrote. He returned Friday evening for a family weekend before Abbie headed for her twentieth college reunion at Mount Holyoke, and Margot, who had a passion for horses, left for riding camp. Van Allen drove to Mount Pleasant that Monday, June 1, where von Braun received an honorary PhD from Iowa Wesleyan College and gave the commencement address.

"I am sure your outstanding graduates are legend," he told the audience. "I must single out one for mention however, whom I have enjoyed visiting again today. He is a very good friend of mine—and a native of Mt. Pleasant—Dr. James A. Van Allen," von Braun said. He recounted their partnership on *Explorer I* and noted the discovery of the Van Allen radiation belts, adding that Van Allen referred to them only as "radiation belts," never by his name.

It was a momentous summer in the physics department as well. Lou Frank successfully completed development of an important new detector called the "low-energy proton-electron differential energy analyzer," a brainteaser of a name soon shortened to LEPEDEA. The detector could track particles at a new range of low energies and could differentiate between protons and electrons as well. LEPEDEA helped fill out a comprehensive range of the energy levels of particles escaping the sun when it was used in missions such as the Orbiting Geophysical Observatory (OGO), NASA's series of compact space laboratories. Frank's LEPEDEA instrument flew for the first time in *OGO 1,* launched on September 4, 1964, and it delivered the first direct measurements of ring currents, confirming them as the cause of magnetic storms that scientists had detected on earth. Chapman hypothesized that solar flares sparked temporary rings of electrical currents that encircled the earth and caused the storm. With results from LEPEDEA, Frank found the actual burst of solar particles that caused the currents.

Frank and Stamatios (Tom) Krimigis, a graduate student from Greece, also worked on finishing touches for the *Mariners 3* and *4* detectors that summer. Van Allen recruited Krimigis from the University of Minnesota where he met the young physics major making detectors as an undergraduate studying with some of Van Allen's competitors.

"He inquired about what I was doing and I explained the detectors that I was building and he asked me if I had considered going to grad school," Krimigis recalled. Van Allen got him a teaching assistantship and also suggested he work on the new solid state detectors for the *Mariners*, a high stakes effort with a technology new to Krimigis. "That was his teaching method. He had an eye for people's abilities, I guess, and he would throw you over board and see if you could swim. If you survived, then you were in great shape," Krimigis said.

Iowa's compact black box of instrumentation developed for *Mariners 3* and *4* weighed only 2.2 pounds but included an advanced package of detectors compared to *Mariner 2*. It had three Geiger tube detectors filtered for different energy levels of particles and a solid state detector to differentiate between electrons radiating from the sun and low-energy cosmic rays piercing the solar system from galactic space. But the solid state detectors, coming from a company in Illinois, couldn't survive the testing for space.

"They would work for a while in testing and then they would malfunction and we were getting fairly close to delivering the instrument to JPL. We finally did deliver the instrument and then the JPL people would call us up to say, 'the counters are going wild.' So I went out there and take a look and I finally looked at the thing under a microscope," Krimigis said. He discovered that contacts were coming loose and he improvised a connection with a special epoxy glue that had conducting properties to seal the contacts but keep power flowing. "I just painted it around and that cured the problem," he said.

Krimigis traveled to JPL to monitor the first data transmission after the launch, scheduled for November 4, 1964. It never took place, halted in mid-countdown when technicians discovered a corroded relay switch as electronic systems fired up to launch. It was just as well, since the revamped fiber glass sheath of the probe proved to collapse when heated in a follow-up vacuum test.

"Apparently such tests in a vacuum have never been done before on the new improved shroud," Van Allen huffed in his journal on November 11. "The implication appears to be that the shroud may very well have collapsed in flight. *Mariner 4* has been delayed indefinitely. May miss Mars window!"

But JPL hastily corrected the problem and "delayed indefinitely" turned out to be only a matter of about three weeks. *Mariner 4* launched for the trip to Mars on November 28, 1964. The rockets functioned perfectly and the solar panel deployed in sixteen minutes after takeoff. *Mariner* was on its way with 28,224 solar cells covering four wings of panels for the long flight. A star-sensor oriented toward the star Canopus helped keep *Mariner* on course, though JPL had to keep a close watch on the system's stubborn tendency to shift to other stars instead.

"Received first substantial batch of data from *Mariner 4* today. Our apparatus seems to be operating perfectly. Got a very good traversal of the radiation belt system and out through the fringe of the magnetosphere," Van Allen wrote in his journal on December 7. He made the most comprehensive mapping yet of the boundaries of the earth's magnetic field as *Mariner* headed into interplanetary space.

Mariner shot toward Mars at speeds exceeding 1 million miles a day, traveling a 325-million-mile trajectory to rendezvous with the red planet at a point when it was 134 million miles away. Then in May and June, the detectors measured a new natural phenomena—electrons in the solar wind at lower energies than had ever been recorded before, extending the observed range of particles radiating from the sun. Once found, the *Injun 4* satellite enabled Van Allen to cross-check for the particles at earth orbit.

Abbie Van Allen drove to Long Island that June with Margot, Tommy, and Peter. As usual, Van Allen stayed home for the first part of the summer. "That's when he got a lot of research done," recalled Peter Van Allen. But it wasn't a usual summer. Sarah and Cynthia stayed with him since Cynthia wanted to take a rhetoric course before starting her freshman year at Western College in Oxford, Ohio, that fall. And Van Allen left for Pasadena in July for the momentous Mars fly-by. Jet Propulsion Laboratory scientists roused the spacecraft with a radio signal at 7:28 A.M. on July 14, 1965, commanding it to warm up the TV camera and instruments. Van Allen joked with reporters as he waited out the last hours before the encounter. He compared the fly-by to watching the baseball scores on a ticker tape. "We can tell if we get any home runs," he told reporters, referring to the ticker tape of the computer printouts of data he would receive from Iowa's detectors. The detectors could register as few as one count a second in space but could also register hundreds of thousands of counts per second if *Mariner 4* encountered a radiation belt.

"The close-up investigation of radiation conditions around Mars will be a crowning achievement to a flight filled with new discoveries about the radiation environment in space," bet the *Iowa City Press-Citizen*.

The planetary operations staff at JPL and scientists visiting for the fly-by uproariously greeted the first messages trickling in from *Mariner* for the July 14–15 fly-by. They had arrived. Van Allen and the Iowa team gathered around the first printouts looking for evidence of a magnetic field. But no change in particle counts occurred even as *Mariner* sliced through the Martian environment to a mere 6,100 miles above the planet.

"If there are any Martian men, they do not use compass needles," Van Allen quipped to reporters. Like Venus, Mars had no magnetic field and no radiation belts. Unlike Venus, Mars rotated rapidly enough to operate as a dynamo but the second piece of the recipe for a magnetosphere was missing, Van Allen noted. "It's a smaller planet, too small to have a molten core."

But the *Mariner*'s TV images provided the main show with the first mesmerizing front-row views of another planet. But the pictures unexpectedly showed a dead, cratered surface, not the red canals of science fiction that people hoped might shelter life. Playback of the images started early on July 15 and continued without interruption for more than eight hours. With *Ranger 7*'s lunar images and the *Mariner*'s spacescapes of Mars, JPL computers recompiled some of the first digital images ever made. The *Mariner* TV camera captured the images in a fifth of a second but it took eight hours and thirty-five minutes to transmit them to Earth, slowly and painstakingly as 4,000 dots per image. *Mariner* sent twenty-one full pictures and a partial landscape, and the computer recovery of the pictures continued nonstop for more than a week. Scientists such as Robert Leighton of Caltech explained that Mars resembled the moon rather than the ruby planet of romance. And the preservation of those visible craters, frozen in time over billions of years, suggested a very thin atmosphere, too thin to sustain life as we know it. Still, several of the scientists conceded, they really couldn't draw definitive conclusions yet. Later mission images indicated that the cratered terrain "was not typical for Mars, but only for the more ancient region imaged by *Mariner 4*," according to NASA.

The *Mariners*, with their exploration of the earth's nearest neighbors, built a fascinated public following and a remarkable success story. *Mariners 1* through *10* cost a total of $554 million, a small fraction of the costs of the *Apollo* missions. But money for the probes became harder and harder to squeeze out of NASA. "Exploring the planets was strictly a sideline business, with about 2 percent of NASA's budget," Van Allen said.

On top of that, Webb considered his grants to universities as "ground-floor" contracts that he believed obligated faculty to contribute "something extra" in terms of time, commitment, and talent. He had a much clearer capitalist vision of industry's relationship to government. The stocky, dynamic attorney who learned the ropes in Washington as director of the Bureau of Budget, wooed the aerospace industry with contracts for *Apollo* that exceeded $4 billion. *FORTUNE* magazine applauded "an intimate new sociology of space, a new kind of government-industrial complex in which each inter-penetrates the other so much that sometimes it is hard to tell which is which." But Webb clearly recognized an invaluable resource that universities could provide. NASA supported graduate students with "traineeships" to develop personnel to go into the space program. "Then Mr. Webb had the idea of supporting the construction of laboratories on a fairly major scale and we were among those who applied for support and did get substantial support from NASA. That contributed importantly to the construction of our new building."

Van Allen applied for a grant in March 1962. He requested and received federal matching funds in the amount of $610,000 from NASA and $750,000 from the National Science Foundation. The attractive commitment convinced the Iowa State Legislature to appropriate $1.4 million for a new physics building and $300,000 for a more powerful Van de Graaff accelerator that Van Allen hoped would rekindle the strength of Iowa's nuclear and theoretical physics programs. Legislators even approved $385,000 for moving expenses and equipping and furnishing the new building.

• • •

The basement of the Iowa physics building had a mystique as one of the most famous laboratories in the world, partially because the cramped, old-fashioned mecca of space science so shocked news reporters, aerospace executives, politicians, and government officials who continued to pour into it for meetings and visits. The space facilities spilled over into the Masonic Hall and rented office space down the street from the physics building by the early 1960s. "Publicized descriptions of the inadequate quarters for achievements of global significance helped prepare the way for a new building," noted James Wells in his 1980 history of the university's Department of Physics and Astronomy.

At first, Hancher approached the state legislature with a plan to expand the existing physics building, with an addition extending across the lawn of the Pentacrest. Such a plan meant virtually sealing off one side of the airy campus landmark and offered an inadequate stopgap in terms of program needs. The strapped state legislature refused this plan in any case before the matching grants saved the day. Now Van Allen considered a location for the new building. The parking lot along Dubuque Street, less than two blocks to the east, seemed like the likeliest prospect for the new building. Van Allen and a building committee on staff began to sketch out a building with 100,000 square feet of working space that provided all the necessities MacLean Hall lacked—a loading dock, a receiving area, elevators, air conditioning, insulated areas for work with high voltage and radiation, and clean rooms. The graduate students begged for office space to improve upon the small desks placed in hallways or storage areas. Architects first estimated a building cost that went $1 million over budget but trimmed back to a simple functional structure, free of ornament, contemporary in design, but built with textured limestone that would mesh with the classic campus structures. The building committee juggled cost estimates and the staff wish list to recommend a seven-story, 83,200-square-foot building, measuring 73 feet by 163 feet with a tower for the new vertical accelerator and a domed observatory. The simplicity of the design brought in the actual construction bids under budget and the Viggo-Jensen firm of Iowa City started construction on October 30, 1963.

But even with plans for the new building underway, Van Allen watched in frustration as several faculty members drifted away. Those who came to Iowa to work with one of the most famous space scientists in the world could command attractive positions elsewhere after a few years' apprenticeship. O'Brien left for Rice University in Houston, Texas, with a lucrative salary boost. The decision frustrated and hurt Van Allen, who had supported his protégé's cost overruns on *Injun 3* ($350,000 versus $100,000 in the original plan). "My own view has been to give him virtually unlimited support and backing through his tenure and to push him along in all regards using my entrée," Van Allen wrote in his journal on January 11, 1963. But O'Brien was frustrated too, over the Starfish results, which he felt he hadn't been allowed to report until they were "stale." He took graduate students Curt Laughlin and Ray Trachta with him to pursue a satellite program at Rice. Later, O'Brien worked with the *Apollo* missions and became a visiting professor at the University of Sydney. He returned to Australia and became the director and chairman of the Environmental Protection Agency in Perth, Western Australia.

Other faculty members left to pursue physics in an environment less focused on space science or for personal reasons. But nothing cut as deep as Max Dresden's resignation in May 1964. "I have been and continue to be very fond of Max and consider him to be one of the best, if not the best professors of physics that the University of Iowa has ever had. I feel reasonably powerless to remedy the deficiencies which he feels exist here," Van Allen noted in a twelve-page passage in his journal. Chief among them was Dresden's inability to attract what he considered first-rank young theorists to teach at Iowa. Dresden was also annoyed at the expectation that he should find grants to finance his research. Van Allen sympathized with the former and dismissed the latter. "I bring in about $1 million a year by [my] own efforts from outside sources not to mention the grants for the new building," he wrote.

Lou Frank helped fill the gap by accepting an assistant professorship as soon as he earned his PhD in June 1964, "the best news of the recent season," Van Allen noted in his journal. Frank had worked closely with Van Allen since he was a sophomore and learned the ropes on the early lunar *Pioneer* missions. Gurnett, another of the trusted inner circle of graduate students, accepted an assistant professorship, too, after earning his PhD in 1965 and giving the university a strong base in the field of plasma physics.

The move to the new Van Allen Hall, the new physics building and research center, began in August 1964 and took a month with the participation of every able body and anything with wheels. Van Allen parked at the dock every day, entered the building through the back service entrance, and visited the instruments shop on his way to the elevator. He had official quarters as department head in the administrative offices on the second floor, but his research haven became Room 701. The spacious corner office overlooked Dubuque Street, his route to and from home, and a half dozen churches. "I'm covered," Van Allen chuckled, taking a rare moment to survey the view in 2004 from the office where he continued to work every day until three months before his death.

The new building enriched the Iowa City community as a whole. The telescope observatory on the roof, the perfect resource for Van Allen's general astronomy class, also offered "clear sky" nights for visits by the public. Often, families would exit the elevators near Van Allen's office and ask at any open door for directions to the stairway to the observatory. "He took us up there to look through the telescope," said Bill Boyd, son of Sandy Boyd and Peter's best friend. "He was so good at explaining the things [in] laymen's terms—things about the solar wind or eclipses."

The Iowa satellite program continued unabated even with the move. *Injun 4* launched in 1964, *Injun 5* in 1968, and *Hawkeye* in 1974, with Van Allen guiding the programs and obtaining funding. Van Allen and Rogers also built a new radio observatory for telemetry near North Liberty in 1969 with a 60-foot antenna dish acquired from the air force as surplus. "We got a gun mount for it from a ship that have been decommissioned and the mount was shipped up the Mississippi River by barge," Van Allen said. In the university's golden space era of the 1970s, the observatory gave Iowa direct ability to track and control the *Hawkeye* satellite during the four years it explored the magnetosphere at high altitudes. "We received all communications from our own receiving station and generated all the commands to the satellite," Van Allen said. North Liberty transmitted the satellite data to a microwave dish on the top of the physics building that fed it directly to computers for read-out in real time. Then in 1990—after twenty-one years of very successful operation—the old antenna was demolished and replaced as part of a national radio telescope network.

Iowa participated in four more *OGO* missions in the late 1960s and in the Imp-D (*Explorer 33*) mission, capturing the dramatic impact of a massive solar flare on July 8, 1968. Imp-D had an earth orbit that extended beyond the moon and Imp-E (*Explorer 35*), launched in 1967, orbited the moon, offering comprehensive study of the outer fringes of the earth's magnetosphere. Van Allen returned to Venus on *Mariner 5* in 1967 as well, but the standards for instrumentation began to tighten. "In one celebrated case, James Van Allen's instrument passed all of JPL's tests. But laboratory personnel opened the box and then tried to reject the experiment because Van Allen had not followed JPL techniques. He was furious. He thought equipment that passed the laboratory's tests should be accepted," JPL historian Clayton Koppes reported. In this case, the instrument was accepted.

But Van Allen's professional focus shifted farther outward now to two of the most exciting and enduring missions of his career—the *Pioneers 10* and *11* missions to Jupiter and Saturn.

16 Pioneers to the Outer Planets

Redondo Beach, California—January 1972. Van Allen could almost catch his reflection in Iowa's glittering gold box fitted with particle counters for *Pioneer 10*'s journey to the outer planets. The instruments on the mission promised the world a front-row seat for new discoveries at Jupiter. Astronomers gave good odds for radiation belts at Jupiter based on the hot glow around the planet that their radio telescopes measured from afar. But no one knew for sure. The gold-cased detector that drew less power than a Christmas-tree light could help unravel such mysteries.

TRW Systems built the spacecraft and Van Allen traveled there while in California in early January. Though he couldn't be sure of the exact date, that's when he believed he made a final check of his "flight unit" as *Pioneer 10* stood anchored in a laboratory bay. The spacecraft itself amounted to a lightweight dish antenna that measured 10 feet across with an aluminum body, six rocket thrusters, and a thermonuclear power generator. Designers kept the design sleek, compact, and simple—their best bet for a trailblazer with a cosmic itinerary and a cargo of ten instruments on the outside surface of the antenna dish. A magnetic field monitor was attached to a long boom to take readings without interference from the craft. The mission and selection of this ark of instruments was fittingly managed by ARC, NASA's shorthand for the Ames Research Center in Moffett Field, California. Van Allen admired the sheer beauty of these devices as they developed with their finely crafted casings. But, for all their style, the instruments stood up to shake tests, heat tests, cold tests, and radiation tests as NASA put them through their paces.

Van Allen wore white cotton gloves to touch the polished gold, hygienic detector fitted with seven miniature Geiger counters. Joe Lepetich, experiments systems manager for Ames, was in and out of TRW and insisted that the scientists wear the gloves. The only evidence of human presence that Ames expected on this mission was the plaque designed by Carl Sagan, then the director of planetary studies at Cornell University. Should *Pioneer 10* ever find its way to a distant star system, Sagan's famous gold plaque carried a message of goodwill along with a galactic map and anatomical drawings of two people who appear sketched for a mural of the Garden of Eden.

Lepetich wore white gloves, too, as Van Allen checked over the detector that he would never see again. Then Lepetich left the room for a minute and Van Allen, the least impulsive of men, couldn't resist. "I took off my glove and, with my forefinger, I left a print on the bottom of the gold plate. And then I put my glove back on before Joe came back," he later admitted to his team of graduate students. *Pioneer 10*, traveling at approximately 250 million miles a year, carries Van Allen's personal talisman as it continues on its journey, more than 8 billion miles from Earth and heading toward the red star Aldebaran, the eye of the Taurus constellation some 2 million years of travel away.

• • •

NASA lost contact with *Pioneer 10* in August 2000. Astronomers at the SETI Institute (the Search for Extraterrestrial Intelligence) tried again with the 300-meter dish at the Arecibo Observatory in Puerto Rico. "They make an attempt to contact it every six months to calibrate their dish. *Pioneer* is the weakest, most reliable transmitter out there so it's been an excellent source of calibration" for a place trying to catch a weak radio signal from some distant voice of intelligent life, Van Allen said. But even with Arecibo's powerful ear—more than three times larger than the antenna dishes in NASA's Deep Space Network—*Pioneer 10* remained lost in space. "There is very important data at this point because of the sun's activity. It just grieves me all over to lose it at this point after it traveled so far—beyond all expectations," Van Allen said in January 2001. His was the only instrument still operating on the craft, but that was enough for supporters to pull out all the stops. Project manager Larry Lasher tapped into a redundant communications system that finally reestablished contact later in 2001. Like Hall, his bet for justifying *Pioneer* remained science—Van Allen's science.

"If we didn't have data from Van Allen's instrument, we couldn't have maintained *Pioneer 10*. His prestige and his constant advocacy is what kept us going," Lasher said. "Headquarters would say, 'Why should we keep this going?' And we said, 'Well, Van Allen has this instrument and he wants the data, okay?' So we were able to convince them, you know. Without Van Allen we couldn't have done it."

Energized by the success of 2001, Van Allen appealed to NASA to formally reinstate the *Pioneer 10* mission, pointing out that it would take years to reach the far corners of the solar system again with any other craft. NASA agreed and stepped up communications with the probe in 2002 for its thirtieth anniversary. Streams of new data poured in, and Van Allen's victory came just in time. Early in 2003, the *Pioneer* called home for the last time, its signal nearly lost in a sea of static some 8 billion miles away. NASA made one last unsuccessful attempt to contact the spacecraft on February 7, 2003, and then cut off communications one final time. Still, that doesn't stop the travels of *Pioneer 10*—a silent ship on an uncharted journey through the cosmos.

The essential mission to Jupiter seemed daring enough. A tiny spacecraft sent off beyond Mars faced dangers that included the asteroid belt that scientists feared might smash it to bits with hurling remnants of an ancient planet. Van Allen and a handful of others secured the ticket for the spectacle of the outer planets by selling NASA and Congress on a vision of potential discoveries that America could make—and make first. But persistent behind-the-scenes badgering won the victory as well. Van Allen bluntly admitted he made a "pest" of himself at NASA in pursuit of the *Pioneer* missions. America beat the Russians to Venus by five years with *Mariner 2* in 1962 and beat them to Mars by seven years with *Mariner 4* in 1964. But a demoralizing sequence of subsequent mishaps on return missions to Mars made Jupiter seem like a long shot. The failed missions led *Time* magazine journalist Donald Neff to speculate about the grip of a "Great Galactic Ghoul" at Mars.

Mishaps or not, the goal was clear to Van Allen. "My own principal thrust of interest is toward planetary missions," he reiterated in his journal early in 1968. He reflected on the fact that the highly successful *Injun* program "may mark the end of an era of building our own satellites. One of the reasons is the dwindling support for university programs in space physics. Perhaps an even more basic one is my feeling that we should not, in a university, be so heavily committed to such a major technological and engineering effort as is now involved in the sophistication of modern spacecraft."

Iowa continued building satellites in the 1970s but, as always, Van Allen hedged his bets and looked ahead to cutting-edge research opportunities. By the time he summed up his thoughts in the 1968 journal entry, he already had proposals submitted for the planned *Pioneer* and *Voyager* missions to the outer planets.

Van Allen began to plan for planetary space exploration with the first lunar probes in 1958. The Space Science Board (SSB) summer session in 1962 gave shape and impetus to planetary missions and the board formed an ad hoc panel on small planetary probes in 1966. Van Allen chaired it. Also that year, goaded by the space scientists, NASA formed its own Lunar and Planetary Missions Board that gave planetary research a direct voice within NASA. Van Allen served as a founding member with Gordon MacDonald of the Institute for Defense Analyses, astronomer Thomas Gold, geneticist Joshua Lederberg, and Joshua Findlay, the chairman of the new board and a staff member of the Space Science Board. As the space agency's budgets plummeted, the Missions Board drafted a planning document that called for a shift to economical robotic space exploration. Lederberg's membership on the board reemphasized a search for life in space over taking human life into space.

The board's formal report, completed in September, recommended that NASA complete the *Apollo* program and then abandon emphasis on human space exploration and a space station. "The interest in long earth orbital missions [justifying a space station] arose historically from interest in long manned planetary missions. We agree with the Space Science Board that this is no longer a timely need," wrote Gold in the report.

The board recommended $350 million annual expenditure on planetary research at a time when Congress slashed NASA's budget. In the late 1960s, the escalating Vietnam War jeopardized even the $75 million per year earmarked for planetary research. Webb quit in disgust over the annual budget bloodbaths. As analyst Daniel S. Greenberg summed it up in *Science* magazine, space exploration extended the hope of fulfilling a global substitute for war. "But now no substitute for war is needed—we have a real war," he wrote.

Despite the turmoil, NASA turned centuries of fantasy into fact when *Apollo 11* reached the moon. Van Allen and millions of Americans watched through the night of July 20, 1969, as Neil Armstrong became the first human being to set foot on the lunar surface. "One small step for man—one giant leap for mankind," he told millions of listeners across the world as Buzz Aldrin followed him out of the lunar module. Van Allen recounted the whole landing with gusto and full technical details in his journal the next morning, starting with the touchdown at 3:17:42 P.M. Central Daylight Time.

"Sally [Sarah] and I watched the entire operations on television at home," he wrote. "The astronauts found it relatively easy to walk about and set up their experiments, despite their awkward space suits," he wrote, outlining each experiment. "[They] also collected over 20 pounds of lunar surface samples. By 12:11 CDT, the two astronauts had re-entered the LM and closed the hatch. I continued to watch the TV show and turned it off at 1:08 A.M. to go to bed. Everything seemed to be going well. An historic human achievement." In

August, he and Abbie attended a state dinner hosted by President Nixon in Los Angeles to honor the *Apollo 11* crew.

With much less fanfare, *Pioneers 10* and *11* took shape, the trailblazers for all missions to come that would head to the outer planets. Jet Propulsion Laboratory and the Goddard Space Flight Center vied for the *Pioneers* but simplicity, a good batting average, and an unassuming engineer named Charles Hall helped cinch the program for Ames Research Center in Moffett Field, California.

• • •

The U.S. Air Force built *Pioneer 1*. It launched and soared to a record height of 72,765 miles from earth before the second-stage booster failed. The pioneer of *Pioneers* fell back toward earth and disintegrated. Even so, Iowa applauded the flight because Iowa's instruments on *Pioneer 1* cut vertically through the radiation belts for the first time. McIlwain designed the detectors with a wide range to prevent saturation and the system's direct vertical measurements verified the mosaic of readings pieced together as the *Explorers* swooped in and out of the belts. A second *Pioneer* mission with the same instrumentation failed when the launch vehicle fizzled out at takeoff.

Pioneer 3, an army/JPL mission carrying Van Allen's Geiger counters, fell back from about 65,000 miles when it failed to reach the speed necessary to escape the earth's gravity. Van Allen turned the partial failure into a coup, confirming the existence of two separate radiation belts with a "slot" between them. *Pioneer 4* carried more Geiger tubes, achieved a fly-by of the moon in March 1959, and went into orbit around the sun. Sky watchers enjoyed the fireworks as the earth's aurorae plumed in a solar storm during the flight and Van Allen measured the intensity of trapped radiation in the outer belt with *Pioneer 4*, another army mission.

The science was solid but the Russians once more stole the show with the three *Lunik* flights to the moon in 1959. *Lunik 1* came closer to the lunar surface than any of the *Pioneers*, became the first craft to escape the earth's gravity, and went into solar orbit two months before *Pioneer 4*. *Lunik 2* crash-landed on the moon itself and *Lunik 3* orbited the moon, photographing the mysterious dark side for the first time. They did it all "with characteristic one-up-manship," Van Allen noted in his journal. *Pioneer 5* restored American self-esteem with its 1960 launch into orbit around the sun. A pair of paddle wheels, lined with solar cells and fitted to the sleek geodesic body of the probe, assured power as *Pioneer 5* proved the basics of technology for long-distant communications and control of a spacecraft.

Ames took over from there to refine *Pioneer* technology for planetary exploration. One of the less glamorous NASA research centers adopted from the old National Advisory Committee for Aeronautics, Ames occupied a military field where the cavernous aircraft hangars housed blimps during World War II. Ames started out the *Pioneer* program with rehearsals for the odysseys to the outer planets. *Pioneers* 6 through 9 made "neighborhood" trips, exploring a few million miles from earth in solar orbits with experiments that tested new craft designs, communications systems, and solar panels to provide electrical power.

NASA awarded Ames the management of the daunting *Pioneer 10* and *11* missions to the outer planets early in 1968. The ingenuity and spirit to bring all the pieces together relied heavily on one man—Charlie Hall, Ames deputy director. Hall worked quietly and competently behind the scenes at Ames for years, an engineer testing aircraft models in the Ames wind tunnel. But the stunning potential of the *Pioneer* missions transformed him into a pragmatic and freewheeling leader. He, in turn, transformed Ames as a major player in NASA's planetary missions. Hall shed Ames's reputation for focusing on research and leaving the action to other places. His famous stand-up meetings and brainstorming sessions reflected the creative momentum that propelled everyone associated with Ames's *Pioneers.*

The lifeblood Hall breathed into Ames reaped results, evident in basics such as the design of the spacecrafts. *Pioneers* evolved from modified satellite designs to the dish-antenna spacecraft focused on long-distance durability and communication. The design became the classic for future space probes. Ames cut weight and strengthened materials wherever possible for *Pioneer 10* and Van Allen participated in the planning for the closest possible encounter with Jupiter. As always with planetary missions, they were shooting at a moving target, aiming *Pioneer 10* at a point in space where it would rendezvous with Jupiter twenty-one months after takeoff.

"Van Allen was interested in all aspects of the planning" from the start in 1966, said David Lozier, who came to NASA and the *Pioneer* project among the earliest computer specialists in the 1960s. The mission finally stamped into the hardware many of the goals he and others had reiterated over and over again since the Space Science Board meeting in Iowa in 1962.

NASA officials hoped the cost-effective *Pioneer* missions would spur support for more journeys through the solar system. Van Allen compared the approach to "sending a number of well-equipped scouting parties to the several planets before we send out the wagon train with all of our women and children and a full set of household furnishings." The wagon train was the huge "Grand Tour," a series of 1970s missions planned to the outer planets.

Ames turned to TRW to construct the spacecraft with the company's veteran team from earlier *Pioneers*. Communication was critical and posed unprecedented challenges. It would need NASA's Deep Space Network, the international system of antennas supporting radio and radar astronomy across the universe, to relay signals to and from *Pioneer 10*. The spacecraft had both a radio receiver to take orders from Ames and a transmitter to send back data and flight information. Ames had a transmitter and receiver for the *Pioneers* as well. The receivers at both ends relied on a lock system to hold onto the signal for the longer time intervals it took to radio *Pioneer 10*.

"When we were about a month from launching the probe, we had a test anomaly with the receiver where it lost lock," said Art Stephenson, a young engineer at TRW when he helped develop the receiver. "We could never reproduce the problem. And we finally sat down with the program manager and said we can't reproduce it. We recommend you just go ahead and launch it because we can't find any way to make this thing fail," said Stephenson, former head of the Marshall Space Flight Center and currently a vice president for space exploration at Northrup Grumman. Their recommendation raised skepticism instead of confidence—everything can be made to fail. But *Pioneer*'s path to the fringes of the solar system vindicated them in the end, Stephenson noted exultantly in an interview nearly thirty years after he took his flight-ready stand.

But the heart of *Pioneer 10* rested on science and every space scientist wanted the chance to make the first discoveries at Jupiter, Saturn, and possibly beyond. In one of his last interviews, John Simpson of the University of Chicago described the competition to win a seat on *Pioneers 10* and *11* as a "fierce, almost bloody competition." The Announcement of Opportunity in June 1968 sparked 150 inquiries by late August. Ultimately, seventy-five research teams vied to be the first to make the trailblazing discoveries at Jupiter and beyond. Simpson, Van Allen, and eleven other teams survived the first, and then the second cut in proposals that were hundreds of pages long. Van Allen streamlined his instrument plans, trimming weight and power for a mission that was tallying up power needs to the milliamp.

His instrument won the cut. "Official notification by TWX from John E. Naugle of NASA Headquarters this morning that my reduced version of [*Pioneer 10/11*] missions proposal has been accepted," Van Allen noted in his journal on March 24, 1969. He could pack his Geiger tubes for Jupiter and Saturn after nearly a decade of political battles to win the flights.

• • •

Van Allen brushed off the Washington politics and concentrated on his science once at his offices in 701 Van Allen Hall. Construction on an addition to the physics building got underway in 1969 while Gurnett and Frank developed another generation of satellite instruments. Both men were on the faculty, pulling in their own space missions. That meant millions of NASA dollars and a growing cadre of faculty, graduate students, and staff involved in building instruments.

A support team of four people now managed the exhaustive process of preparing space proposals and travel arrangements. Administrative tasks mounted in Iowa where the physics and astronomy department faculty and staff had grown to about a hundred people. Van Allen's personal secretary, Mrs. Evelyn Robison, deftly smoothed out all the jagged edges from Room 702, next door to Van Allen's office. She screened visitors and phone calls with grace, organized the downpour of reports in Van Allen's office, typed PhD theses, and tactfully relegated letters about space aliens and brain implants to her "Unfortunate Persons" file. She managed press relations, personnel files, and budget deadlines. She arrived by 8 A.M. and she also made it clear she intended to leave every day at 5 P.M. But slowly and graciously, the weight of many day-to-day tasks shifted from Van Allen to Robison's reliable shoulders.

Van Allen commuted regularly between Ames and Iowa for intensive negotiations and meetings that hammered out weight, power, and telemetry limitations for *Pioneer* instruments. He raced against the July 1970 deadline to build the design verification model of the Iowa experiment for *Pioneer 10* with research associate Bruce Randall, project engineer Roger Randall (no relation), and graduate students Daniel Baker, Davis Sentman, and Mark Pesses. Another graduate student, Michelle Thomsen, joined the team when Van Allen recruited her right out of her entrance interview.

"I had a degree in physics and I've always been really fascinated by space. So I was really excited when Professor Van Allen mentioned that he had a little project that he'd like some help with," said Thomsen, now a research fellow at Los Alamos National Laboratory. The "little project" turned out to be the analysis of the first direct measurements of the radiation belts at Jupiter and Saturn.

More women and minority students signed up for physics classes now, a welcome sign to Van Allen who kept close tabs on the numbers of students in the department, routinely reporting the count in his journals. A good student was a good student. Race and gender didn't matter to him and hadn't mattered, even twenty years before when he welcomed Bob Ellis on board among the very first African American physics PhD candidates anywhere in the country.

Van Allen had noticed as the hair of his students grew longer and shaggier on both sexes during the late 1960s. T-shirts replaced collegiate sweaters and sandals replaced penny loafers. The undergrads in his Introduction to Astronomy class and classes across campus cut lecture on the days of peace rallies at the Pentacrest. Van Allen didn't sympathize with the anti-American sentiment on campus and refused to write letters that might excuse students from the draft. But the campus unrest had little impact on physics and astronomy or on his activities until Monday, May 4, 1970, when the National Guard opened fire on a demonstration at Kent State University in Ohio and killed four students. The smoldering violence erupted on campuses across the country including the University of Iowa.

On the morning following the shootings, four gruesome figures covered in bandages and splattered in red paint lay on both sides of the front steps at 508 River Street, the home of University of Iowa President Sandy Boyd. He, his wife Susan Boyd, and their three children awoke to the sight. "Death joined the group. I saw him park his Volkswagen a block down on a side street and then don his black robe and mask. He carried a flag. He returned the next morning and then the next," former newspaper reporter Susan Boyd wrote in her diary account summarized here and published in her book *The Wide-Brimmed Hat*.

Worse was to come. Queues of demonstrators made vigils to the River Street home as students vandalized the venerated Old Capital, commandeered the Pentacrest for nonstop demonstrations, and demanded a student strike to close the university. The Boyds stayed with neighbors and then, on Saturday, May 9, they settled in with the Van Allens. Abbie welcomed them with a cheery hubbub of making up temporary beds and a big, friendly dinner. Susan was her best friend and kindred spirit, a like-minded promoter of education and of independent lives for women, especially women with workaholic husbands. But the work the two men pursued that May took on a surreal quality.

Due to escalating demonstrations and threats, Boyd called for twenty-four-hour fire watches at every building on campus. Van Allen barely left the physics building. In his journal, he summarized the "massive window breaking fray through the downtown business district, obstruction of ROTC meeting, burning of the old armory temporary building" and another small fire. The space program and its links to the military made the physics buildings a possible target. He walked his own halls that first weekend of the fire watch until he could recruit volunteers. He arrived home after midnight on the Saturday the Boyds came to stay.

"From the time of our arrival, Abbie always answered the phone and screened callers. We went to bed about 1:30 A.M., exhausted. My eyes burned almost constantly, awake or asleep," Susan Boyd wrote.

A week into the confrontation, Sandy Boyd, like presidents of campuses across the country, took measures to clear the campus. He offered students the opportunity to take grades as earned up to that time. They could also take a simple pass grade in any course they were passing or an incomplete. The Pentacrest emptied and the campus quieted until June, when proud graduates returned with their families for commencement. "There were various peace signs and some sloppy footwear, but the mood generally was one of pride," Susan Boyd wrote. She summed up her husband's graduation speech: "It was so short there was no time for demonstrating."

Meanwhile, construction on the physics building—halted that spring by labor strikes rather than student unrest—resumed, with Van Allen pressing for completion by the start of school in September. He and his team also completed the *Pioneer* design model. Roger Randall, chief engineer for the project, took charge of designing and assembling a particle detector system durable enough for a 640-day journey to Jupiter but sensitive enough to measure radiation levels expected to be far stronger than those at earth. They filtered and assembled a series of Geiger tubes to measure for various energy ranges. The whole experiment required less power than any other experiments on the craft—and that ensured it life support for thirty-one years.

Van Allen planned to deliver the design model himself for the August 12, 1970, verification model deadline, but he didn't go. On August 11, his mother died at Mercy Hospital in Iowa City following a brief illness. She was eighty-six years old. Seven months later, his younger brother William died of heart failure at only forty-nine years old. William's longtime employer, Hughes Aircraft Company, had sent him to London to represent the company in NATO countries. He was vacationing in the Canary Islands with his wife, Kay, when he died.

"His health has been a matter of considerable concern for about a year," Van Allen wrote. To his consternation, Kay planned to bury her husband on the islands rather than shipping his remains home. But after a discussion with Maurice, the brothers decided they should accept her judgment "under the circumstances."

Van Allen put the tragedies behind him as his team completed, calibrated, tested, and shipped the flight units for *Pioneers 10* and *11*. The *Pioneer 10* flight unit "came home" again briefly in fall for final testing. Van Allen used the most intense radiation source on campus—the 600-curie cobalt source used for cancer treatment at the university hospital—to make final calibrations before shipping the instrument back to Ames in December. "As soon as they got the cancer patients out of there, we would step up with our instruments for the evening," Bruce Randall said.

• • •

The *Pioneer* mission held special pride for Van Allen as the March 1972 launch approached. Not only did Iowa have an instrument on board, but several physicists he had mentored as space scientists also had experiments, including Frank McDonald at Goddard and Carl McIlwain and R. Walker Fillius, both at the University of California at San Diego.

Roger Randall went to the cape for the *Pioneer 10* launch. Van Allen went to Ames. At opposite ends of the country, they sweated out three days of launch delays. *Pioneer 10* finally rose into the evening sky on March 2. Van Allen watched the counts as the instruments switched on at Ames. "Looks great!" he wrote in his journal on March 10. Within eleven hours, the mission crossed the orbit of the moon, reaching there three times faster than *Apollo* had.

On May 24, *Pioneer* crossed the orbit of Mars and then stepped into uncharted territory and the ominous hazards of the asteroid belt. Mini planets such as Ceres and Vesta occupied the belt. *Pioneer* could avoid them but the countless asteroid pebbles and particles—as small as grains of sand—could pelt or pierce the spacecraft and its instruments with an impact "at a speed 15 times higher than a rifle bullet," Van Allen and his colleagues wrote. Scientists expected a storm of static transmissions from the collisions but *Pioneer* hummed through the asteroid belt registering barely a touch. Instruments detected no concentration of the tiny, fearsome particles. But on August 2 Van Allen and the other scientists gathered direct readings from *Pioneer* that documented the greatest solar storm ever recorded. It took another fifteen months for the spacecraft to reach Jupiter. Thomsen, Daniel Baker, and Davis Sentman all headed to Ames with Van Allen in late November 1973 for the historic encounter choreographed by thousands of commands transmitted to *Pioneer 10* from Ames, long-distance calls that took forty minutes, one-way.

"Since the *Pioneer 10* spacecraft had nothing more than a very limited data storage capacity, practically everything it had to tell Earth about Jupiter during its swift fly-by had to be transmitted in real time, as quickly as possible," noted Mark Wolverton in his book *The Depths of Space*. "There would be no second chance at any observations; whatever *Pioneer* missed seeing the first time around, it would never see again."

Lozier provided the critical computer support for the PIs, the "principal investigators," with instruments aboard *Pioneer 10*.

"Charlie Hall says to me: 'Hey these PIs are coming here and they have computer programs they want to put on the mainframe so they can process their data in real time,'" Lozier said. Van Allen handed him Iowa's "deck of cards," hand-punched computer cards that Lozier loaded on the IBM 7094 at Ames. The mainframe was in the computer buildings, separate from mission control and the science area. The scientists used remote terminals linking to the main system that would deliver their raw data. With all sorts of bugs to be worked out of this interface and a steep learning curve for the scientists in processing data on an unfamiliar system, Lozier had his hands full.

"Some of the PIs—my God you couldn't do anything to please them," he said. "I sat down day after day with Dr. Van Allen processing his data and running his program because he wasn't familiar with the operating system. His stature was already established but, as a person, that's when I realized this man really treats people right."

The computer printouts gave raw results, but didn't offer analysis in response to the rapidly unfolding drama of the magnetosphere at Jupiter. Analysis still had to be done back in Iowa in an era before the Internet allowed the digital transport of computer files. "Some things would work and some wouldn't, so it was a matter of getting all the data read correctly," said Bruce Randall, who stayed in Iowa City. Thomsen and Van Allen hand-graphed the raw data nonstop until others took over the shift. Then Van Allen shipped everything to Randall to run through Iowa's analysis programs.

"We'd get the data tape from the Cedar Rapids airport in the afternoon and run it about midnight and go through the night. I'd generate plots and printouts—we had until about five or six in the morning to work," Randall said. Then Randall loaded everything in a briefcase so a student could taxi it back for the morning flight from Cedar Rapids back to California. Randall's three kids were just getting up when he'd get home to catch a nap and then head back to the physics building to lead a discussion section for a class, filling in for the team in California.

The drama at Jupiter began to unfold November 26 when *Pioneer 10* and Iowa's detectors suddenly and unexpectedly slammed into Jupiter's churning magnetic field more than 4 million miles above the planet. Caught by surprise, Van Allen was still at the university while the Iowa expeditionary force of Sentman and Baker lounged on the California coast drinking Irish coffees. They stopped at Ames later that day and found the place in an uproar over the new discovery, millions of miles beyond where anyone expected to find it. The two graduate students stayed up all night "plotting points and getting our story in order for Dr. Van Allen," Baker recalled. Van Allen arrived for the Jupiter encounter on November 29 with Thomsen and another graduate student and the entire Iowa team worked feverishly to map Jupiter's huge magnetic field. "It was the first time we were really seeing inside the magnetosphere [of Jupiter]. We were getting these fifteen minute points of data. It was kind of like a slow motion picture filling in—this was such an enormous place," Thomsen recalls. "So you'd kind of wander around the science area and see what other people were doing. I remember being exhausted but it was really exciting."

But excitement gave way to an imminent sense of doomsday. As *Pioneer* moved in closer, the instruments flared with radiation readings as intense as levels following the detonation of a large nuclear bomb. "Everything on the spacecraft was being fried," noted Robert Kraemer, NASA's deputy for Planetary Programs. "TRW engineers said flatly that *Pioneer 10* was about to go off the air."

"The radiation got real intense, ten thousand-fold more intense than at the earth, so intense that we were quite concerned about the survival of the instrument. And then it started to finally go back down again and it was still working," Van Allen said. He knew from the pattern of readings that he had located Jupiter's radiation belt, a momentous discovery confirming astronomical expectations. At first, lots of scientists held their findings "close to the chest because they wanted to be first. But they couldn't each stand alone without the others. Van Allen—he got it. 'Now if you look at this like this, what does your instrument say we have?' He was always willing to share," Lozier says.

Hall built the process of sharing data in his morning meetings during the course of the fly-by. "We would get there at about eight o'clock and stagger in and all the PIs would walk into Hall's 9 A.M. meeting from their various little cubicles—walk in and give five to ten-minute presentations on what they found the previous day or even that morning," Van Allen said. They kept their reports short and sweet since Hall kept everyone standing for the brief exchange.

A 10 A.M. press conference followed each day. Van Allen's detectors and those of Fillius, McIlwain, and John Simpson of the University of Chicago began to sketch the immense contours of a radiation belt of unimagined force. The JPL magnetometer backed them up with overall magnetic field readings.

"I think we've got a different ball game here than we do on Earth," Van Allen told the press on November 30.

"The Van Allen radiation belts on earth reach out like a tear-shaped doughnut. The Jovian belts of trapped [charged] particles, Dr. Van Allen said, seem to shoot out like a long narrow tongue," reported John Noble Wilford of the

New York Times. The spin of the planet seemed to rock the radiation belts up and down. Van Allen watched the radiation counts from the Geiger tubes rise and fall every ten hours, Jupiter's equivalent of a twenty-four-hour day on Earth. "I immediately interpreted that as being due to the tilt of the planet and I was the first to identify that, actually," he said. The tilt changed the angle of the belts as the planet rotated on its axis. The tilt and the tongue-like shape of the radiation belts—more of a disc than a belt—saved *Pioneer 10*. It veered out of the belt—and the staggering radiation intensity—as the planet rotated on its axis, a key finding of the mission even before the actual fly-by.

The fly-by was fast approaching as the Iowa team worked round the clock in the small cubicles, snatching a couple of hours sleep here, catching irregular meals there, and stripping off the Iowa data from the streams of printouts. Couriers brought by printouts every quarter hour.

On December 3 at 6:25 P.M. Pacific Time (PT), *Pioneer 10* flew within 82,000 miles of Jupiter, the closest point, approaching the planet near the equator and sending back extraordinary photographs of a turbulent gas-ball of a planet covered in multicolored clouds of ammonia. *Pioneer 10* finally confirmed an explanation for the signature horizontal bands visible across the face of Jupiter and noted by astronomers for centuries. Here was a planet ten times the diameter of Earth, yet spinning through ten-hour days. The rapid spin swirls heat rising from the planet into the bands of hot air alternating with bands of cold, the scientists explained at a press conference. The spin also generated the massive electrical currents that accounted for the intensity of the magnetic fields. And what about Jupiter's giant red spot, glaring from the Southern Hemisphere like a wounded eye? Even *Pioneer 10* didn't have an explanation for that.

The scientists sat behind a table with name cards in front of them for a press conference in one of the old blimp hangars the morning after the fly-by.

"They looked really tired. They weren't spit-polished but they were wearing sports coats," recalls *Chicago Tribune* editorial writer Pat Widder, a young reporter for the *Pleasanton Times* who covered the press conference at the time of the encounter. "There was this sense of reaching for the stars, a sense that all things were possible."

An elated James Fletcher, Webb's successor as head of NASA, came to Ames and welcomed the press. "This is more than we ever dreamed of," he said simply. Ames scientist John Wolfe laid it on the line: "We're still alive."

"Charlie Hall got up with a lightbulb and told the press the amount of energy we have to run the spacecraft and all of the experiments would operate this lightbulb. That impressed people in 1973 with the energy crisis," recalled Ed Smith, the principal investigator for JPL's *Pioneer 10* team.

Reporters asked detailed questions of the scientists because, by this time, the space program had spawned a whole new breed of writer. "That's the first time I met what you would call space reporters—that was a new category back then—and they were asking very technical stuff," Widder said.

Van Allen wanted to help reporters visualize the magnificent radiation belts and called the shop in Iowa. "Could you make me a model of Jupiter and send me a bunch of flexible wires?" He told them where to drill the holes for him to insert the wires. Van Allen received the model on the morning flight out of Cedar Rapids. It was painted to look like Jupiter and even motorized to rotate. Exhausted but exhilarated, Van Allen worked through the night after his shift in the science area and fashioned the strips of wire into a crosshatch of Jupiter's radiation belts. He set his motorized radiation belts spinning to the glee of reporters at a subsequent press conference.

Van Allen and the Iowa team back home simulated the intensities of the radiation readings from *Pioneer 10* at Jupiter by testing their spare instrument with streams of particles generated by the old Van de Graaff accelerator, now reassembled in the basement of the new building. They also used the more powerful vertical accelerator in the tower and the cobalt cancer treatment facility at the hospital. The modeling allowed Randall to estimate the tilt of the planet at 9.5 degrees, although instruments flown by JPL and Goddard estimated the tilt anywhere between 7 and 15 degrees. Randall had to wait for the *Pioneer 11* fly-by at closer range to confirm the simulation and prove that the 9.5-degree tilt, estimated by the simulation, was correct.

Pioneer 11, launched April 5, 1973, carried the same suite of instruments as *Pioneer 10,* plus one additional instrument to measure the high magnetic field strengths. It offered scientists the opportunity to confirm and expand on their initial discoveries at Jupiter and resolve a few controversies such as the degree of tilt of the planet.

Pioneer 11 dropped within 27,000 miles of the swirling, turbulent ammonia clouds of Jupiter, close enough for the scientists to see the polar caps of the moon Callista—so close they could make more precise observations of the circulating bands caused by the immense heat radiating from the planet.

At one point, *Pioneer 11* changed orientation, as though it had developed a mind of its own, and scientists realized an electron storm had actually mimicked a command from Ames. They righted the probe but, at its close proximity to the planet, the real challenge was to free it. As *Pioneer 11* accelerated, the mighty arm of Jupiter's gravity acted as a kind of slingshot to propel the spacecraft on its way at 108,000 miles per hour, fifty-five times "the muzzle velocity of a high-speed rifle bullet," according to Van Allen and his coauthors of the 1980 book on the *Pioneers*. *Pioneer 11* proved that the "slingshot technique" of gravitational assist could boost the speed of a craft on a long-distance space mission. The spacecraft streaked toward Saturn, but scientists would have to wait nearly five years for it to get there.

Pioneer 11 navigated the billion-mile journey to Saturn and then, on August 31, 1979, crossed the bow shock, the turbulent boundary where the solar wind hit Saturn's magnetic field on August 31, 1979. Distance cloaked Saturn from even the most powerful radio telescopes and now *Pioneer* removed the veil and made a string of momentous discoveries over the next ten days. Van Allen and four other scientific teams discovered the magnetic field and radiation belts of Saturn, the most fundamental discovery of the mission. The belts had contours more closely resembling those at the earth than at Jupiter. Saturn's moons attracted particles from the belts and the signature absorption pattern led to the unexpected discovery of another moon. Saturn's rings also pulled in particles, nearly freeing the environment of radiation close to the planet, and the absorption pattern led to the discovery of another ring at Saturn—F Ring. The pass came within 13,000 miles of the visible rings and then the "slingshot technique" of gravitational assist propelled *Pioneer 11* beyond Saturn.

"This was one of the most exciting times of my life," said Van Allen, recalling the luxurious and continuing streams of data compared to the stressful enigmas created by the spotty transmissions from *Explorer I*.

New data from both probes continued to pour in, painting in the picture of the dynamic increase in cosmic rays as the solar wind slowly subsided with distance from the sun. Most people thought of Pluto—3.5 billion miles from earth—as standing sentinel at the boundary of the solar system. But *Pioneer 10* flew past the orbit of Pluto more than a decade after the Jupiter encounter and showed that the solar wind blows billions of miles beyond it. The *Pioneers* and the *Voyagers* that followed pushed the front gate of the solar system billions of miles farther out into space.

Hall later estimated that fifty students owed PhDs to *Pioneers 10* and *11* at a pay scale that probably amounted to "less than $1 an hour." Sentman, Baker, and Thomsen were among the *Pioneer* PhDs from Iowa. Thomsen developed a model to explain the dramatic regeneration of the radiation belts as particles constantly leaked from them. As he had from the start, Van Allen relied on his best graduate students as trusted collaborators. He ironed out one conflict in his schedule by sending Thomsen in his place to report on the *Pioneer 10* findings at Jupiter for a solar wind conference.

"I had never even been to a meeting like this before and here I was on stage with John Simpson and Frank McDonald and all these really big names. And we were having this debate because our [radiation] results were somewhat different from theirs—theirs were higher. It was quite an experience—these two grand old men of the field and little me," Thomsen said. But Thomsen made a convincing case for the lower readings, especially when she showed the audience how the Iowa team had subtracted out the effect of the radiation coming from the spacecraft's own thermonuclear generators.

As *Pioneer 11* sped to Saturn, Iowa space scientists designed and constructed another satellite—*Hawkeye 1.* The name no longer conflicted with the now obsolete Hawk missile. The satellite, also known as *Explorer 52*, launched on June 3, 1974, and explored the high latitude polar regions for the next four years. It carried one of Gurnett's VLF receivers, one of Frank's low-energy electron and proton analyzers, and one of Van Allen's magnetometers. The mission assessed new features of radiation and the solar wind in the auroral zones and surveyed the earth's magnetic field and the boundary of the magnetic field in the polar areas. Gurnett extended his observations of the earth as a radio source and of solar radio bursts during the mission.

By the time *Hawkeye 1* launched, NASA's budget had dropped steadily for nine years—plunging from a high of $5.25 billion in 1965 to $3 billion in 1974. That meant cutbacks for many university space science programs. But the $3.6 million *Hawkeye* program had dual support from NASA and the Naval Research Laboratory. The construction of the *Hawkeye* satellite along with instruments for other *Explorers* and *Voyager 1* brought the Iowa space program to a zenith in the mid-1970s, with about one hundred full-time employees.

Though the *Pioneers* continued to pour in data and Gurnett, Frank, and others participated in a continuing pipeline of space missions, *Hawkeye* marked Van Allen's final launch of a new mission. For the *Voyager* missions, he chaired the committee that developed the scientific goals scaled down from the long, painful saga of the Grand Tour. Van Allen proposed his original Grand Tour experiments in 1965. But with Richard Nixon's election as president in 1968, Congress quickly reined in the space program, and the precarious position of the Grand Tour became dismally evident in 1970. Van Allen's term of service as a charter member of the Space Science Board "has now been terminated as part of the reorganization of the Board," he wrote in his journal in July 1970. At the same time, NASA placed the Lunar and Planetary Missions Board on a "stand-by basis," shutting down a key forum for the "pesky" scientists who pushed for planetary research, he added. After that, Van Allen determined he would partner with another research team rather than lead a mission. "I decided last fall not to spend my time on the politics required to develop such a group myself," he wrote in his journal on February 14, 1971.

But, with NASA's priority set on the space shuttle, Congress scrubbed the Grand Tour early in 1971. Plans for two missions to Jupiter, Saturn, and Pluto in 1977 and two to Jupiter, Uranus, and Neptune in 1979 were scaled back to a total of two missions—the twin *Voyagers 1* and *2* spacecraft would be sent to fly by as many planets as possible.

Essentially the same scientists flew instruments on both *Pioneers 10* and *11*, and NASA gave the *Voyager* seats to several new investigators, including Gurnett and Krimigis, now a physicist with the space program that had been reestablished at the Applied Physics Laboratory at Johns Hopkins. McDonald, one of the few scientists to participate in both the *Pioneer* and *Voyager* missions, paired with Ed Stone of Caltech (later the director of JPL) on a cosmic ray system. The *Voyagers* launched within weeks of each other in August and September of 1977 and reached Jupiter a year later.

For years in the foyer of Van Allen Hall, blue push pins for *Pioneers 10* and *11* and green ones for *Voyagers 1* and *2* mapped the travels of the space probes. NASA lost contact with *Pioneer 11* on September 30, 1995. *Pioneer 10* sailed on, falling behind the speedier *Voyager 1* in terms of distance from the sun. But it remained in the race to discover the termination shock, the inner boundary of the solar system where the solar wind and incoming cosmic rays crash and begin to mix. Van Allen's cosmic ray detectors on board the *Pioneer* and several instruments, including Don Gurnett's VLF radio receiver on board *Voyager 1*, could detect the boundary, predicted to be located more than 8 billion miles from earth. Gurnett's open door is practically visible from Van Allen's longtime office just around the corner, though their instruments on *Voyager 1* and *Pioneer 10* headed toward opposite ends of the solar system, billions of miles apart in 2003 when the intrepid *Pioneer* fell silent.

The fact that *Pioneer 10* had survived so long represented something of a technological and political miracle. No sooner had *Pioneer 11* reached the mission objective of Saturn when Hall and the *Pioneer* scientists shouldered the battle to convince NASA to keep *Pioneers 10* and *11* operating.

"'You finished the job—what's the problem with turning this thing off?' That was NASA's point of view," Van Allen said. "Our point of view was simple: You can't spend a million dollars a year to keep this thing going when it's the farthest thing out in space? A million dollars a year, but it was a continuing fight on the part of the investigators. I'm amused to say it was a bitter fight in 1979 and was still a fight over 20 years later."

To conserve power as the *Pioneers* probed farther into space, Charlie Hall made the tough calls to turn off instrument after instrument over those years. A few instruments, including Van Allen's, continued to operate on *Pioneer 10*, and Hall parlayed them into justification to keep *Pioneer* operating. NASA insisted on retiring the mission in 1997 as the escalating costs of the International Space Station created another budget crunch. But even as he celebrated the twenty-fifth anniversary milestone at a party at Ames, Van Allen "worked the hallways" of NASA to give the mission a second life.

A dedicated corps of scientists including Lozier and Larry Lasher, the project manager who succeeded Charlie Hall, volunteered time to maintain the mission at Ames. Lasher garnered funding for the mission to test a chaos theory study involving weak transmission signals. Scientists at the Deep Space Network insisted it needed to stay in contact with *Pioneer 10* for projects such as the advanced concept studies of communication technology. Space fans across the globe flocked to the *Pioneer* Web site. "We got 20,000 hits a day," said Lasher.

17 Space Politics

Jet Propulsion Laboratory, Pasadena, California—January 24, 1986. Van Allen joined colleagues at a Planetary Society banquet in Hollywood before heading to Pasadena on Thursday, January 23, for the big day at JPL. Friday brought *Voyager 2* within 50,000 miles of Uranus.

Scientists had waited eight years and 2 billion miles for this day. Project scientists Norm Ness of the Goddard Space Flight Center invited Van Allen to the encounter as a guest of his magnetometer team. "I was thereby admitted to the investigators' 'inner sanctum' and was able to circulate fully in the working areas, attend science group meetings, etc.," Van Allen wrote in his journal. Friday morning, just a minute off cue, *Voyager 2* made its closest pass at Uranus, delivering mesmerizing images of a hazy, windblown planet. Ness's magnetometer discovered the planet's magnetic field and other discoveries included ten new moons and new rings. "An immensely exciting day," Van Allen wrote of the new mapping of Uranus. Now it was time for *Pioneer 11* and the Ames Research Center to get into the act. *Pioneer 11* also happened to be in the neighborhood of Uranus, reaching there at just about the same time as the *Voyager*. Van Allen flew to San Francisco, Saturday, and Bruce Randall picked him up in a rented car for the drive to Ames. Van Allen, Charlie Hall, and the other investigators lobbied hard for a period of unbroken transmissions from *Pioneer* during the *Voyager*'s Uranus encounter. Here was the unparalleled opportunity to obtain dual mission coverage. Their request involved some delicate finessing, however. The *Voyagers* had long since bumped the *Pioneer* missions from top priority at the Deep Space Network, giving the *Pioneer* teams only periodic data streams. But NASA finally agreed that twenty-four-hour coverage of *Pioneer 11* from January 26–31 could back up *Voyager*'s finds. By Monday night, January 27,

Van Allen, Randall, Walker Fillius, and *Pioneer* mission manager Richard Fimmel reviewed some early findings over dinner at Fimmel's home.

Randall met Van Allen for an early breakfast on Tuesday and they drove to Ames, anticipating the bustling arrival of other *Pioneer* investigators. Instead, hushed corridors greeted the pair beyond the front entrance. The space shuttle *Challenger* had just exploded, Van Allen learned. All seven astronauts were dead. Glued to NASA's broadcasts and to the television news, Van Allen watched colleagues give interviews on the disaster. As a key opponent of human spaceflight, calls flooded in for him in Iowa but he steered clear of any interviews at first. "I didn't want to be quoted as a critic during this tragedy. I didn't want to in any way to use this disaster to justify my position," he said.

Not that anyone needed much clarification on his position, given his widely read eight-page article on the proposed International Space Station published in *Scientific American* that month. "With no difficulty I can think of a billion dollar space mission before breakfast any day of the week and a multibillion dollar mission on Sunday. Ordinarily, I do not inflict such visions on my fellow citizens" as the space station promised to do, Van Allen wrote. But tragedy, like triumph, only made it more important for human space programs to proceed.

• • •

Computer technician Bob Brechwald unlocked a door in 2004 to a bunker in a subbasement of the Van Allen physic buildings, a room with reinforced concrete walls and a door heavily insulated against radiation leaks. But the need for such protection passed long ago with the retirement of the old Van de Graaff accelerator. Now the room housed space artifacts including Iowa's *Plasma Diagnostic Package,* a satellite launched from the space shuttle *Challenger* in 1985 to analyze the solar wind and then plucked back for the journey home by one of the shuttle's robotic arms. "Touch it," urged Brechwald. "It's been in space."

The romanticism of objects drawn from space and the vision of inevitable next steps in human space exploration conflicted sharply with the $10,000 a pound price tag (in current dollars) of launching anything into space. "It is only the presumed capability of the shuttle for recovering expensive spacecraft from orbit that gives it any hope of economic advantage," Van Allen told Congress during hearings on the NASA budget in 1971, already in a bitter debate that divided scientists and politicians alike.

The *Apollo* moon landings set the agenda at NASA for a "natural progression" of potential space milestones: a space laboratory, reusable space shuttles to provide transportation to a space station orbiting the earth and an astronaut expedition to Mars. Von Braun already was promoting the space shuttle in the works when his crowning achievement, the Saturn V, took the *Apollo* spacecraft to the moon. But for many in Congress and for the public, the moon landing was the final victory before life—and NASA funding—went back to normal.

Von Braun transferred to NASA headquarters in Washington in 1970 to head space planning. There, he promoted the shuttle as a space ferry for people and equipment, a launching pad for interplanetary probes and even a tanker to refuel space vehicles on long-term flights. The move seemed to signal a push for more aggressive space programming. But, in Huntsville, where aerospace jobs had been declining since the 1965 peak, kicking him upstairs appeared ominous. The suspicions proved well founded. Von Braun watched helplessly as rocket scientists and engineers, who had been with him since World War II, lost their jobs in Huntsville. Feeling marginalized and isolated in his new position, he retired from NASA in 1972 to join Fairchild Industries where company president Edward Uhl wanted to expand the aerospace firm into satellite programs.

Van Allen and other university space scientists felt both the budget cutbacks and increased competition from within NASA's own space centers. By 1976, full-time personnel in the space program at Iowa numbered sixty-three, with thirty part-time employees, compared to nearly a hundred full-time faculty and staff just two years earlier.

Van Allen already economized wherever he could on hotels, meals, and transportation, even taking the Greyhound bus home from meetings in Chicago. Economics professor Peter Alonzi recalled sitting down next to a nice man with a friendly smile on a 1971 bus ride from Chicago to the University of Iowa where he was in graduate school. Alonzi introduced himself and the man shook his hand and said, "Hi, I'm James Van Allen."

"Two questions burst in my mind. What are those radiation belts? And why is the world-renowned James Van Allen taking the Greyhound bus to Iowa City?" Alonzi said. "As the bus rolled out of the Chicago terminal, I knew Iowa City was no longer the destination. I was on a journey to the stars."

A four-hour conversation on astrophysics and space missions followed and Alonzi, who teaches at Dominican University just west of Chicago, finally got around to asking about the Greyhound bus.

"If I wasn't on a bus back to Iowa then I would miss the opportunity to soar with a wonderful Iowa student like you," Van Allen said.

This warmhearted manner combined with Van Allen's scientist stature made him the perfect ambassador for Iowa-based Amana Refrigeration. When the Consumer's Union warned of radiation hazards from microwave ovens,

Amana gathered leading scientists, including Van Allen, to launch a counterattack on March 26, 1973. More than one hundred reporters gathered at the company's plant, right across the street from the home of founder and president George Foerstner in the tiny village of Middle Amana. Experts from Bell Labs, the federal Bureau of Radiological Health, and Underwriters Laboratories backed the safety of Amana's Radarange and other microwave ovens. Then Van Allen stole the show.

"I am personally prepared to sit on top of my Amana Radarange for a solid year while it is in full operation, with no apprehension as to my safety," Van Allen said. "As I stated a few years ago, in my judgment its hazard is about the same as the likelihood of getting a skin tan from moonlight." Newspapers buzzed his comments across the country.

"George Foerstner was smiling from ear to ear when he heard that," recalled Alex Meyer, retired vice president of Amana and the late Foerstner's son-in-law. But Van Allen's equally outspoken opinions on the follies of NASA's human space program often found deaf ears or vocal dismay.

• • •

Space scientists worked as a close-knit fraternal group through the early *Explorer, Pioneer*, and *Mariner* missions, but *Apollo* escalated the space program and focused the program on a dramatic national effort. As the field got bigger, and as the *Apollo* program took on its grandiose scale, Van Allen escalated his battle on behalf of small satellites and probes with economical instrumentation and a long life in space. While the *Apollo* program took NASA into the stunning business of creating the behemoth Saturn V, Van Allen's philosophy helped guide the robotic program toward compact *Mariner, Pioneer*, and *Voyager* spacecraft that pierced across the solar system.

In this financial juggernaut, plans for the $8 billion space shuttle program went forward and space scientists realized that budget cuts would once again sacrifice planetary missions instead of rewarding them for their low cost. Anything but the shuttle could fall to the hit list. Even the destruction of *Skylab*, America's apartment-size space station meant as the destination of the shuttle, didn't stop plans to build it. *Skylab*, compromised by overheating from the start and finally crippled by solar flares, plunged back into the atmosphere on February 8, 1974, after a mere 171 days in orbit. Three crews had occupied the $2.5 billion space laboratory. The shuttle was promoted now as a cost-effective reusable launch vehicle for civilian, military, and research satellites and space probes. The shuttle seemed to promise that the one-time-use rockets for these launches soon would go the way of the horse and buggy.

Van Allen assessed the glaring irony in the new math, a catch-22 in which the costs of the shuttle would leave but a trickle of funds for the satellites and probes that now justified its existence. As an ambassador of space science, Van Allen stepped into the tempest, a witness at congressional hearings where he tried to stem the tide of cuts for robotic space exploration. He promoted the message that Congress could reap a rapidly accelerated space exploration program on a shoestring budget with economical satellites and probes and still have plenty of money left over for the conventional launch of communications satellites.

Van Allen had watched with pride and admiration as Neil Armstrong stepped onto the moon. He had no quarrel with the "collective adventure" of sending humans into space in world of unlimited resources. But in the world of congressional budget hearings, even New York Mayor John Lindsay chided Congress for sending money into space instead of into urban slums. In this world, Van Allen carried a pragmatic, oft-repeated case for the cost benefits and scientific successes of robotic space science.

"The junior partner in space—the unmanned/automated/commandable spacecraft—has provided at least 80 percent of the new scientific results in space and almost all of the utilitarian applications, both civil and military," Van Allen advised the Congressional Committee on Aeronautical and Space Sciences in April 1973. He pointed to examples in telecommunications, weather surveys, and the system of flight navigation satellites serving both civilian and military air travel. Other scientists offered arguments similar to Van Allen's but Congress greeted them to some degree as narrow specialists unable to grasp national policy with a Cold War defense agenda. The scientists grasped it all too well, however, predicting correctly that military operations would preoccupy the shuttle. The science community flinched at the idea that the Defense Department would get economy seats on the shuttle to launch its satellites, essentially subsidizing the military with the pinched funds earmarked for the civilian space program.

"The military use of the shuttle is going to be dominant, while civilian uses will be minor," Van Allen later told the *New York Times* in 1981. "NASA is going to be trampled to death by the Defense Department on shuttle use so why not be honest about it and call it a military program?"

The Soviet Union called for a halt to the shuttle as the military emphasis became all the more apparent and NASA set a 30 percent limit on military use of the flights. The military ultimately claimed about half of the flights at a 32 percent discount compared to commercial users in the early years, however. But the military benefits and the idea of a recyclable space vehicle—plus an ailing aerospace industry as the 1972 election year approached—gave the shuttle the political support it needed. "*Apollo* was a matter of going to the moon and building whatever technology could get us there; the Space Shuttle was a matter of building a technology and going wherever it could take us," wrote McDougall.

In justifying the shuttle, Congress soon accepted and developed the larger concept of a permanent human presence in space. As a direct result, the International Space Station was proposed as a way of sharing the cost of space exploration internationally and testing long-term endurance in space, a prerequisite for any mission beyond the moon. The space station, far more costly than the shuttle, justified the expensive investment in the shuttle and rescued national prestige in the latest round of catch up. Once again the Soviet Union was leading the space race with the pioneering Mir (meaning "peace") space station launched into orbit in 1985.

• • •

Van Allen laced a 1983 address at an American Astronomical Society banquet with homespun anecdotes that became mainstays for his end of the continuing debate about the shuttle and the space station.

He told the story of balloon pioneer Ed Ney opening a lecture up for public discussion. "A woman in the front row stood up to ask a question. 'Professor Ney, please tell me: Is there anything that a man can do in a balloon gondola that an instrument cannot do?' Ed's answer, after only a moment's hesitation, was: 'Yes, madam. Yes, there is. But why would anyone wish to do it at such a high altitude?'"

Van Allen's 1986 *Scientific American* article emphasized cost overruns, the delays, and the "slaughter of innocents" mentality that kept the shuttle running at the expense of space science and automated space missions. Van Allen predicted more of the same for the space station, expected to cost $8 billion in 1984 dollars (an estimate raised to $20 billion when construction began in 1988 and to about $130 billion currently).

In his *Scientific American* article, Van Allen reviewed the record of the shuttle, which was expected to supplant all other launch vehicles with fifty shuttle flights per year. "Each flight would deliver 50,000 pounds into low earth orbit at a cost of $100 per pound of the 50 annual flights at least four would carry spacecraft for the exploration of other planets," he wrote. "There is a striking disparity between those claims and the present situation. In 1985 only 10 shuttle flights were carried out at a true launching cost of about $5,000 per pound."

The cost today is about $10,000 per pound. For Van Allen, as he noted in lectures and articles in the spring of 1986, the *Challenger* tragedy was all the more distressing because the key purpose of the mission was to deliver a communications satellite into orbit. "This function has been performed successfully and much less expensively for many years with minimal risk to human life," he wrote in *Science* magazine. The military—temporarily grounded after the tragedy since the shuttle had been launching all its satellites—essentially agreed with him. The military stand secured renewed support for rocket launches of satellites and probes.

But Van Allen's stand became a lightning rod for supporters of human spaceflight both inside and outside of NASA. Author Ben Bova of the National Space Society pointed out that driving the NASA budget lower with reductions in human space endeavors would do nothing to ensure increased funding for space science. Van Allen understood the drive of the human mind to explore the world through science, noted James R. Arnold, of the California Space Institute at the University of California at San Diego. "Why does he reject so sharply, in this context, the drives to travel, to explore, to build and develop, and to create wealth, which are just as basic to our nature and which have built the resources that support science, music and all other cultural endeavors?" Arnold questioned in *Science* magazine in August 1986.

Van Allen argued that all those endeavors could be done and had been done with robotic missions. Robotic missions had made initial explorations of every planet in the solar system except Pluto. As for wealth and commercialization of space, Van Allen pointed to automated telecommunications satellites as the most obvious and profitable application of space technology thus far. "My own view is that our national predicament is the result of the clash between the mythology of manned space flight and the real achievement of space technology in practical applications to human welfare and the expansion of human knowledge," he wrote in *Science*.

Van Allen went before Congress in 1987 to challenge Reagan's claims that "a space station will permit the quantum leaps on our research in science, communications and in metals and life-saving medicines which can be manufactured only in space." He quoted a massive two-year National Academy of Sciences study that he had helped prepare. It blueprinted space science from 1995–2015 and concluded that "apart from biomedical research on human subjects, this study has again found few first order scientific objectives that require human crews in space, much less a permanently manned space station."

But Van Allen found his position unpopular and openly unwelcome for the first time when he left Ames to attend the Space Science Board (SSB) planning retreat in January 1989 at the futuristic Beckman Center on the University of California's Irvine campus. Such retreats became an outgrowth of the Iowa summer study session in 1962 and the goal of this session was to hammer out an updated long-range plan for 1995–2015. Though Van Allen was no longer on the board, he was on the steering committee that brainstormed to map out the future of lunar and planetary research. The retreat and the planning had taken an unexpected turn, however, as the focus shifted to microgravity research, the insider phrase for research on how long-term residency at the space station would affect the human body. The SSB took the field of microgravity under its wing for the first time that year, a signal that human spaceflight had become the primary issue for upcoming board strategy sessions. Amid the boosterism for a new round of astronaut missions to the moon and Mars during the first George Bush administration, Van Allen resolutely approached the podium on Wednesday, January 18.

The SSB had written some excellent strategy documents to guide national science programs well into the twentieth century, Van Allen started out. But then he stated that the board had failed to prevent any of NASA's major policy disasters of the last twenty years. He listed among the disasters almost the entire philosophical foundation of NASA: the *Apollo* mind-set; the "big project" emphasis; the shuttle development; tying all satellite and probe launches exclusively to the shuttle; and unsubstantiated expectations about the commercial opportunities of space.

Among the current and upcoming disasters, he listed the shuttle, the space station, and the goals of human missions to Mars, Titan, and even Pluto. Old-timers in the audience applauded his candor. But many who had come of age professionally during the *Apollo* era left the talk almost speechless. The SSB had turned a "deaf ear" to his warnings, Van Allen wrote in his journal a few days later. "On Thursday, the SSB and NASA representatives held an exclusive session, to which I was not invited. I flew home on that day."

The space station went forward but the aging shuttle fleet and the tragic loss of the shuttle *Columbia* and its seven-member crew on February 1, 2003, ignited the debate over human spaceflight again. The half-completed space station now carried an estimated price tag of about $130 billion. "Yet for all that money and effort, NASA and Congress are still struggling to define a purpose for the orbiting lab," said Michele Norris, as she opened the July 11, 2005, program of NPR's *All Things Considered*. Astronaut John Phillips, interviewed from the space station itself, shifted tactics. "I'd like to emphasize that up in here on the International Space Station, we are the experiments," he said. The crew-as-experiment remains the main focus of the space station in terms of "the equipment on it, our operations concept, our mission control operations, and even our international partnership," he said.

Senator Kay Bailey Hutchison (R-Texas), interviewed for the show as well, reiterated the argument that the space station is the tool for further and farther space exploration. She held hearings on retiring the shuttle as NASA pushed for a new generation of reusable vehicles, and President George W. Bush dusted off his father's plans for a new suite of human missions to the moon, Mars, and beyond. The trips could cost an estimated $1 trillion, requiring a rapid acceleration of NASA's fiscal year 2007 $16.8 billion budget.

Van Allen, still actively pursuing research in his late eighties, took the occasion of a ceremony honoring him with a service award at the March 25, 2004, meeting of the National Space Grant Foundation to update his position on the debate. The title of the talk laid his position on the line—"Is Human Spaceflight Now Obsolete?"—a speech adapted from an article of the same title.

"I am an unqualified admirer of the courageous individuals who undertake perilous missions in space," he stated in the speech. "Few persons doubt that the *Apollo* missions to the Moon as well as the precursory *Mercury* and *Gemini* missions had a valuable role in America's Cold War with the Soviet Union and lifted the spirits of mankind. Also, the returned samples of lunar surface material fueled important scientific discoveries. But the follow-on space shuttle program has fallen far short of the *Apollo* program in its appeal to human aspirations." Van Allen highlighted shuttle achievements such as launching and repairing the Hubble Space Telescope. But he argued that, after forty years of human spaceflight, almost all of the space program's important advances in scientific knowledge have been accomplished by hundreds of robotic spacecraft.

"The visions of the 1970s and 1980s look more like delusions in today's reality," he concluded. "The promise of a spacefaring world with numerous commercial, military and scientific activities by human occupants of an orbiting spacecraft is now represented by a total of two persons in space—both in the partially assembled International Space Station—with barely enough time to manage the station, never mind any significant research."

Van Allen's crusade to promote robotic space exploration marked many of his most quotable public speeches and statements over the last twenty-five years of his life. But Van Allen focused most of his efforts and energy during this period as a working scientist, documenting the continued findings of *Pioneer 10* in ongoing popular and scientific articles. With no new space missions to prepare for in the 1980s, he also began writing and editing books. He collaborated with Fimmel at Ames and science writer Eric Burgess to produce the lavish 1980 book *Pioneer: First to Jupiter, Saturn and Beyond.* Stunning photographs anchored the book. The chapters read like a series of articles in *Popular Mechanics*—easily accessible to amateur space enthusiasts but grounded with technical details for the science community.

Van Allen spent eight months in 1981 on sabbatical at the National Air and Space Museum in Washington, D.C., where a fellowship supported his authorship of his book, *Origins of Magnetospheric Physics,* republished by the University of Iowa Press in 2004. Despite the daunting title, Van Allen said he wrote the book to explain "how I got into this business" from the V-2 rocket days through all the *Explorer* and lunar *Pioneer* missions that established the contours of the radiation belts.

He also pulled together key papers by Scott Forbush that laid the groundwork for space science and edited them into a 1993 book published by the American Geophysical Union, *Cosmic Rays, the Sun and Geomagnetism.* "Nearly every cosmic-ray paper of Forbush's relatively short bibliography is a landmark in the subject," Van Allen wrote in the Introduction. "Forbush either discovered or put on a firm basis for the first time" cosmic ray shifts in intensity caused by twenty-seven-day cycles on earth, solar flares, recovery effects when the flares subsided, and the eleven year cycles of solar storms. "So much for Merle Tuve's disdain," Van Allen wrote in a parting shot in the Introduction, reflecting back on the old-timers at DTM and their tiresome charts that ultimately inspired Van Allen's career.

• • •

After thirty-four years at the helm of the physics department at the University of Iowa, Van Allen retired when the school year closed in 1985, the same year his brother Dr. Maurice Van Allen retired from his position as head of the neurology department of the University of Iowa Hospitals. Friends, colleagues, family, political leaders, and scientists from across the world wrote letters for the occasion that Mrs. Robison mounted in a series of three leather-bound scrapbooks, among the treasured volumes that Van Allen kept on the bookshelves nearest his desk. President Reagan thanked him for helping to "formulate a space program every American can be proud of" in his letter. Iowa Governor Terry Braustad declared June 28, 1985, as Van Allen Day in Iowa. "Jim has always been generous with his time and patient with the thick-headedness of reporters in explaining his work," wrote Walter Sullivan of the *New York Times.*

"I (clearly) remember how in the spring of 1958, I did not believe your interpretation of the radiation belts," wrote Russian space physicist A. Y. Chudakov. "But I remember also that [the late Professor Vernov] believed you from the beginning."

Paul Engle, founder of Iowa's Writers' Workshop, wrote him a poem that begins: "Head in the startled stars, feet on the ground, mind in far orbits, but common-sense sound."

Frank penned a hilarious seven-page account of some of the pranks and foibles of physics that carried instruments from the famous cramped basement to the vaulted realm of outer space. He hailed all the respected tenants of the basement labs including the "two bats that flew up and down the hallway after midnight on summer nights." He recalled the mysterious appearance of a colleague's small car on the roof of the new physics building. He noted his personal use of a radiation badge to calibrate an X-ray machine, a shortcut resulting in radiation exposure readings. As a warning, the physics department sent him on "a trip to see radiation damage to rabbits in the medical research labs."

But it was left to Cynthia Van Allen Schaffner to sum up Van Allen's career in her letter. "I remember one day being very upset that no room was left on the frieze that surrounds the outside of the [old] building for your name along with the other scientists of the past. And now you have a new physics building with your name!"

Two of Van Allen's greatest awards followed in quick succession soon after his retirement. President Reagan hailed "the heroes of the modern age" as he bestowed the National Medal of Science on Van Allen and nineteen other recipients on June 25, 1987, in a ceremony in the Rose Garden. Reagan said Van Allen earned the medal for his "central role in the exploration of outer space, including the discoveries of the magnetospheres of Earth, Jupiter and Saturn." Abbie, the Van Allen children—now all grown—and grandchildren joined him for the ceremony.

Van Allen treated the whole family to plane tickets to Stockholm in 1989 for his next award, the Crafoord Prize, bestowed on only one scientist in the world annually by the Royal Swedish Academy of Sciences. His Majesty King Carl XVI Gustaf awarded the heavy gold medallion engraved with the faces of Anna-Greta and Holger Crafoord, who endowed the prize in 1980 to promote basic research. Her Majesty Queen Sylvia shepherded Van Allen's grandchildren Andrew and Elizabeth Cairns right in front of her and the king as the family posed for pictures.

Despite retirement, Van Allen returned to work full-time without pay as professor emeritus in the fall of 1985. "I think that made me a research associate emeritus," quipped Bruce Randall, who continued to work with Van Allen along with the indomitable Mrs. Robison who computerized all his activities. Colleagues continued to check in with Van Allen for advice on new space missions and data. Unusual inquiries often required him to revisit old missions. He continued to receive requests for updated insights on findings of the *Explorers*, the *Pioneers*, Argus, and the role of artificial radiation belts. The restoration of the USS (later USCGC) *Glacier*, where Van Allen had first heard *Sputnik*'s beeps, added to the growing mystique of the rockoons.

Van Allen continued his close, professional relationships with many of his former graduate students who had taken positions of leadership in the space programs across the world. Meredith, McIlwain, and Cahill stopped in periodically, and Ludwig visited with the latest chapters of his memoir and filled in more details in a day-by-day time line. Van Allen wrote technical papers, encyclopedia entries, and accounts of space history. He continued to serve on advisory committees for NASA, JPL, and the National Academy of Sciences. After retirement, he agreed to serve on the board of directors of IE Industries, an energy company in Iowa, and became a member of the Governor's Science Advisory Council for Iowa. In addition to analyzing the *Pioneer* data that continued to pour in, he mentored other space missions. The *Galileo* mission that he had helped shepherd and plan with JPL launched to Jupiter in 1989. Van Allen didn't have an instrument on the mission but Gurnett and Frank both did.

• • •

In transition, the physics department Van Allen molded remained strong, with a vital mix of theorists and experimental scientists in space physics and a growing emphasis on other disciplines, including quantum physics and semiconductor research. As at other universities, the plasma physics program grew rapidly at the University of Iowa. But a brutal mass murder robbed the department of several of its most brilliant minds on November 1, 1991, the morning space scientist John Fix arrived and stepped into Dwight Nicholson's office on the second floor. Nicholson, the kindly and gregarious physicist who succeeded Van Allen as chairman of the department, casually chatted with Fix and other faculty about their research and their families. Almost everyone had a habit of stopping by to talk with Nicholson on a daily basis. After their visit, Fix headed up to his own office on the seventh floor, just down the hall from Van Allen.

As the day took a routine start at Van Allen Hall, physics doctoral candidate Gang Lu, 28, cleaned his Jennings J-22 automatic pistol and gathered his .38 caliber Taurus gun at his apartment at 515 East Jefferson Street. He wrote letters explaining his anger and pain at being passed over for a prestigious physics dissertation award for which Linhua Shan, 27, of the People's Republic of China was nominated instead. Then Lu, also from the People's Republic, walked three blocks west on Jefferson Street to his office in Room 513 of Van Allen Hall. He loaded the guns and left the black plastic cartridge boxes behind. He laid out his thesis, his résumé, his passport, his international student ID, and personal photographs. Then he joined seven or eight other people, including Shan and his adviser, theoretical physicist Christoph Goertz for a 3:30 meeting in Room 309 on the third floor.

Lu excused himself briefly from the meeting, returned about 3:40, pulled his guns out, and started shooting before anyone had time to move. He fired at point-blank range, killing Shan and Goertz immediately. Professor Robert Smith survived and called for help as Lu raced down one floor, walked through the open door to the chairman's office, and shot and killed Nicholson. Lu returned to the third floor, ordered those helping Smith to leave and fired a final fatal shot at the stricken physicist. He had killed all three professors who participated in the dissertation committee and fled down the east stairway, bumping into Bruce Randall.

Randall sprinted up the east stairway of Van Allen Hall, his speed heightened by the shell casings strewn across the landing where the gunman had reloaded. On the seventh floor, he pulled Van Allen's door shut, and locked it, and told Van Allen not to come out. Randall also ran into John Fix's office with the warning. Police responding to frantic calls arrived on the scene, racing through the hallway, and ordering people in the building to turn out the lights in their offices and lock the doors. But the gunman headed out of the building and west two blocks to Jessup Hall where he shot T. Anne Cleary, the associate vice president of academic affairs and a professor of education, who died the next day. He seriously injured Miya Sioson, a student working for her. The entire rampage took ten minutes. Then Lu turned the gun on himself and died at the scene.

A single gold wreath hung outside Van Allen Hall the following Monday, with seven black ribbons—one for each of the victims and for Lu. Several hundred students and faculty gathered around it for a moment of silent prayer. "It is midnight for this university right now," one engineering student told reporters.

"We can only trust that the Father of us all will give the survivors the strength to reassemble their lives," Van Allen told the 500 mourners gathered for a department memorial service the following Wednesday. But shock waves rippled through the university and a larger memorial service was arranged for the next night at the Hawkeye-Carver Arena. There, Van Allen shared with 4,000 people the role the three physicists had played in the space program. Their research in solar physics focused on efforts to produce abundant energy through nuclear fusion, he told them.

"Each of these (professors) was at the height of his professional career. Each was devoted to teaching and advising students at all academic levels ranging from undergraduate freshman to PhD candidates and postdoctoral associates. Each was of national and international stature," Van Allen said. Smith, forty-five, came to the university just two years before from the Naval Research Laboratory. The warmhearted Nicholson, forty-four, joined the University of Iowa in 1978, a few years after earning a PhD from the University of California at Berkeley. He walked to work every morning from a restored Victorian home three blocks north of the physics building.

Goertz grew up in Germany and earned his PhD in South Africa where Van Allen recruited him to come to the University of Iowa in 1973 to guide the space scientists with theoretical research and models. The term *theorist* meant magician in Renaissance times, and Goertz had the magic touch to cross the divide by working closely with theorists and experimenters. Van Allen knew him best, working with him almost daily from 1973 until 1980 for the *Pioneers 10* and *11* missions. Van Allen often joined Goertz at the *Voyager 2* encounters, for Goertz partnered with MIT on the plasma physics experiment on board *Voyager*. Goertz was in the middle of a four-year rotating term as editor of the *Journal of Geophysical Research-Space Physics,* an editorship Van Allen had previously held. Now, Van Allen sadly took the job back to complete Goertz's term.

As though the cosmos joined in recognizing the lives and work of the three men, Van Allen noted that a bright and unusual aurora settled over Iowa City on Friday, November 8. The murders marked a difficult phase for the university as physicist Gerald Payne hastily took over as chairman of the department and faculty scrambled to cover classes, graduate research programs, and NASA mission contracts.

Even with the added challenges, the space program at the university continued to pursue a steady stream of space missions and findings. Iowa's twenty-first-century space explorers comprise a small team compared to the large and growing space staff during the heyday in the 1960s and 1970s. But this group takes Van Allen's legacy of research with satellites, probes, and sounding rockets into new arenas of ongoing and future space missions.

As one of the scientists who helped pioneer the field of plasma physics, Gurnett continues to pursue this research in a steady stream of space missions. He recorded radio sounds across the solar system with his radio receivers and antennas on the *Voyager, Galileo, Cassini, Cluster* (the European Space Agency's mission to study the solar wind with a cluster of interrelated satellites), and other missions. The Kronos Quartet relied on Gurnett's cassette-tape collection of space sounds he has found and on composer Terry Riley to turn the recordings of magnetic fields and plasma flow into *Sun Rings,* a musical work in ten movements, which debuted at the University of Iowa in October 2002.

With NASA's *Galileo* mission to Jupiter (1989–2003), Gurnett's plasma wave experiment picked up sounds that first marked the signature of a magnetic field surrounding Ganymede, Jupiter's largest moon. NASA's *Cassini* spacecraft arrived at Saturn in 2004 and began to deliver its breathtaking spacescapes of the planet. En route in 2003, Gurnett's radio and plasma wave instrument on board recorded the sounds of two of the most powerful solar flares observed in decades as they generated radio signals moving outward from the sun.

"These solar flares are the biggest events of their kind ever seen," Gurnett said.

The successes offered a welcome verdict for a mission with a troubled start. Gurnett's experiment for *Cassini* navigated a seemingly endless barrage of added engineering demands from mission partner JPL, and *Cassini* itself came under attack by environmentalists threatening violence to prevent the launch of a spacecraft with a nuclear power source similar to the ones on *Pioneers 10* and *11* and the *Voyagers*.

As *Cassini* toured to Saturn, the *Mars Express* spacecraft (sponsored by the European Space Agency) proceeded on schedule to Mars for a December 2003 rendezvous. Even during a limbo period in 2001, when he wasn't sure the Mars mission would fly, Gurnett and the Iowa team readied their low-frequency radar device designed to search for subsurface water. During a federal funding freeze in 1999 and 2000, Gurnett sweated out receipt of an initial NASA contract to cover costs, already nearing $300,000, as he and his staff continued to work to meet the tight launch schedule. He worried about the repercussions if Iowa were left to pay the tab for these initial development expenditures. "I've always had this fear that the university would make me work for ten years for free if NASA didn't come through with the money," he said. But NASA came through with the full $6 million contract to build the instrument. "This was the last large project that I planned to lead," said Gurnett as he approached retirement. The continuation of his research group is assured by the recent success of veteran Iowa research physicist Bill Kurth in winning a $12 million contract to lead the *Juno* mission that will study the aurora at Jupiter. A plasma wave instrument will be the focus of Iowa's research for the mission, expected to reach Jupiter in 2016.

In addition to Earth-orbiting satellites and planetary probes, Van Allen's "hands-on legacy" continues with sounding rockets as well, noted University of Iowa space scientist Craig Kletzing. "On the rocket side, we have a strong sounding rocket program that is training the next generation of instrumenters. Students get to build, test, fly, and then analyze data from experiments that they themselves do the work on. We have one rocket flying this coming winter from Alaska to make measurements of Earth's aurora—also a topic of early research by Professor Van Allen—and two more rockets flying as a pair from Norway in the fall of next year to investigate the physics of Earth's polar cusps. We don't have to use balloons to get the needed altitude anymore, but the spirit of giving students hands-on experience has not changed," Kletzing said at Van Allen's memorial service on September 10, 2006. He noted that sounding rockets often offer students a more ready opportunity to participate in space exploration than satellites and probes, where missions require years of planning and lead time before they ever launch. Kletzing, Kurth, and assistant research scientists Scott Bounds and George Hospordarsky recently received a $21 million NASA contract to provide instrumentation for a mission to explore how solar storms reshape the Van Allen radiation belts and impact communications, satellites, and global navigation systems.

The mission, called the Radiation Belt Storm Probes, pursues science at the crossroads of Van Allen's legacy. It takes a new look at geospace, the region reaching from the upper atmosphere of the Earth to the boundaries of the planet's magnetic field. This region includes the radiation belts, communication systems, and other technologies affected by the volatile weather of solar flares and solar storms. The mission is part of NASA's Living with a Star program that the space agency initiated to explore the dynamic interactions of the sun and geospace and their influence on human technology and society.

The radiation belts continue to pose risks to astronauts and space vehicles, for instance. The Radiation Belt Storm Probes will study the forces that create the belts, alter them, and cause them to dissipate—information needed for models. Modeling the radiation belts "will be used by engineers to design radiation-hardened spacecraft, while the physics-based models will be used by forecasters to predict geomagnetic storms and alert both astronauts and spacecraft operators to potential hazards," according to NASA.

The mysteries and the applications echo Van Allen's concerns and curiosity since the belts that bear his name were first discovered, as he pushed his students to make discoveries with him.

Van Allen mentored other space explorers through forty-seven master's degrees and thirty-four PhDs spread out across the world. Van Allen's global space science family journeyed to Iowa City from Australia, Italy, and across the United States to celebrate when the Department of Physics and Astronomy threw a ninetieth birthday bash for him. They shared photo albums, memories, and technical papers amid one of the fiercest debates in space history.

18 Journey to the Edge of the Solar System

Iowa City, Iowa—October 9, 2004. The debate simmered as top space scientists from across the world gathered for James Van Allen's ninetieth birthday colloquium. Amid the cocktail parties, personal reunions, and technical papers, Tom Krimigis of APL insisted *Voyager 1* had crossed the inner boundary of the solar system and Frank McDonald of the University of Maryland said *Voyager* hadn't crossed yet. Both had instruments on the mission, both stood at the peaks of their profession, and both painted competing pictures of realms of the solar system never before explored.

The debate between Krimigis, one of Van Allen's former graduate students, and McDonald, one of his earliest space science professors, focused on a tumultuous boundary in space—a boundary called the *termination shock*. The shock zone encases the solar system like the skin of a bubble billions of miles beyond Pluto. Here, the solar wind hits the incoming stream of interstellar gas in crashing shock waves as the solar stream slows down in speed. No one really knew where or what this frontier of space might hold. But scientists speculated that the gust of the solar wind pools and mixes with interstellar gas at the termination shock, and Krimigis said he had found that point about 8 billion miles from the sun. McDonald adamantly placed the point farther out in space.

Krimigis's low-energy particle detector and McDonald's cosmic ray detector system involved teams of scientists who could all see rapid declines in the solar wind as predicted at the termination shock. But McDonald said the cosmic rays and magnetic field measurements remained well below the stormy environment expected at the termination shock. "The magnetic field would jump by a factor of three," McDonald said. The charged particles of the solar wind slow down like cars piling up in dense rush-hour traffic, and a denser mass of charged particles increases the magnetic field, he said.

Ed Stone, former head of JPL, discussed the controversy during the public lecture he gave later in the day but echoed McDonald's conclusions when approached about it afterward. "The magnetic field readings just aren't there," he said. "The particle counts aren't enough." From his data, Krimigis sketched the concept of a volatile termination shock—not a nice, neat magnetic landmark but a heaving, pulsing zone of space. "*Voyager* moves at the rate of 61,000 km/hr (about 38,000 mi/hour) but the termination shock can move ten to twenty times as fast. It's like a weather front and *Voyager* keeps moving in and out of it. That's exactly what we think is happening. The weather front has lasted for nine months. Now we think we may be out of it for good based on the particle counts. The intensity went up by a factor of one hundred. We've never seen anything like this in twenty-seven years with *Voyager*," Krimigis said.

• • •

Like a father with two sons competing in the family business, Van Allen kept mum on his position. But he was preparing a paper. The decades of readings from *Pioneer 10* through the Bastille Day solar storm of July 14, 2000, showed clearly that the spacecraft was approaching the termination shock, though Van Allen couldn't say definitively where it would be found. Wherever it was, let *Voyager* soar beyond it and continue to explore, McDonald said. "People keep saying we should turn off the *Voyager* when we get to the termination shock. Our answer is 'gee'—this is like calling Lewis and Clark back when they got to Kansas."

Everyone agreed on that and on the eventual discovery of an even more elusive boundary in space. Beyond the termination shock, a gentler breeze of the solar wind continues to push outward. It blows ever weaker, until incoming streams of cosmic rays cut it off at the final boundary—a boundary that encases the solar system like the surface of a bubble.

Space luminaries from across the world—and across fifty years of space history—gathered in Iowa City October 8–10, 2004, to celebrate sixty years of space exploration and Van Allen's ninetieth birthday. The space science homecoming brought former Iowa associate professor Brian O'Brien and his wife, Avril, from Perth, Australia. It brought Guido Pizzella, a former graduate student of Sydney Chapman's and Van Allen's, from Rome.

Ludwig, McIlwain, Gurnett, Gurnett's administrative assistant Kathy Kurth, and several others organized the weekend that started with a cocktail party at Iowa City's Sheraton hotel Friday night, October 8. Retired University of Iowa electronics technicians Lowell Swartz and Michael Nowack came from towns in the area and introduced themselves by their old nicknames of Ugger 1 and Ugger 2. In a space science environment, where the impossible just took a slight bit longer, "we'd go *ugh* and tackle the job," Nowack said. They teamed up once again at the party to go over Swartz's album from the 1970s and the glory days of the *Hawkeye* satellite.

"Well, there's Bertha," Nowack said, smiling at the portrait of Bertha the truck, customized transportation and communications hub for the *Hawkeye* as it traveled across the country for testing and, later, launching at the cape. "So much for security. We had 'University of Iowa's *Hawkeye* Space Mission' plastered all over the truck in letters 10-inches high," Swartz said. Bertha housed an entire control room that Lowell manned to get early results from *Hawkeye*, said Nowack. The satellite launched from a five-stage rocket to give it a kick into high polar orbit.

Instrument maker Ed Freund, from Iowa City, reminisced about the frantic months when George Ludwig raced to the tool shop with drawings of the latest design for his tape recorder for the early *Explorers*. Freund tooled it into shape, figuring out the lightest-weight metals and parts he could use and improvising techniques to keep parts together during tests on the shake table that unfastened tightly fixed screws as though they were paper clips.

Van Allen's space exploration provided the vital link between some two hundred guests and spouses who arrived for the weekend of events. Scientists, recognizable from old photographs in Van Allen's office, arrived one by one. Some of their faces were fifty years older now. Tall, gangly Les Meredith was still tall and gangly with eyes as playful as ever. Feisty Agnes McLaughlin, who warded off even federal agents on busy days, reminded eminent space scientists of their student pranks as though they were eighteen again. Ludwig buttonholed colleagues, trying to pin down dates for his memoir from the chaotic period during the discovery of the radiation belts. "There's hardly anything written from that period—neither Van or I had time," he said remorsefully.

Ludwig and O'Brien exchanged notes on the memoirs they both are writing as they greeted old friends at the cocktail party. People lingered long after the party officially ended at nine, catching up with colleagues some of them hadn't seen since they left Iowa for space science labs across the world. They talked with fondness about the rusted-out Pinto Van Allen used to drive, as recognizable an emblem of the great scientist as his battered, brown leather briefcase and his complete oblivion to status symbols.

"He taught by one example—hard work. He was there at all hours," said Pizzella, the suave, soft-spoken Italian space scientist who still fretted that Van Allen was "a little annoyed" that he left Iowa.

Van Allen didn't attend the cocktail party. As a mild concession to turning ninety, he saved his energy for the packed schedule of events planned for Saturday. He arrived with Abbie and Peter at Van Allen Hall, parking his sedan in his usual spot by the dock of the building. Abbie came in through the side door to the vestibule of the auditorium, exchanging lots of hugs and taking a seat at the back where she sat with Ed and Marcella Freund.

Van Allen and Peter walked through the maze of old computers and electronics gear in the storage rooms to reach the doorway closest to the front row of seats in the auditorium. They brushed past a glass case with a Loki cosmic ray detector from the rockoon days and a spare of Iowa's instrument for *Explorer I*. Well-wishers crowded into the auditorium from the foyer with Van Allen's arrival. As the program began, Van Allen sat in the center of the front row, flanked by three generations of his former graduate students. He turned to greet the crowd, resplendent in a navy-blue blazer, button-down shirt, and red-patterned tie and beaming as friends and colleagues gave him his first standing ovation of the day. He saluted them all and, of course, saluted Abbie Van Allen among the circle of explorers who had shared with him the first steps into outer space.

"I want to say welcome and I want to thank everybody for coming," said Abbie, wearing a tailored sky-blue jacket and slacks. "Happy birthday. Let the games begin." As master of ceremonies, Gurnett paid the opening tribute to the "father of space science" and noted the challenge of finding a fall college weekend near Van Allen's September 7 birthday that didn't conflict with a football game and packed hotels. Then he introduced the morning chairperson, *Voyager* scientist Norm Ness, who recalled coming to Iowa for those Midwest cosmic ray conferences. One baffled Bell Labs scientist, expecting to focus on satellite telecommunications rather than satellite detectors pondered, "What the hell am I doing here," Ness recalled.

Ludwig opened the sessions with the story of *Explorer I*, and although a familiar tale to this audience everyone listened raptly—like people of ages past gathering as the village storyteller turned oft-repeated adventures into legends. Ludwig revisited the cross-country trek of America's first space experiments in the trunk of his car, the pink potting foam that hardened to cushion sensitive instruments and the rapid-fire firsts of the early *Explorer* missions. Out of more than seven hundred boxes in the Van Allen archive, Ludwig located the chart, dated April 2, 1958, where Van Allen used a drugstore ruler to graph the continuous plots resulting from Ludwig's tape recorder on *Explorer III*—and discovered the radiation belts.

"I marvel to this day that I, as a largely untested graduate student, was given so much freedom and responsibility for such an important endeavor," Ludwig said. "My admiration for James Van Allen, with his willingness to take great risks with his graduate students, is unbounded."

McIlwain, the flutist who decided to study physics, told the audience about the grant proposal that ushered him into the harmony of space science. "How long was the proposal? One page. And you can see the amount—$2,000" for a rockoon experiment in the freewheeling era of early space science. He explained how he saturated the *Explorer I* instrument with the output of a simple X-ray machine, reinforcing those graphs Van Allen drew in the hotel room. Van Allen called himself a scoutmaster but "he was a mentor and more than a mentor," McIlwain said. Here was a man who taught students how to make instruments that would fly in space with 10¢ parts from the local hardware store, he noted.

At the morning break, everyone streamed into the golden light of a beautiful warm fall day, weather Gurnett promised had been negotiated at the price of two minus 20-degree days in January. Van Allen signed copies of the new paperback release of his book, *Origins of Magnetospheric Physics,* republished by the University of Iowa Press. Everyone gathered for a group picture and then returned to the auditorium for the next episode on *Pioneers 10* and *11*. "When we discovered the new satellite [at Jupiter] we decided to call it Bob—named after my husband. As grad students, we were too tired to believe we had discovered anything real," said Michelle Thomsen, now a scientist at Los Alamos National Laboratory. The name didn't stick. "It's too bad that in the Greek pantheon there wasn't a Bob," she said.

Thomsen laid out all the remarkable trailblazing of the *Pioneers* but revealed as well the emotional enormity of the first Jupiter encounter. "I will treasure that for the rest of my life," she said. "I discovered that space physics was really exciting. I learned that nature is truly amazing and sometimes makes sense." She closed with the story of Van Allen pressing his fingerprint to the instrument loaded on *Pioneer 10*. "That little spaceship, that little [fingerprint] will be working its way through to the stars. That's about as close to immortality as I can imagine."

Who could think of food at such a moment? But it was time for lunch. Alums revisiting campus often can't resist a visit to the Hamburg Inn, two blocks from the physics building, for grilled sandwiches and fries served in plastic baskets. But The Cottage bakery and deli just down the street drew a large luncheon crowd, too. Conversations there turned to McIlwain's mountain biking, new grandchildren, and some of those missing from the gathering. Ludwig recalled von Braun and the Huntsville science team's trip to Pasadena after they lost the *Explorer I* satellite project to JPL. "Von Braun told his science team, 'No grumbling or I will turn this plane around,'" Ludwig recalled.

Ed Smith, a coinvestigator on *Pioneers 10* and *11,* introduced McDonald and Krimigis as the afternoon speakers. They took the audience to the opposite sides of the termination shock with their differing interpretations of where *Voyager* now stood. But while they didn't agree on space boundaries, they agreed on Van Allen's inspiration, leadership, and pivotal role in space science. Krimigis placed Van Allen at the forefront of the new map of the solar system that filled in radiation belts and magnetospheres—or lack of them—at the planets. And he raised the compelling speculation that perhaps the solar system has radiation belts yet to be discovered.

"Van was the leader of the pack in getting *Pioneer 11* to Saturn" when others were content to keep the mission to a more modest itinerary to back-up *Pioneer 10* at Jupiter, McDonald said in his talk that closed the colloquium.

Then everyone paraded to the neoclassical auditorium of Macbride Hall where more than five hundred friends, neighbors, and Iowa City space enthusiasts filled the ground floor and sweeping balcony for a public lecture in honor of Van Allen.

"Van Allen is the only person at the university who has a natural phenomenon and a building named for him," said Linda Maxson, dean of the College of Arts and Sciences. At one point in the 1960s, Iowa had "sent up more satellites than any nation except the United States and the former Soviet Union," she said.

University President David Skorton spoke next. "Please indulge me as I share a vignette about a very important lesson I learned when I first became vice president for research, an important lesson that Dr. Van Allen taught me," he said. "I got a call from Dr. Van Allen and I tell you the truth, when I picked up the phone, I stood up. I really did. And I said, 'What can I do for you?' Van Allen requested a small additional equipment grant of a few hundred dollars to augment a large NASA grant," Skorton recalled. "Well I thought, I'm a red hot administrator. I'll see what this guy's made of. And I said, 'You know we have criteria for this. And the main criterion is if we think the person to whom we're giving the funds has some promise for an academic career. So do you have anything to help me out on this?' So I thought, I'd like to see what he says now. Well, in less time that it took me to draw my next breath, he said, 'Well, you know that's a very good point and if you'll just take a chance on me, maybe you won't go wrong.' Well, I still stand up every time he calls me on the phone and I don't mess around when anyone from physics and astronomy calls for a grant match anymore."

Skorton presented Van Allen with a permanent appointment as an "outstanding Iowan" on behalf of Iowa Governor Thomas Vilsack. "Professor Van Allen's stellar achievements in revealing the secrets of nature have brought great pride, recognition and honor to our state," the award read and it brought Van Allen his second standing ovation for the day.

"I understand a previous person to receive this award—perhaps the first one—was a man by the name of Herbert Hoover," Van Allen said. "And it's quite humbling to be put in that same company." Van Allen shared his vivid memory of meeting Hoover during his high school trip to Washington, D.C., with his brother George and his shock in seeing the president's drawn face in the midst of the Great Depression. "I've never seen anyone as lifeless as the president was at that time. I didn't expect him to live through the afternoon. But he lived on for another thirty years."

Next, Ed Stone took the podium. The former head of JPL and a native son of Iowa had studied under Van Allen's colleague John Simpson at the University of Chicago in the 1960s. Stone unfolded Van Allen's role in jump-starting the space age with the discovery of the radiation belts. Afterward, well-wishers lined up for pictures with Van Allen and children vied for a taste of the planets frosted onto the cake served at the reception. Then the core group of visitors headed for cocktails and the banquet at the Iowa Memorial Union.

Dan Baker, the graduate student who had missed *Pioneer 10*'s first encounter with Jupiter's magnetosphere, spoke at the banquet with a script straight out of *Comedy Central*, despite his solemn title as Director of the Laboratory for Atmospheric and Space Physics at the University of Colorado. He was also a NASA veteran who delivered an irreverent, rollicking roast of space-age infighting and politics. Then he turned to personal matters. Baker showed the pattern of solar storms since 1900—and showed Van Allen's birth during an auspicious rise in solar activity. He traced the receipt of his college degree to a low point in solar activity and also joked about scientists coining monster-movie slang terms such as "killer electrons" for very high energy electrons from solar flares. "Are they really killers or just misunderstood? Did they start out bad or just had a bad home life?" he quizzed.

It was left to Van Allen to close the day. He said he was happy to report he had beat the expectations of a college classmate who pegged him as good material—possibly for a high school science department. He told the story of the all-knowing young woman who regaled him at a reception about some contentious matter. "She concluded, 'Well, you'd have to be a rocket scientist to understand that one.' And I said, 'My dear lady, actually, I am a rocket scientist.' About a week later I got in the campus mail this T-shirt." Van Allen held it up. "Actually, I am a rocket scientist," it read.

But he saved his concluding remarks to acclaim Abbie. "We're just coming up on our fifty-ninth anniversary next week. I owe so much to her it's almost impossible for me to put it into words. She's first of all been my most-trusted supporter over the years. Simultaneously, she's been my most cogent and constructive critic," he said. He noted that they agreed at the beginning of their marriage that he would make the decision on really important matters. And he did so when the first matter came up involving placement of the paper-towel dispenser in their first kitchen. "I wish to report that in the subsequent fifty-nine years there hasn't been a single other matter of sufficient importance to require my decision," he said.

Spared of such domestic labors, he reached for the stars instead.

• • •

Van Allen and I met in his office Monday morning after the colloquium for what he liked to call a "postmortem" of an event. He marveled that so many had come so far for a reunion of many of the founders of the space age—the men and women behind the *Explorers*, the *Injuns*, the *Mariners*, the *Pioneers*, the *Voyagers, Galileo, Cassini,* and beyond—the men and women who touched the first 8 billion miles of space right along with him.

Then the conversation fell to the discord over the inner boundary of the solar system billions of miles away. "I didn't see any fist fights but all the principal protagonists were here," Van Allen said. "Embedded is the deep rivalry between the APL group and the JPL group on *Voyager*. They're at each other's throats competing for new missions."

He pointed to a set of two graphs pinned to the wall near his desk. He and Bruce Randall prepared them for an article in the works. One graph charted the force of the solar wind as it rose and fell with the eleven-year cycles of solar storms. Against all odds, *Pioneer 10* witnessed three great solar storms across its voyage of thirty-one years. Each time, the strength of the solar flare temporarily forced back the streams of cosmic rays. But by 2001, the force of the solar wind at *Pioneer 10* began to plateau.

"There was only a 3 percent dip in the cosmic rays at that distance from Earth," Van Allen said. If previous trends prevailed, the cosmic ray counts would have been way down. Yet the solar wind measured by Iowa instruments on the Imp-8 satellite orbiting Earth remained at a constant force with each cycle of flares.

"The meeting contributed to my feeling that Krimigis is right," Van Allen said. "*Pioneer 10* supports his position."

"So you say they found the termination shock?" I asked.

"It's hard to avoid that conclusion," Van Allen said, wishing once again that *Pioneer 10* was still providing data for added context. But several months later, Ness's magnetometer on *Voyager 1* registered the heightened magnetic field that marked the termination shock as predicted at 8.7 billion miles from Earth instead of about 8 billion where the Krimigis team reported they had found it. The difference is about twice the distance between the earth and the sun.

"*Voyager* has entered the final lap on its race to the edge of interstellar space, as it begins exploring the solar system's final frontier," said Stone, speaking as the *Voyager* project scientist at JPL in announcing the discovery.

Always open to new thinking emerging from solid scientific observations, Van Allen agreed that Ness's findings were definitive. Now the search was on for the heliopause, the final boundary of the solar system. Gurnett discovered the first direct evidence of the heliopause with intense low-frequency radio emissions picked up by his antennas on *Voyager 1* in the 1990s. He had detected powerful radio signals generated by the shock wave from a solar flare as it slammed into the frigid interstellar gases at the heliopause. From the speed of the shock wave, he estimated that this boundary fences in the solar system at about 14 billion miles from the sun. It could take another fifteen years for *Voyager* to reach that point, Van Allen noted in 2005.

"There's only an outside chance it can last that long," he added with pragmatic, professional detachment—and the twinkle of discovery in his eye.

• • •

On March 8, 2006, the National Air and Space Museum of the Smithsonian Institution honored James Van Allen with a lifetime achievement award and a trophy accepted on his behalf by his daughter Cynthia Schaffner at a black-tie awards reception and dinner. Van Allen stayed in Iowa but addressed the gathering by phone link. His remarks summarized his sixty-year career in two minutes and then tackled NASA with a tongue-in-cheek bombshell.

"I have recently moved over to the slow lane, making the transition from being an active participant in space research to being a fascinated bystander. As an irresponsible bystander and in the lighthearted spirit of this occasion, I decided to offer an iconoclastic suggestion for your consideration," he said. "In brief, I propose that the United States sell to China the partially completed International Space Station plus its remaining components plus the three surviving shuttles plus the portfolio of related obligations to other countries.

"I submit that this would be a win-win deal. On the one hand, it would enable China to realistically calibrate its declared aspiration to engage in world-class human spaceflight. On the other hand, it would help reduce our trade deficit with China." He noted "that many details of such a deal must be worked out. But, in its omission of details, my proposal is in the good company of the president's January 2004 proposal, briefly entitled the Moon-Mars Initiative."

Van Allen offered a few closing words on his career. "As a 10-second summary of my life as a space scientist, I have adopted and adapted a writer's recent characterization of a fictional private detective: 'He had curiosity, patience, persistence *and* a good time.'"

Van Allen also received the annual Goddard Trophy from the National Space Club at the annual Robert H. Goddard Memorial Dinner in Washington, D.C., later that month. Peter Van Allen accepted that award. He talked lightheartedly about his father's scientific precision, frugality, and spirit of improvisation in everything he did. "For many years, he drove a '62 Volkswagen Bug. To deal with the cold Iowa mornings, he rigged a rope with a handle so that he could wrap the rope around the belt drive and start the car—much like you would a lawn mower. AAA was not an option," he said. "Around the house and in the family cars were small notebooks with an attached string and tiny pencils, sharpened to a point with a pocket knife. These were the 'logs.' When you got gas, you wrote down the date, the gallons, the price, the location. When you traveled, you wrote down the mileage, time you left, destination, time of pit stops, arrival, motel check-in, tolls, money spent, etc. Science wasn't just about observation; it was about keeping track."

• • •

On Monday, May 15, 2006, James Van Allen drove to his office as usual. On a routine day, he arrived at about 9:30 A.M., read the *New York Times*, answered correspondence and e-mails, and fielded requests for interviews and forewords to books. He logged every telephone call, ate lunch at his desk, and kept his door open to colleagues and former graduate students who stopped by whenever they were in town. He remained an active participant in space science and completed two technical papers in spring 2006. Before he left on May 15, he penned a quick note in his journal about his back surgery the next day. He left the journal open to the page, but he never returned to finish the passage. It was his last day in the office he had occupied for decades.

The back surgery at the University of Iowa Hospitals went well. Doctors removed a cyst that had been causing increasing pressure on his spine and an increasing shuffle in his walk. Van Allen and his doctors believed the surgery would improve his mobility. But he suffered a heart attack and the onset of pneumonia compounded his condition as he recovered. Even while on a feeding tube, he was determined to return to work. Bruce Randall visited him on the Fourth of July and came away with instructions for making 701 accessible to Van Allen from a Scooter Store power chair. Randall moved Van Allen's computer to the desk at the front of the office and finally relegated the antiquated 1972 graphics plotter to the back of the office.

But Van Allen's heart and breathing continued to drain his strength. Doctors put him on oxygen and he rebounded, improving enough to go off the oxygen on Friday, August 4, while he convalesced at an extended-care facility. Abbie was there every day and noticed his breathing becoming more and more strained over the weekend. Even though he had seemed so much better, Abbie was called to the University of Iowa Hospitals emergency room at 2 A.M. Monday, August 7. Later that morning, he had another heart attack and slipped into a coma. With his family gathered around him, he lingered for two more days. At 6:15 A.M. on Wednesday, August 9, James Van Allen died. The Department of Physics and Astronomy shrouded its Web site in black and set up a scholarship fund in Van Allen's honor. Letters and e-mails poured in to the department and to Abbie from friends and colleagues around the globe. Scientists who had never met Van Allen wrote to say that his discoveries had inspired them to become physicists and astrophysicists.

The machine shop at Van Allen Hall designed a box for Van Allen's ashes. On September 10, the family and hundreds of friends gathered for a memorial service in Hancher Auditorium to celebrate Van Allen's life and accomplishments.

"Jim Van Allen was also simply a kind, generous man, and I am sure he will be remembered most fondly in that way," said acting university president Gary Fethke in opening remarks for the service. "He was a superb model of an Iowa native achieving world-renowned excellence, someone who, from humble beginnings, realized revolutionary accomplishments."

Van Allen's close friend Sandy Boyd summed up humble childhood beginnings and early influences that eventually brought him back to Iowa City in 1951. He always found the person needed to settle problems and expedite solutions with a brief chat. "Woe unto me, if I as president or provost, had the temerity to suggest that I should be consulted. And, of course, that modus operandi worked well in advancing the university," Boyd said. "Jim was both intellectual and modest—a rare combination. He listened more than he spoke. He was an astute observer of his environment and judicious in his commentary. When he spoke, he had something worth hearing. He never raised his voice to make a point, and we paid attention. He had a wry and insightful sense of humor. In discussing the evils of smoking with [law professor emeritus] Charles Davidson, he commented that he had never heard of a pipe smoker who was convicted of murder."

Gurnett and Krimigis spoke of the influence Van Allen had had on their lives and space science overall. Kletzing spoke of Van Allen's space legacy, carried on through current Iowa space programs involving new probes to other planets, satellites, and sounding rockets. Dr. Margot Van Allen Cairns spoke of her father's sense of genius as teamwork—the collective passion of the team working on a project. Cynthia Schaffner read a poem about her father.

Then Peter Van Allen took the podium and captured the soul of James Van Allen, a man who could bring the sophisticated wisdom of physics and the homespun art of frugality to almost everything he did. Peter spoke of the lessons his grandfather imparted in reading to his four sons from the encyclopedia for entertainment and the lessons his father imparted with the rigorous discipline brought to any project, even if it was to replace the mailbox at the bottom of the hillside that led up to their home.

"The word in vogue now is *grit,*" Peter said. "Grit is what I really identify with my dad. Everyday, he left the house at 7:45 A.M. Everyday, he was home at 6 P.M. and, after dinner, he retired to his home study to work some more. He followed that schedule six days a week," though Sundays were different. "We had this ancient waffle iron that would steam and hiss and you'd plug it in and it would throw off sparks and it was a fire hazard and so on. But he made these amazing waffles on it every Sunday morning and bacon. But then, after he put the waffle iron away and got it cleaned up and stuff, he was off to work and he would work for at least half a day on Sunday. And the work he did was really that chipping away—you know—the chip, chip. You know—that tenacity you demonstrate over a period of years.

"And that really meant, if there was any kind of inspiration it came after many long hours and, if there was genius, it was the result of many hours of toil applied to the same problem. Grit was not something my dad wore on his tattered shirts or work gloves. That was simply his frugality. Rather, grit was something that showed in the papers that he published, and in the classes he taught at the University of Iowa, and in the doggedness of making sure the mailbox was plum and center."

Notes

Quotes from interviews with the author are generally not cited in the chapter notes below. Clarity necessitated some exceptions. A complete list of interview sources and dates is included in the bibliography.

The transfer of papers from his office to the existing James A. Van Allen Papers and Related Collections at the University of Iowa Special Collections was still underway as this book went to publication. This applies to any documents cited as "Van Allen office papers." The author recommends inquiry at Special Collections and University Archives. University of Iowa Libraries, Iowa City, to check for documents in the future. Documents cited as "Van Allen family papers" are in family collections.

Introduction

"The cosmic ray ceiling": James Van Allen, "Exploratory Cosmic Ray Observations at High Altitude by Means of Rockets," *Sky and Telescope* 7 (1948): 171–175.

1. Frontier Roots

"If you would have your machine": George Clinton Van Allen, February 24, 1848 journal entry, Van Allen family papers.

"One term was all that I dared hope": George Clinton Van Allen, 1851–1852 journal, "Introduction," September 10, 1851.

"Getting lazy": journal, October 1, 1851.

"The same routine of study": journal, September 12, 1851.

"The Lord's presence": letter, September 2, 1851.

"O! God, parent of all good": journal, September 25, 1851.

"I have $2.55 and Mart has 38 cents": letter, September 2, 1851.

"It cost me about $10,": George Van Allen, undated letter.

"I have let him have control": Martin Van Allen, undated letter.

Van Allen left home before dawn: journal, December 26, 1851.

"I am afraid if I go home": journal, June 2, 1853.

"You will shorten your life": George Van Allen, undated letter.

"Sarah too feared the trap": George Van Allen, letter, October 24, 1853.

"One of my little girls": George Van Allen, letter, January 1, 1854.

"He returned to Falley": *Henry County Biographical Review*. Chicago: Hobart, 1906, 16.

"The building boom": George Van Allen, letter, May 23, 1856.

"The disaster": *Henry County Biographical Review*.

Near the county courthouse: *Portrait and Biographical Album, Henry County, Iowa*, Chicago: Acme Publishing, 1888.

"One of the novel improvements": *Mount Pleasant Daily News*, July 26, 1886.

"Standardized title search procedures": Alfred Van Allen, journal, undated inserted typewritten essay, Van Allen family papers.

"Some fishermen hang out the hook": George Van Allen, letter, October 1992.

"The ring rule": *Mount Pleasant Daily News*, September 20, 1897.

"Mr. AMVA called me": Alma Olney journal, January 19, 1907, Van Allen family papers.

2. Heartland Boyhood

The girls of Phi Mu sorority: Alfred Van Allen journal, May 29, 1923.

"George furnished two small cedar trees": Alfred Van Allen journal, typed insert.

"Exactly in the center": James Van Allen, elementary school notebook at the Van Allen House Heritage Center, Mount Pleasant, Iowa.

"There was much beating of the bushes for students": Thomas Poulter, unpublished memoir, Van Allen office papers.

"Sardines . . . Little America lived, ate": C. J. V. Murphy account of Byrd rescue, "Night Journey," chapter in Richard Byrd, *Discovery, The Story of the Second Antarctic Expedition*, New York: G. P. Putnam Sons, 1935, 217.

"Please don't ask me crank": Ibid., 236.

"Come on down": Ibid., 237.

3. The Making of a Scientist

"A sedate game of croquet": Stow Persons, *The University of Iowa in the Twentieth Century*, Iowa City: University of Iowa Press, 23.

McLean revamped: Ibid., 26.

Alexander Ellett came on board . . . "Atoms à la Bohr": James Wells, *Annals of a University of Iowa Department: From Natural History to Physics and Astronomy*, Iowa City: Department of Physics and Astronomy, University of Iowa Publication 80-19, 1980, 119.

But he brought luster to the physics department . . . multidisciplinary themes such as "Physics and Society": Ibid., 124.

"He took me on . . . a purchase order": James Van Allen interview with David De Vorkin, and Allan Needell, February-August, 1981, Smithsonian Institution National Air and Space Museum, The Space Astronomy Oral History Project, transcript, 70.

"There are 21 students in the physics class": Van Allen office papers.

He "found that the trend . . . reaction chamber was faulty": James Van Allen, "What Is a Space Scientist?" *Annual Review of Earth and Planetary Sciences*, June 1989, reprinted in *The James A. Van Allen Papers and Related Collections*, Iowa City: University of Iowa Archives, 1993, 19.

"My appointment": Van Allen office papers.

While the "young Turks" at DTM: James Van Allen, editor, *Cosmic Rays, the Sun and Geomagnetism: The Works of Scott E. Forbush*, Washington, D.C.: American Geophysical Union, 1993, viii.

"Passionate and solitary": Ibid., viii.

"He regarded the old timers": Ibid., viii.

Millikan found that he: William Pickering, draft of a speech in celebration of the seventy-fifth anniversary of Millikan's Nobel Prize Award, copy from Pickering.

The 1933 Century of Progress Exposition: Information about the competition to settle Compton's and Millikan's differing premises about cosmic rays is drawn from David De Vorkin, *Race to the Stratosphere*, New York: Springer-Verlag, 1989, 55–82.

"Vic Neher and I": William Pickering, Millikan anniversary lecture, June 27, 1998, Jackson County Historical Society, Maquoketa, Iowa, 19, copy from Pickering.

4. Physicists to the War Effort

Winston Churchill wrote: Quoted in Ralph Baldwin, *The Deadly Fuze*, San Rafael, California: Presidio Press, 1980, 4.

"The funny fuze": General George Patton quoted in David Colley, "Deadly Accuracy," *Invention and Technology*, Spring 2001, 50.

"I think that when all armies": General George Patton quoted in Baldwin, *The Deadly Fuze*, 303.

"Broke the back": *The Deadly Fuze*, produced by PBS affiliate WGVU, Grand Valley State University, Allendale, Michigan, based on *The Deadly Fuze* by Ralph Baldwin.

The proximity fuze: For further details on the early development of the fuze at DTM, see Allan Needell, *Science, Cold War and the American State: Lloyd V. Berkner and the Balance of Professional Ideals*, Amsterdam: Harwood Academic Publishers with the Smithsonian Institution, 2000, 69–78.

Rumors of a tiny, rugged glass tube: *Sunday Globe*, January 5, 1964.

"The exuberant response": Van Allen, "What Is a Space Scientist?"

"I don't want": Quoted by Van Allen, interview with the author.

But duds and premature firings: January 19, 1943, report to BuORD, Van Allen office papers, navy files.

Parsons told them . . . "The fleet is very short": Van Allen office papers, navy files.

Apply for his uniform reimbursement: July 14, 1943, uniform gratuity, Van Allen office papers, navy files.

Thanksgiving at sea: Van Allen office papers, navy files.

Proximity fuze from Deke Parsons: Van Allen office papers, navy files.

"A very close burst": Deke Parsons February 23, 1943, proximity fuze report, Van Allen office papers, navy files.

BuORD ordered him back: COMSOPAC order, June 9, 1943, navy files.

In nearly 250,000 projectiles: Mark 32 inventory, January 31, 1944, Van Allen office papers, navy files.

Dispatch: COMSOPAC memo to the *Helena*, February 15, 1944, Van Allen office papers, navy files.

"From the beginning . . . Russian fleet at Tsushima in 1905": Harry Gailey, *The War in the Pacific*, Novato, California: Presidio, 1995, 308.

“Lee’s antiaircraft gunners”: Gailey, *The War in the Pacific,* 313.

“The haze of battle”: “My Life at APL,” *Johns Hopkins APL Technical Digest*, 18 (1997) 175.

“The gunners’ effectiveness”: Gailey, *The War in the Pacific,* 313.

“I could say that . . . Japan was defeated technologically”: Ralph Baldwin. *They Never Knew What Hit Them,* Naples, Florida: Reynier Press, 1999, 175.

“Almost every night”: Hudler quoted in Baldwin, *They Never Knew*, 176.

After his return from the Pacific: naval order, December 21, 1944, Van Allen office papers, navy files.

5. Enter Abigail Fithian Halsey

He soon owned 100 acres of land . . . “sell it together”: Jacob Halsey and Edmund Halsey, *Thomas Halsey of Hertfordshire, England and Southampton, Long Island, 1591–1679, with his American Descendants to the Eighth and Ninth Generations,* Morristown, N.J.: Yankee Peddler Book Company, 1971, 19–20.

“From the town records”: Ibid., 23.

“A party of young people making the ascent”: “Lake Placid Campers Busy,” *New York Times,* July 12, 1908, 4.

“Abigail and I have decided to get married”: July 31, 1945, letter home, Van Allen family papers.

“The Rev. Dr. and Mrs. Jesse Halsey”: *New York Times,* August 28, 1945.

“My old Mercury”: Van Allen office papers.

“Am hoping to get”: August 31, 1945, letter, Van Allen office papers.

“I regard his condition as dangerous”: George Van Allen, August 31, 1945, letter, Van Allen office papers.

“Seventeen more days”: Abbie Van Allen, letter to James Van Allen, September 26, 1945, Van Allen office papers.

“Marriage is wonderful”: James Van Allen, October 25, 1945, letter home, Van Allen family papers.

“Every morning, I think of something pleasant”: Babbie Halsey, to Alma Van Allen, July 30, 1946, Van Allen family papers.

6. The Dawn of Space Exploration

The V-2 rocket gleamed . . . rocket and payload: Van Allen and Stuhlinger interviews; Homer Newell, *Sounding Rockets,* New York: McGraw Hill, 1959, 237–238.

Later, he proposed: Frank Winter, “Planning for Spaceflight,” in *Blueprint for Space,* eds. Frederick Ordway III and Randy Liebermann, Washington, D.C.: Smithsonian Institution Press, 1992, 104–105.

“Successful Firing”: Susan Enscore, *Operation Paperclip at Fort Bliss: 1945–1950,* Fort Bliss, Texas: United States Army Air Defense Artillery Center, 1998.

Von Braun is known to have worn: Michael Neufeld, *The Rocket and the Reich: Peenemünde and the Coming of the Ballistic Missile Era,* New York: The Free Press, 1995, caption to photo with Riechsfuhrer-SS Heinrich Himmler, photos following 210.

A staggering twenty thousand: Ibid., 264–265.

The OSI singled out: Ibid., 187.

Von Braun had no control: Ernst Stuhlinger and Frederick Ordway III, *Wernher von Braun: Crusader for Space*, Malabar, Florida: Krieger Publishing, 1996, 43–53.

"He was really not interested": Wernher von Braun and Frederick Ordway III, *History of Rocketry and Space Travel,* New York: Thomas Y. Crowell, 1966.

"Had the V-1 and its": Doris Kearns Goodwin, *No Ordinary Time,* New York: Touchstone, 1995, 520.

Von Braun's most ardent: Tom Crouch, *Aiming for the Stars,* 90.

Goddard to analyze them: Stuhlinger, 158.

"You have a man in your country": Paul Dickson, *Sputnik,* New York: Walker and Co., 2001, 60.

Krause moved forward with a crazy idea: Van Allen interviews; James Van Allen, "Early Days of Space Science," *Journal of the British Interplanetary Society,* 41, 1, January/February 1988, 11–17.

"Every agency in the United States": Homer Newell, *High Altitude Rocket Research*, New York: Academic Press, 1953, 13.

"Weekly trips . . . earth's atmosphere": *Popular Science*, 145, 1, July 1946.

"The immense opportunity for finally being able": James Van Allen, *Origins of Magnetospheric Physics,* Washington, D.C.: Smithsonian Institution, 1983, 19–20 (Reprinted, University of Iowa Press, 2004).

"A post flight review showed that": Homer Newell, *Beyond the Atmosphere,* 43.

Then hit the same plateau: James Van Allen, and H. E. Tatel, "The Cosmic Ray Counting Rate of a Single Geiger Counter from Ground Level to 161 Kilometers Altitude," *Physical Review,* 73, February 1948, 245–251.

"Dips at about 100 seconds": Van Allen field log, August 31, 1947, Van Allen office papers.

Balloon research continued: De Vorkin, *Race to the Stratosphere,* 297.

"The camera encased in": "Rocket Camera 65 Miles Up," *Washington Post*, November 21, 1946, 12.

"He was almost our ombudsman": De Vorkin, National Air and Space Museum, 1981 series of Van Allen interviews, transcript, 40–41.

Astronomer Jesse Greenstein: David De Vorkin, *Science with a Vengeance,* New York: Springer-Verlag, 1992, 117.

Greenstein wasn't the only: Ibid., 117.

7. The Mighty Little Aerobee

It was Van Allen's baby: Homer Newell, *Guide to Rockets, Missiles and Satellites*, New York: McGraw Hill, 1961, 22.

"It's a hazard": Conversation quoted by Van Allen, Van Allen interview, January 23, 2001.

"The stop order caused a substantial": James Van Allen, et. al., "The Aerobee Rocket," in *Sounding Rockets,* ed. Homer Newell. New York: McGraw Hill, 1959, 59.

"The least controversial": De Vorkin, *Science with a Vengeance,* 170.

Van Allen and APL physicist Fred Singer: *APL News,* November 1948, archive of the Applied Physics Laboratory, Johns Hopkins University.

Van Allen stressed: "The Rockets Report," *Johns Hopkins Magazine,* October 1950.

"V-2 helps swell": De Vorkin, *Science with a Vengeance,* 249.

He would revisit Millikan's: I. S. Bowen, A. Millikan and H. V. Neher, "New Light on the Nature and Origin of Incoming Cosmic Rays," *Physical Review,* 53 (1938): 855–61; Van Allen, *Origins,* 14.

Rossi referred to: De Vorkin, *Science with a Vengeance,* 249–252.

Rear Admiral A. G. Noble: *APL News,* March 1949, archive of the Applied Physics Laboratory, Johns Hopkins University.

"If any small detail of the rocket": James Van Allen, "The Rockets Report," *Johns Hopkins Magazine,* October 1950.

American Rocket Society Medal: *APL News,* December 1949.

"The North Atlantic much like": Van Allen field log for Aerobee launches, January 1950, Van Allen office papers.

Van Allen spent time: Van Allen field log for Aerobee launches, January 1950, Van Allen office papers.

"The equipment used for launching Aerobees": Homer Newell, *High Altitude Rocket Research,* New York: Academic Press, 1953, 24–25.

"Over-all, the potential biological": James Van Allen, "The Nature and Intensity of the Cosmic Radiation," in eds. C. S. White and O. O. Benson, Jr., *Radiation Physics and Medicine of the Upper Atmosphere,* Albuquerque: University of New Mexico Press, 1952, 262.

The report, written by panel founding members: UARRP Panel Report, No. 26, September 7–8, 1950, James A. Van Allen Papers, Box 234-4.

The report essentially: De Vorkin, *Science with a Vengeance,* 172–182.

"In response to the very kind letter": James A. Van Allen Papers, Box 232-14.

After World War II. . . should march hand in hand: Needell, *Science, Cold War and the American State: Lloyd V. Berkner and the Balance of Professional Ideals,* 130–131.

Berkner, whose research: Ibid., 163.

"His former mentors": Wells, *Annals of a University of Iowa Department: From Natural History to Physics and Astronomy,* 160.

"Tyndall was very frank about it": Van Allen interview with De Vorkin, transcript, 174.

Van Allen was the unanimous choice: Wells, *Annals of a University of Iowa Department: From Natural History to Physics and Astronomy,* 161.

8. It's a Rocket! It's a Balloon! It's a Rockoon!

"The small living room": James Van Allen, "What Is a Space Scientist? An Autobiographical Example," *Annual Review of Earth Planet Science,* 18, 1990, 1–26.

"You don't understand": Ellis quoted by Van Allen, interview with author.

Meredith based his: Les Meredith, "A Measurement of the Vertical Cosmic Ray Intensity as a Function of Altitude," master's thesis, June 1952.

"Already I can tell by looking": Les Meredith, "*Northern Travels and Travails of 1952,*" unpublished journal, photocopy in Van Allen office papers.

"We double did everything": Meredith journal, August 23, 1952.

"I remember one day": Lou Frank, letter to James Van Allen, May 17, 1985, correspondence scrapbooks from Van Allen's retirement, Van Allen office papers.

"The 1953 expedition": Van Allen, "What Is a Space Scientist?" 1993 reprint, 27.

"Frank McDonald and I": Carl McIlwain, "Music and the Magnetosphere," in Gillmor and Spreites, eds., *Discovery of the Magnetosphere,* Washington, D.C.: American Geophysical Union, 1997, 130.

During the two-stage assembly: Ibid., 131.

Van Allen later: Van Allen, *Origins of Magnetospheric Physics*, 26; James Van Allen, "Energetic Particles in the Earth's External Magnetic Field," in Newell, *Discovery of the Magnetosphere*, 238; Van Allen chapter in *Sounding Rockets*; Van Allen interview with De Vorkin; Van Allen interview with the author, February 2001.

9. *Sputnik* and the Space Race

"Presently, [the Huntsville scientists] have": Van Allen journal entry, November 18, 1956, Van Allen office papers.

"Boehm-Van Allen line-up": Von Braun's desk log, December 7, 1956, archive of the U.S. Space and Rocket Center, Huntsville, Alabama.

They had found a way "to bootleg": Herbert York, in *James Van Allen: Flights of Discovery*, video documentary produced by the University of Iowa Center for Media Production, narrated by Tom Brokaw, 2000.

The report: "Preliminary Design": Douglas Aircraft Company, Inc., Santa Monica Plant Engineering Division, Report No. SM-11827, May 2, 1946.

Popular Science reported: Martin Mann, "Going up for Keeps," *Popular Science,* March 1947, 150, 66–71.

"As we were all sipping brandy": Conversation quoted in interview with Van Allen, February 21, 2001.

"Cosmic radiation does not represent an overwhelming": James Van Allen, "The Nature and Intensity of Cosmic Radiation," in *Physics and Medicine of the Upper Atmosphere,* eds. C. S. White and O. O. Benson, Jr., Albuquerque: University of New Mexico Press, 1952, 239–263.

"Connie, go to": Stuhlinger and Ordway, *Werner von Braun*, 112.

After a meeting at the Jet Propulsion: Stewart interview transcript for *James Van Allen: Flights of Discovery.*

Durant approached the Huntsville team: Stuhlinger, *Wernher von Braun: Crusader for Space*, 123.

Minimum Orbital Unmanned Satellite: Constance Green and Milton Lomask, *Vanguard: A History,* Washington, D.C.: Smithsonian, 1971, 17.

"Van Allen sat": Stuhlinger, 125; for Van Allen description of the event, see *Origins of Magnetospheric Physics,* 49.

"It was obvious that only": Van Allen, *Origins of Magnetospheric Physics,* 49.

Rocket panel . . . satellite briefing: UARRP correspondence and minutes, James A. Van Allen Papers, Box 232-1.

"The purpose of the meeting was": Ibid.

Whipple spoke next: Ibid.

Hoover detailed plans . . . "ONR is engaged": Ibid.

"Outline of a Proposed Cosmic Ray Experiment for Use in a Satellite (Preliminary)": James A. Van Allen Papers, Box 84-1.

"With tongue in cheek": Stuhlinger, 128. Also see Wernher von Braun, *This Week* magazine, "The Explorers," *Des Moines Sunday Register*, April 13, 1958, 9.

Boundary involving "freedom of space": Herb York interview transcript for *James Van Allen: Flights of Discovery*; testimony of Don Quarles, Preparedness Investigating Subcommittee hearings chaired by Lyndon Johnson, 1957–1958; declassified Stewart Committee Papers, JPL archive.

Kaplan's letter to Alan Waterman, favoring the Vanguard: Michael Neufeld, "Orbiter, Overflight and the First Satellite: New Light on the Vanguard Decision," in *Reconsidering Sputnik,* eds. R. Launius, J. Logsdon, R. Smith, Amsterdam: Harwood Academic Publishers, 2001, 242.

The Ad Hoc Advisory Group: The Stewart Committee Papers, Jet Propulsion Laboratory, archive, Pasadena, California.

Stewart endorsed the Orbiter: Ibid.

"What I disliked": Stewart interview transcript, *James Van Allen: Flights of Discovery.*

The Stewart Committee issued: Stewart Committee Papers, JPL archive.

Rosen, as technical: Green, *Vanguard: A History*, 54.

"Unanimity prevailed on one point": Ibid., 51.

Clifford Furnas, chancellor of the University of Buffalo: Now the State University of New York at Buffalo.

"I think they were optimistic about the Vanguard": Stewart interview transcript, *James Van Allen: Flights of Discovery.*

"Political insistence" on a civilian satellite: Walter McDougall, *The Heavens and the Earth,* New York: Basic Books, 1985, 203.

Michael Neufeld: "Orbiter, Overflight and the First Satellites."

In a *Life* magazine: Clifford Furnas, *Life*, Oct. 21, 1957, 22–25. For more details, Senate Preparedness Investigating Subcommittee hearings chaired by Lyndon Baines Johnson, 1957–1958.

The Orbiter encountered prejudice: Stuhlinger interview; Fred Whipple, "Recollections of Pre Sputnik Days," in *Blueprint for Space*, eds. F. Ordway III and R. Liebermann, Washington, D.C.: Smithsonian, 1992.

"Look at this von Braun": "Journey into Space," *Time* magazine, Dec. 8, 1952.

The army once again: Stewart Committee Papers, JPL archive.

That gave Huntsville adequate "cover": Van Allen interview, Stuhlinger interview, Pickering interview, von Braun, *Des Moines Sunday Register*.

"Backyard scientist": Green, *Vanguard: A History*, 115.

Van Allen edited: James Van Allen, ed., *Scientific Uses of Earth Satellites*, Ann Arbor: University of Michigan Press, 1956.

For one 1956: Van Allen travel journals, James A. Van Allen Papers, Box 233-2.

"I located and tested subminiature": George Ludwig, unpublished manuscript and time line, "Opening Space Research at the University of Iowa." Contact ludwiggh@visuallink.com.

"We traded in our 1950 Ford station wagon": Van Allen journal August 8, 1956.

"On 10 August, 1956, Abigail, Cynthia,": Van Allen journal.

Stewart's ashes: Van Allen journal, October 27, 1956.

"Public announcement yesterday": Van Allen office papers.

"Re Van Allen wire": Van Allen office papers.

So Van Allen and Ludwig lived double lives: von Braun's desk logs.

Ludwig to visit Huntsville: Ludwig time line, July 9–12, 1956.

10. Countdown to *Explorer I*

The Boston Navy Yard: now called the Boston Naval Shipyard.

"Radio reports said [*Sputnik*]": Van Allen field log, October 5, 1957.

Additional field notebooks, James A. Van Allen Papers, Boxes 383–388.

"Yesterday night—the 4th—was very exciting": Van Allen field log, October 5, 1957.

IGY scientists sipped: George Ludwig, unpublished manuscript, "Opening Space Research at the University of Iowa"; Clayton Koppes, *JPL and the American Space Program*, New Haven: Yale University Press, 1981, 83.

Reporters poured into the embassy: Paul Dickson, *Sputnik: The Shock of the Century*, New York: Walker and Company, 2001, 13.

Von Braun turned to McElroy: John Medaris and Arthur Gordon, *Countdown for Decision*, New York: Putnam, 1960, 155.

"He was therefore stunned": Herbert York, *Making Weapons, Talking Peace*, New York: Basic Books, 1987, 140.

"The American loves his car": "*Sputnik*'s Week," *Time* magazine, October 28, 1957, 51.

"One thing we . . . flight-testing for a considerable time?": Stuhlinger, *Wernher von Braun: Crusader for Space*, 133.

Physicist Edward Teller: Koppes, *JPL and the American Space Program*, 84.

"The famed beeps": "Sputnik's Week," *Time*, October 21, 1957, 50–51.

Eberhardt Rechtin at JPL: Ludwig time line, October 23, 1957.

"When a big pot is won": Koppes, *JPL and the American Space Program,* 86.

"With a system that owed more to ": Dickson, *Sputnik: The Shock of the Century*, 170.

"He asked if I had authority": Ludwig time line, October 28, 1957.

"Would you approve transfer": William Pickering, Western Union telegram via U.S. Naval Dispatch, October 30, 1957, affixed in Van Allen field log, October 18, 1957–November 3, 1957.

"Present planning": William Pickering, Western Union telegram, via U.S. Naval Dispatch, November 2, 1957, affixed in Van Allen field log, October 18, 1957–November 3, 1957.

"Pickering now has me": Van Allen field log, November 1, 1957. The entry responds to both of Pickering's wires, even though it appears to be dated before receipt of the second message.

JPL would build: Hall qtd. in Koppes, *JPL and the American Space Program*, 288.

"Von Braun swallowed hard": Stuhlinger, *Wernher von Braun: Crusader for Space,* 136.

"You don't say": Van Allen and others quoted this comment, attributed to von Braun.

Ludwig, his wife: Ludwig time line, November 13, 1957; Ludwig interview.

"George quite happy": telegram, November 13, 1957, Van Allen office papers.

Low-budget rocketry: James Van Allen, "The Inexpensive Attainment of High Altitudes with Balloon-launched Rockets," in R. L. P. Boyd, *Rocket Exploration of the Upper Atmosphere*, London: Pergamon Press, 1954, 53–64.

Reedy spelled out: Robert Divine, "Lyndon B. Johnson and the Politics of Space," in *The Johnson Years,* ed. Robert Divine, Lawrence: University Press of Kansas, 1987, 217–248.

"Whew, at first": McDougall *The Heavens and the Earth: A Political History of the Space Age*, 167; originally in the *Washington Post*, November 21, 1957.

"The military was still in charge": Walter Cronkite, interview with the author. See details about covering early space missions in his chapter 12 of his autobiography *Walter Cronkite: A Reporter's Life*, New York: Alfred A. Knopf, 1996.

Von Braun described: Preparedness Investigating Committee hearings transcript, 618.

"Received Nuclear Corporation": Ludwig time line. January 27, 1958.

The January 29: Ludwig time line.

Hydyne fuel: von Braun, *Des Moines Sunday Register*.

"Within two minutes": Ludwig time line, January 31, 1958.

Van Allen, von Braun, and Pickering gathered in the War Room: Van Allen, *Origins of Magnetospheric Physics,* 58.

"Boys, this is just": Medaris Collection, Evans Library, Florida Institute of Technology, Orlando, Florida.

"We are out of coffee": Pickering log, JPL archive.

11. Celebrity Scientist and the Birth of NASA

York became a key advisor: Herbert York, *Making Weapons, Talking Peace,* New York: Basic Books, 1987, 137–138.

"The army exploited the fact": Ibid., 115.

The air force claimed space: Ibid., 115.

The term blitzed: Ibid., 122.

His success "demonstrated clearly": John Naugle, *First among Equals*, Washington, D.C.: NASA, 1991, 16.

A National Mission to Explore Outer Space: A Proposal by the Rocket and Satellite Research Panel, November 21, 1957; *National Space Establishment: A Proposal of the Rocket and Satellite Research Panel,* December 27, 1957, Van Allen Archive; discussion in Newell, *Beyond the Atmosphere: Early Years of Space Science.*

Van Allen, Pickering, and Whipple lobbied for their proposal: Van Allen Archive, RSRP minutes, December 6, 1957.

"We do not wish to see the satellites of Mars": Letter submitted to Lyndon Johnson as chairman of the Special Committee on Space and Astronautics, Hearing Transcript on S.B. 3609, 380.

"The military side of space technology": McDougall, *The Heavens and the Earth,* 173.

The Eisenhower administraton: Naugle, *First among Equals,* 18–21; Van Allen interview with De Vorkin, transcript, 284–285; Van Allen interviews with the author.

"I have a vivid memory of it": Naugle, *First among Equals,* 32.

Glennan moved into: Ibid., 45.

"With NASA in the driver's seat": Newell, *Beyond the Atmosphere,* 205.

"Decisions concerning the space science program": Ibid., 204.

The multiple roles York: *Making Weapons, Talking Peace,* 113–115, 143–144.

One of NASA's first reports . . . now called the Saturn: *Report of the Special Committee on Space Technology,* in Stuhlinger, *Wernher von Braun: Crusader for Space,* 149, 151–153.

"Among other things that I learned": Van Allen, "What is a Space Scientist?"

"Math department has decided": Van Allen journal, May 20, 1958.

"We were like the heroes": Van Allen interview with De Vorkin transcript, 268.

Time magazine covered: "Reach into Space," May 4, 1959, 64–70.

"One that lingers in my mind": James Van Allen, "Energetic Particles in the Earth's External Magnetic Field," in *Discovery of the Magnetosphere,* Washington, D.C.: American Geophysical Union, 1997, p. 243.

Blasted the bomb shelter myth: Wells, *Annals of a University of Iowa Department: From Natural History to Physics and Astronomy,* 200.

"For $30, I built": Wells, Ibid., 200.

12. Discovery of the Radiation Belts

"I pointed out that another possibility": McIlwain, "Music and the Magnetosphere," 137.

"A masterpiece of miniature electronics": Walter Sullivan, *Assault on the Unknown*, New York: McGraw-Hill, 1961, 121.

A satellite with a memory: Ibid., 59.

Ludwig started working . . . to operate for a full forty-four days: Ludwig memoir.

Because of the design differences: Ibid.

"A beautiful majestic sight": Van Allen journal, March 22, 1958.

"My reaction to that was to": Ludwig quoted in *James Van Allen: Flights of Discovery*.

"At 3 A.M., I packed my work sheets": Van Allen, *Origins of Magnetospheric Physics*, 66.

"So we knew at once": McIlwain, "Music and the Magnetosphere," 139.

"The work of Størmer, at his desk": Sullivan, *Assault on the Unknown,* 116.

Throughout his career: Van Allen, *Origins of Magnetospheric Physics,* 13.

"This was equivalent to": Sullivan, *Assault on the Unknown,* 132.

The "leaky bucket" model of the radiation belts: James Van Allen, "Radiation Belts around the Earth," *Scientific American,* March 1959, 46.

"Radiation Belt Dims Hope of Space Travel": Walter Trohan, *Chicago Tribune*, May 2, 1958, 4.

"In particular, the second page of the translation": James A. Van Allen Papers, Box 33; reprinted in Van Allen, *Origins of Magnetospheric Physics,* 128.

"The first public announcement": Ibid., 129.

But the Russians continue to revisit: Ivan Vladimirovich Zavidonov, "Sputniks, Explorers and Propoganda: The Discovery of the Earth's Radiation Belts," in *History and Technology, an International Journal*, December 2000, 99–124.

Vernov focused the paper: S. A. Vernov, et al., "A study of the soft component of cosmic rays outside the atmosphere," Zavidonov, ibid., *Iskusstvennye sputniki Zemli*, article censored October 22, 1958, 117, 121.

But in 1960: S. N. Vernov and A. E. Chudakov, "Terrestrial Corpuscular Radiation and Cosmic Rays," *Space Research I*, 1960, 751.

Zavidonov dismisses any confusion: Zavidonov, "Sputniks," in *History and Technology, an International Journal,* 99–124.

"The Soviet Cosmic Rocket 'Mechta'": Reprinted in Van Allen, *Origins of Atmospheric Physics*, 129–130.

"In retrospect, these firings represent": James Van Allen, "The Radiation Environment of the Earth," *General Semantics Bulletin,* Nos. 24 and 25, 1959, 33–34.

13. Space Shield for the Cold War

"Nick was more strongly moved": York, *Making Weapons, Talking Peace*, 130.

"Only a few weeks after *Sputnik*": Ibid., 130–131.

"Nick always thought in terms of big numbers": York interview for *James Van Allen: Flights of Discovery.*

"A-bomb explosions must be carefully designed": Nicholas Christofilos, "The Argus Experiment," in *The Exploration of Space,* ed. Robert Jastrow, New York: MacMillan Co., 1960.

"Agreed [Iowa] will coordinate payload assembly!": James A. Van Allen Papers, Box 89-3.

The Iowa team took up the task: James A. Van Allen Papers, Box 89-3.

"This job is so": Carl McIlwain has quoted this sign, as did Sullivan, *Assault on the Unknown*, 129.

"We had no idea what was up there": McIlwain, "Music and the Magnetosphere," 140.

Texas Instruments provided: With Van Allen's permission, Texas Instruments used a photo of one of the disk-shaped pink-foam pancakes of electronics from *Explorer IV* as insets in company ads.

"Continued hard work in the laboratory": Ludwig time line, May 31–June 5, 1958.

"We sang variations": Ludwig time line, July 16, 1958.

"Once, curious about a Redstone rocket": McIlwain, "Music and the Magnetosphere," 140.

"You are the important ones": Wernher von Braun quoted by McIlwain; McIlwain, Ibid., 140; interview with the author.

"Mrs. Annabelle Hudmon": James A. Van Allen Papers, Box 89-7.

"When I paid another visit": Sullivan, *Assault on the Unknown,* 129.

Despite their yield: Van Allen, *Origins of Magnetospheric Physics*, 78.

"He chuckled merrily": "Buttoned-Up Spaceman," *Time,* September 21, 1959, 72–75.

"Blagonravov entertained our children": Van Allen, *Origins of Magnetospheric Physics,* 84.

14. Space as a Cottage Industry

North American Air Defense Command: NORAD is now called the North American Aerospace Defense Command.

And recently published: Donald Gurnett and A. Batacharjee, *Introduction to Plasma Physics,* Cambridge, U.K.: Cambridge University Press, 2005.

"We had a lovely, friendly reception": Brian O'Brien, August 2, 2002, memo to the author.

"Let's call it *Injun*": The name of the satellite raised one letter of complaint in the 1970s for its slang reference to Native Americans and, by then, the coast was clear for Van Allen to switch back to his first-choice name of *Hawkeye*.

"Injun Cookbook": James A. Van Allen Papers, Box 97–11.

Van Allen swiftly: Van Allen's challenge to the *Telstar* data: James Van Allen, "Geomagnetically Trapped Radiation Produced by a High-Altitude Nuclear Explosion" on July 9, 1962, *Nature,* 195, September 1962, 939–943.

"The telling points up two things": "No Ivory Tower for Iowa's Van Allen," *Des Moines Sunday Register,* December 8, 1965.

"I well realize that these offers": Van Allen journal, February 21, 1960.

"Abbie and I and the children": Van Allen journal, November 17, 1960.

Old mentor Tyndall retired: Van Allen journal, February 22, 1960.

Sentinella retired: Van Allen journal, June 29, 1962.

New York Times science reporter Jim Glanz: Glanz lectured on his reporting tour of duty in Iraq while at the University of Iowa for the September 7–9, 2005, ceremonies honoring him and four other alumni fellows of the College of Liberal Arts and Sciences.

The home cost: Van Allen journal, December 1, 1960.

15. The *Mariners*

Earth's cloud-shrouded neighbor: At its closest alignment with Earth, Venus is about 26 million miles away but the trajectory of the journey to rendezvous with the planet made the total distance much farther.

Newspaper articles speculated: *New York Times* editorial, December 20, 1962, 6.

"I was very keen on this": Van Allen interview with De Vorkin, transcript, 318.

"It is most important for the university . . . will be carefully heeded": "A Review of Space Research: The Report of the Summer Study conducted under the auspices of the Space Science Board of the National Academy of Sciences at the State University of Iowa, Iowa City, Iowa, June 17–August 10, 1962," *Publication 1079,* NAS, Washington, D.C., 1962, 1–2.

"The search for extraterrestrial life . . . such questions have little meaning": Ibid., 1–13.

"One crew member of each *Apollo* mission": Ibid., 11–14.

And scientists wanted that crew member: Ibid., 11–14.

Space Science Board, February 11–12, 1961, meeting minutes, James A. Van Allen Papers, Box 269-2.

"Deeply felt doubts . . . wasteful of resources": Needell, *Science, Cold War and the American State*, 360–361.

But Berkner . . . national priority: Ibid., 362.

"I believe that": It took another thirty-four years for the House of Representatives to endorse a plan (383–15) to send astronauts to another planet—Mars, based on a 2005 proposal from President George W. Bush.

"Will doubtless stand": Van Allen journal, April 12, 1961.

"We are convinced": Summer Study Session report, 4.

As for the other planets: Ibid., 4–7.

Goddard proposed a probe: Koppes, *JPL and the American Space Program*, 165–166.

"Worked with George Derbyshire": Van Allen journal, May 29, 1964.

"Received first substantial batch of data": Van Allen journal, December 7, 1964.

"We can tell if we get": *Iowa City Press-Citizen,* July 14, 1965, 3.

"The close-up investigation": *Iowa City Press-Citizen,* July 14, 1965, 3.

"If there are any Martian men": John Wilford, *New York Times*, July 15, 1965.

"It's a smaller planet": Ibid.

Later images indicated: http://nssdc.gsfc.nasa.gov/database/MasterCatalog?sc+1964–077A.

University "ground floor" contracts: Koppes, *JPL and the American Space Program*, 145.

"An intimate new sociology of space": *Fortune*, July 1969, 86.

"Then Mr. Webb had the idea": Van Allen interview with De Vorkin, transcript, 304.

"Publicized descriptions": Wells, *Annals of a University of Iowa Department: From Natural History to Physics and Astronomy*, 190.

Details on the building description: Ibid., 214.

Van Allen journal, January 11, 1963.

"I have been and continue to be very fond": Van Allen journal, May 9, 1964.

Lou Frank helped fill the gap: Van Allen journal, May 9, 1964.

"In one celebrated case:" Koppes, *JPL and the American Space Program*, 198.

16. Pioneers to the Outer Planets

"Great Galactic Ghoul": Quoted in Mark Wolverton, *The Depths of Space,* Washington, D.C.: Joseph Henry Press, 2004, 41.

"My own principal thrust . . . modern spacecraft": Van Allen journal, January 29, 1968.

Drafted a planning document: Hand-printed Space Science Board planning document, July 1968, Van Allen office papers.

The board's formal report: Thomas Gold, draft Space Science Board report, September 15, 1968, Van Allen office papers.

"But now no substitute": Daniel Greenberg, *Science,* 153, September 9, 1966, 1222.

"Sally and I watched . . . *An historic human achievement*": Van Allen journal, July 21, 1969.

"With characteristic one-up-man-ship": Van Allen journal, April 5, 1966.

"Van Allen was interested": C. W. Hall, Technical Memorandum 391–221, August 16, 1971; *Pioneer F* Jupiter Periapsis Working Group Meeting, 9, 1971, James A. Van Allen Papers, Box 138-5.

"Sending a number of well-equipped scouting parties": James Van Allen, *Science*, August 11, 1967, 659.

In one of his last: John Simpson quoted in Wolverton, *The Depths of Space*, 63.

"Death joined the group . . . no time for demonstrating": Descriptions and quotes regarding May 1970 campus demonstrations. Susan Boyd, "Death on the Lawn, a Diary of May 1970 in Iowa City," in *The Wide-Brimmed Hat,* Iowa City: The Long River Press, 2002.

"His health has been": Van Allen journal, March 30, 1971.

"At a speed 15 times higher than a rifle bullet": Richard Fimmel, James Van Allen, and Eric Burgess, *Pioneer: First to Jupiter, Saturn and Beyond,* Washington, D.C.: NASA (SP-446), 1980, 75.

Pioneer 10 . . . long-distance calls: Ibid., 93.

"Since the *Pioneer 10* spacecraft": Wolverton, *The Depths of Space*, 103.

The two graduate students: Daniel Baker, Van Allen Day banquet speech, October 9, 2004, Iowa City, Iowa.

"Everything on the spacecraft ": Quoted in Wolverton, *The Depths of Space*, 113.

"The Van Allen radiation belts": John Wilford, *New York Times,* November 30, 1973, 68.

"Charlie Hall got up": Ed Smith, University of Iowa colloquium for Van Allen Day, October 9, 2004.

"The muzzle velocity": Fimmel, Van Allen, Burgess, *Pioneer*, 244.

"Slingshot technique": Ibid., 22.

Belts and rings at Saturn: James Van Allen, et al., "Saturn's Magnetosphere, Rings, and Inner Satellites," *Science,* 207, January 25, 1980, 415–421; Fimmel, Van Allen, et al., 244.

Hall later estimated: http://quest.arc.nasa.gov/sso/cool/pioneer10/general/am twotxt.html.

Van Allen's term: Van Allen journal, July 16, 1970.

17. Space Politics

"An immensely exciting day": Van Allen journal, February 9, 1986.

"With no difficulty": James Van Allen, "Space Science, Space Technology and the Space Station," *Scientific American,* 254, January 1986, 32–39.

Potential space milestones: McDougall, *The Heavens and the Earth*, 421; Stuhlinger, *Wernher von Braun: Crusader for Space*, 233 and 298–299.

Economics professor Peter Alonzi . . . Van Allen said: Peter Alonzi, letter to the author, January 26, 2002.

"The military use": *New York Times,* March 29, 1981.

But the military benefits: Koppes, *JPL and the American Space Program*, 251.

"*Apollo* was a matter of": McDougall, *The Heavens and the Earth*, 421.

"A woman in the front row": James Van Allen, speech transcript to the Division for Planetary Sciences, Ithaca, New York, October 18, 1983.

Van Allen, *Scientific American*, January 1986.

"This function has been performed": James Van Allen, "Myths and Realities of Space Flight," *Science*, 232, May 1986, 1075–1076.

Ben Bova: *Science,* 233, August 8, 1986, 610.

James R. Arnold: *Science,* 233, August 8, 1986, 610.

"My own view is that our national predicament": James Van Allen, "Is Human Spaceflight Now Obsolete?" *Science,* May 2004, 822.

He quoted a massive: "Space Science in the Twenty-First Century," 1986 study of the Space Science Board of the National Academy of Sciences.

The SSB took the field of microgravity: Van Allen journal, January 21, 1989.

"On Thursday, the SSB and NASA": Van Allen journal, January 13, 1989.

She held hearings on retiring the shuttle: Science and Space Subcommittee of the Senate Commerce, Science and Transportation Committee, Senator Kay Bailey Hutchison, Chair. April 20, 2005.

The title of the talk: Van Allen, "Is Human Spaceflight Now Obsolete?" May 2004, 822.

Wrote letters for the occasion: Van Allen office papers, 1985.

A bright and unusual aurora: Van Allen journal, November 13, 1991.

Two of the most powerful solar flares: Listen for the sounds of the flares at http://www.-pw.physics.uiowa.edu/space-audio/.

Modeling the radiation belts: http://lws.gsfc.nasa.gov/missions/geospace/geospace.htm.

18. Journey to the Edge of the Solar System

University President: Skorton left the University of Iowa to take the presidency of Cornell University in 2006.

He and Bruce Randall: "Projected disappearance of the 11-year cyclic minimum of galactic cosmic ray intensity in the antapex direction within the outer heliosphere," *Geographical Research Letters*, April 15, 2005, v.32, L07102.

"Around the house": Peter Van Allen, speech to the National Space Club, Washington, D.C., March 17, 2006.

Completed two technical papers: James Van Allen, "Encounter of an asteroid with a planet," *American Journal of Physics*, 74(8), 2006, 717–719; James Van Allen, "Inference of magnetospheric currents from multipoint magnetic field measurements," *American Journal of Physics*, 74(9), 2006, 809–814.

Selected Bibliography

Interviews

Peter Alonzi, January 26, 2002
Ralph Baldwin, August 16, 2002
Bill Boyd, November 9, 2002
Susan Boyd, October 30, 2002
Thomas Boyd, November 9, 2002
Willard Boyd, November 9, 2002; July 20, 2004
Larry Cahill, July 8, 2003
Fritz Coester, July 20, 2004
Walter Cronkite, March 28, 2001
Louis Frank, December 18, 2001
Edward Freund, October 8, 2004
Donald Gurnett, May 21, 2001; May 22, 2001; October 8, 2004; September 12, 2005
Sophie Haroutunian, April 24, 2003
Annabelle Hudmon, February 23, 2001
Craig Kletzing, September 19, 2001
Stamatios Krimigis, August 29, 2002; October 9, 2004
Larry Lasher, April 26, 2003
William Littlewood, July 12, 2003
David Lozier, May 1, 2002
George Ludwig, March 13, 2002; October 8, 2004; October 9, 2004
Carl McIlwain, January 2, 2002; October 10, 2004
Agnes McLaughlin, April 28, 2001
Les Meredith, February 16, 2001; October 8, 2004
Mike Neufeld, January 8, 2008
Edwin Norbeck, July 8, 2004
Michael Nowack, October 8, 2004
Brian O'Brien, August 2, 2002
William Pickering, August 13, 2001
Guido Pizzella, October 9, 2004
Bruce Randall, August 21, 2001; July 23, 2004
Alan Rogers, April 7, 2003
Cynthia Van Allen Schaffner, March 23, 2001
Richard Schwoibel (interview with Van Allen in which the author participated), April 23, 2001
Art Stephenson, August 3, 2001
Edward Stone, October 9, 2004
Ernst Stuhlinger, March 28, 2001; July 11, 2001

Lowell Swartz, October 8, 2004
Michelle Thomsen, June 4, 2001; October 9, 2004
Abigail Halsey Van Allen, November 28, 2000; June 26, 2001; October 10, 2001; May 29, 2002; November 15, 2005
David Van Allen, March 19, 2001
James Van Allen, October 19, 2000; November 27, 2000; November 28, 2000; November 29, 2000; December 14, 2000; December 15, 2000; January, 23, 2001; January 24, 2001; February 20, 2001; February 21, 2001; March 20, 2001; March 21, 2001; April 24, 2001; May 22, 2001; May 23, 2001; June 25, 2001; June 26, 2001; June 27, 2001; September 17, 2001; September 18, 2001; September 19, 2001; October 15, 2001; October 16, 2001; October 17, 2001; December 18, 2001; December 19, 2001; January 14, 2002; January 16, 2002; February 20, 2002; February 18, 2002; February 19, 2002; February 20, 2002; March 25, 2002; March 26, 2002; April 22, 2002; April 23, 2002; April 24, 2002; May 25, 2002; March 25, 2003; November 11, 2003; July 6–13, 2004; October 11, 2004; June 6, 2005; September 12, 2005; November 15, 2005; February 8, 2006; March 24, 2006
Janet Hunt Van Allen, March 19, 2001
Peter Van Allen, April 22, 2002; October 9, 2005; January 22, 2006
Sarah Van Allen, July 26, 2001
Pat Widder, March 26, 2001

Transcripts of Interviews from *James Van Allen: Flights of Discovery,* University of Iowa Film, 2000

Larry Cahill
George Ludwig
Carl McIlwain
William Pickering
Homer Joe Stewart
Edward Stone
Ernst Stuhlinger
Abigail Halsey Van Allen
Cynthia Van Allen
James Van Allen
Sarah Van Allen
Herbert York

Archives and Other Unpublished Sources

Applied Physics Laboratory Archive, Johns Hopkins University, Silver Spring, Maryland
Dwight D. Eisenhower Library, Totowa, New Jersey
Evans Library, Florida Institute of Technology, Orlando, Medaris archive

Flower Memoral Library, Genealogy Department, Watertown, New York
Jet Propulsion Laboratory Archive, NASA Center managed by the California Institute of Technology Pasadena, California
Ludwig, George. "Opening Space Research at the University of Iowa." Contact ludwiggh@visuallink.com
Lyndon Baines Johnson Library and Museum, Austin, Texas, space policy and history files
Meredith, Les. Journals of the 1952 and 1953 rockoon missions
Mount Pleasant City Hall, Mount Pleasant, Iowa, meeting minutes
Mount Pleasant Historical Society, Mount Pleasant, Iowa
NASA, John F. Kennedy Space Center, Cape Canaveral, Florida
NASA Archive, Marshall Space Flight Center, Huntsville, Alabama
National Air and Space Museum, Smithsonian Institution, Washington, D.C., Space Astronomy Oral History Project, interview transcripts
Poulter, Thomas. *Over the Years*, unpublished autobiography, Van Allen Office Papers
U.S. Space Center, Huntsville, Alabama, von Braun Archive
James A. Van Allen Papers and Related Collections, University of Iowa Library Special Collections and University Archives, Iowa City, including more than 700 boxes of documents, journals, correspondence, mission specs, photographs, and logs
Van Allen Office Papers. The transfer of papers from Van Allen's office to the *James A. Van Allen Papers and Related Collections* was still underway as this book went to publication
Van Allen House Heritage Center, family papers and photographs

Books

Baldwin, Ralph. *The Deadly Fuze*. San Rafael, CA: Presidio Press, 1980.
Baldwin, Ralph. *They Never Knew What Hit Them.* Naples, Florida: Reynier Press, 1999.
Beon, Yves. *Planet Dora: A Memoir of the Holocaust and the Birth of the Space Age.* Boulder, CO: Westview Press, 1997.
Bille, Matt, and Erika Lishock. *The First Space Race: Launching the World's First Satellites.* College Station: Texas A & M University Press, 2004.
Boyd, Susan Kuehn. *The Wide-Brimmed Hat.* Iowa City: Long River Press, 2002.
Brandt, E. N. *Growth Company: Dow Chemical's First Century*. East Lansing, Mi: Michigan State University Press, 1997.
Bulkeley, Rip. *The Sputniks Crisis and Early United States Space Policy: A Critique of the Historiography of Space*. Bloomington: Indiana University Press, 1991.
Byrd, Richard Evelyn. *Discovery: The Story of the Second Byrd Antarctic Expedition*. New York: G. P. Putnam, 1935.
Clarke, Arthur. *The Promise of Space.* New York: Harper and Row, 1968.

Cronkite, Walter. *A Reporter's Life*. New York: Alfred Knopf, 1996.
Crouch, Tom. *Aiming for the Stars: The Dreamers and Doers of the Space Age*. Washington, D.C.: Smithsonian Institution Press, 1999.
Dallek, Robert. *Lyndon B. Johnson: Portrait of a President*. New York: Oxford University Press, 2004.
De Vorkin, David. *Race to the Stratosphere: Manned Scientific Ballooning in America*. New York: Springer-Verlag, 1989.
De Vorkin, David. *Science with a Vengeance: How the Military Created the U.S. Space Sciences after World War II*. New York: Springer-Verlag for the Smithsonian Institution, 1992.
Dickson, Paul. *Sputnik: The Shock of the Century*. New York: Walker and Co., 2001.
Enscore, Susan. *Operation Paperclip at Fort Bliss: 1945–1950*. Fort Bliss, TX: United States Army Air Defense Artillery Center, 1998.
Frank, Louis, with Patrick Huyghe. *The Big Splash*. Secaucus, N.J.: Carol Publishing Group, 1990.
Friedman, Herbert. *Sun and Earth*. New York City: Scientific American Books, 1986.
Gailey, Harry. *The War in the Pacific*. Novato, California: Presidio Press, 1995.
Gilmore, Stewart C., and John Spreiter, eds. *Discovery of the Magnetosphere*. Washington, D.C.: American Geophysical Union, 1997.
Glennan, T. Keith, and J. T. Huntley, eds. *The Birth of NASA: The Diary of T. Keith Glennan*. Washington, D.C.: NASA, 1993.
Goodwin, Doris Kearns. *No Ordinary Time*. New York: Touchstone, 1995.
Green, Constance McLaughlin, and Milton Lomask. *Vanguard: A History*. Washington, D.C.: Smithsonian Institution Press, 1971.
Gurnett, Donald, and Amitava Batacharjee. *Introduction to Plasma Physics: With Space and Laboratory Applications*. Cambridge, U.K.: Cambridge University Press, 2005.
Halas, Christine. *Guide to the James A. Van Allen Papers and Related Collections*. Iowa City, Iowa: The University of Iowa Archives, 1993.
Harford, James. *Korolev: How One Man Masterminded the Soviet Drive to Beat America to the Moon*. New York: John Wiley & Sons, 1997.
Jastrow, Robert, ed. *The Exploration of Space: A Symposium of Space Physics (April 28–30, 1959)*. New York: Macmillan, 1960.
Killian, James. *Sputnik, Scientists, and Eisenhower: A Memoir of the First Special Assistant to the President for Science and Technology*. Cambridge, MA: MIT Press, 1977.
Koppes, Clayton. *JPL and the American Space Program*. New Haven, CT: Yale University Press, 1982.
Kranz, Gene. *Failure Is Not an Option*. New York: Simon and Schuster, 2000.
Kuhn, Thomas. *The Structure of Scientific Revolution*. Chicago: The University of Chicago Press, 1962.

Launius, Roger. *Frontiers of Space Exploration.* Westport, CT: Greenwood Press, 1998.

Launius, Roger, and Howard McCurdy. *Presidential Leadership in the Development of the U.S. Space Program:* Washington, D.C.: NASA, 1994.

Launius, Roger, John Logsdon, and Robert Smith, eds. *Reconsidering Sputnik: 40 years Since the Soviet Satellite.* Amsterdam: Harwood Academic Press, 2000.

Ley, Willie. *Rockets: The Future of Travel Beyond the Stratosphere.* New York: Viking Press, 1944.

Ley, Willie. *Watchers of the Skies: An Informal History of Astronomy from Babylon to the Space Age.* New York: Viking Press, 1963.

Logsdon, John M., ed. *Organizing for Exploration, Exploring the Unknown: Selected Documents in the History of the U.S. Space Program.* Washington, D.C.: NASA History Series, National Aeronautics and Space Administration, 1995.

McDougall, Walter. *The Heavens and the Earth: A Political History of the Space Age.* New York: Basic Books, 1985.

Montoya, Earl, and Richard Fimmel. *Space Pioneers.* Washington, D.C.: NASA, 1987.

Naugle, John. *First Among Equals: The Selection of NASA Space Science Experiments.* Washington, D.C.: NASA, 1991.

Needell, Allan. *Science, Cold War and the American State: Lloyd V. Berkner and the Balance of Professional Ideals.* Amsterdam: Harwood Academic Publishers in Association with the National Air and Space Museum, Smithsonian Institution, 2000.

Needell, Allan, ed. *The First 25 Years in Space.* Washington, D.C.: Smithsonian Institution Press, 1983.

Neufeld, Michael J. *The Rocket and the Reich: Peenemünde and the Coming of the Ballistic Missile Era.* New York: The Free Press, 1995.

Newell, Homer. *Beyond the Atmosphere: Early Years of Space Science.* Washington, D.C.: NASA, 1980.

Newell, Homer. *High Altitude Rocket Research.* New York: Academic Press, 1953.

Newell, Homer, ed. *Sounding Rockets.* New York: McGraw-Hill, 1959.

Nicks, Oran. *Far Travelers: The Exploring Machines.* Washington, D.C.: NASA, 1985.

Odishaw, Hugh, et al, eds. *A Review of Space Research: The Report of the Summer Study conducted under the Auspices of the Space Science Board of the National Academy of Sciences at the State Univesity of Iowa, June 17–August 10,* Publication #1079, Washington, D.C. National Academy of Sciences, 1962.

Ordway, Frederick, III, and Randy Liebermann, *Blueprint for Space: Science Fiction to Science Fact.* Washington, D.C.: Smithsonian Institution Press, 1992.

Ordway, Frederick, III, and Mitchell Sharpe. *The Rocket Team.* New York: Crowell, 1979.

Piszkiewicz, Dennis. *Wernher von Braun: The Man Who Sold the Moon.* Westport, CT: Praeger, 1998.

Schlain, Leonard. *Art & Physics: Parallel Visions in Space, Time and Light*. New York: William Morrow, 1991.
Stuhlinger, Ernst, and Fred Ordway, III. *Wernher von Braun: Crusader for Space: A Biographical Memoir.* Malabar, FL: Krieger Publishing Co., 1996.
Stuhlinger, Ernst, and Fred Ordway, III. *Wernher von Braun: Crusader for Space: An Illustrated Memoir.* Malabar, FL: Krieger Publishing Co., 1996.
Sullivan, Walter. *Assault on the Unknown*. New York: McGraw-Hill Book Co., 1961.
Van Allen, James, ed. *Cosmic Rays, the Sun and Geomagnetism: The Works of Scott E. Forbush.* Washington, D.C.: The American Geophysical Union, 1993.
Van Allen, James. *924 Elementary Problems and Answers in Solar System Astronomy*. Iowa City: University of Iowa Press, 1993.
Van Allen, James. *Origins of Magnetospheric Physics.* Washington, D.C.: Smithsonian Institution, 1983. Iowa City: University of Iowa Press, 2004.
Van Allen, James, Richard Fimmel and Eric Burgess. *Pioneer: First to Jupiter, Saturn and Beyond.* Washington, D.C.: NASA, 1980.
Van Allen, James, ed. *Scientific Uses of Earth Satellites.* Ann Arbor: University of Michigan Press, 1956.
von Braun, Wernher, and Frederick Ordway III. *History of Rocketry and Space Travel*. Chicago: J. G. Ferguson Publishing Co., 1966.
von Braun, Wernher. *Space Frontier*. New York: Holt, Rinehart and Winston, 1971.
Webb, James. *Space Age Management*. New York: McGraw-Hill, 1962.
Wells, James. *Annals of a University of Iowa Department: From Natural History to Physics and Astronomy*. Iowa City: Department of Physics and Astronomy, University of Iowa Publication 80-19, 1980.
Whipple, Fred. *Orbiting the Sun: Planets and Satellites of the Solar System.* Cambridge, MA: Harvard University Press, 1981.
Winter, Frank. *Prelude to the Space Age: The Rocket Societies: 1924–1940.* Washington, D.C.: National Air and Space Museum, Smithsonian Institution Press, 1983.
Wolverton, Mark. *The Depths of Space: The Story of the Pioneer Planetary Probes.* Washington, D.C.: Joseph Henry Press, 2004.
York, Herbert. *Making Weapons, Talking Peace: A Physicist's Odyssey from Hiroshima to Geneva*. New York: Basic Books, 1987.

Selected Articles, Chapters, and Web Sites

Arnold, James. Letters. *Science* 233 (August 8, 1986): 610.
Baker, D., and J. Van Allen. "Energetic Electrons in the Jovian Magnetosphere." *Journal of Geophysical Research* 81 (1976): 617–632.
Bova, Ben. Letter. *Science* 233 (August 8, 1986): 610.
Bowen, I., et al. "New Light on the Nature and Origins of Incoming Cosmic Rays." *Physical Review* 53 (1938): 855–861.
Bradt, H., and B. Peters. "Investigation of the Primary Cosmic Radiation with

Nuclear Photographic Emulsions." *Physical Review* 74 (December 15, 1948): 1828–1837.

Cahill, Laurence, Jr., and James Van Allen. "High Altitude Measurements of the Earth's Magnetic Field with a Proton Precession Magnetometer." *Journal of Geophysical Research* 61 (1956): 547–558.

Christofilos, Nicholas. "The Argus Experiment." In Robert Jastrow, ed. *The Exploration of Space*. New York: Macmillan, 1960.

Ellis, R. J., et al. "Low-Momentum End of the Spectra of Heavy Primary Cosmic Rays." *Physical Review* 95 (1954): 147–159.

Frank, L., et al. "Mariner II: Preliminary Reports on Measurements of Venus Charged Particles." *Science* 139 (1963): 905–907.

Frank, Louis, and James Van Allen. "Survey of Magnetospheric Boundary Phenomena." In Hugh Odishaw, ed. *Research in Geophysics, Volume 1: Sun, Upper Atmosphere, and Space*. Cambridge: Massachusetts Institute of Technology, 1964. 161–187.

Fraser, L., et al. "Methods in Cosmic-Ray Measurement in Rockets." *Physical Review* 72 (1947): 173.

Furnas, Clifford. *Life* (October 21, 1957): 22–25.

Gangnes, A., et al. "The Cosmic-Ray Intensity above the Atmosphere." *Physical Review* 75 (1949): 57–69.

Gehrels, T., and J. Van Allen. "New Ring and Satellites of Saturn." *Circular No. 3417, Central Bureau for Astronomical Telegrams, International Astronomical Union*, October 25, 1979.

Krimigis, Stamatios, and James Van Allen. "Observations of the February 5–12, 1965, Solar Particle Event with *Mariner 4* and *Injun 4*." *Journal of Geophysical Research* 72 (1967): 4471–4486.

Ludwig, George, and James Van Allen, "Instrumentation for a Cosmic Ray Experiment for the Minimal Earth Satellite." *Journal of Astronautics* 3 (1956): 59–61.

Mann, Martin. "Going up for Keeps." *Popular Science* 150 (March 1947): 66–71.

Meredith, Les, et al. "Direct Detection of Soft Radiation above 50 Kilometers in the Auroral Zone." *Physical Review* 97 (1955): 201–205.

Meredith, Les, et al. "Cosmic-Ray Intensity above the Atmosphere at High Latitudes." *Physical Review* 99 (1955): 198–209.

Neufeld, Michael. "Orbiter, Overflight and the First Satellite: New Light on the Vanguard Decision." In R. Launius, J. Logsdon, and R. Smith, eds. *Reconsidering Sputnik*. Amsterdam: Harwood Academic Publishers, 2001.

"Reach into Space." *Time* magazine cover story profiling James Van Allen, May 4, 1959.

Sullivan, Walter, and James Van Allen. "Discoverer of Earth-Circling Radiation Belts Is Dead at 91." *New York Times*, Obituaries, August 10, 2006, C14.

Tatel, H. and J. Van Allen. "Cosmic-Ray Bursts in the Upper Atmosphere." *Physical Review* 73 (1948): 87–88.

Van Allen, James. "Apparatus for Measuring Young's Modulus at Small Strains." *Proceedings of the Iowa Academy of Sciences* 44 (1937): 152.

Van Allen, J., et al. "Cross Section for the Reaction $H^2 + H^2 \rightarrow H^1 + H^3$ with a Gas Target." *Physical Review* 56 (1939): 383.

Van Allen, J., and N. F. Ramsey, Jr. "A Technique for Counting High Energy Protons in the Presence of Fast Neutrons." *Physical Review* 57 (1940): 1069–1070.

Van Allen, James, and Nicholas M. Smith, Jr. "The Absolute Cross Section for the Photo-Disintegration of Deuterium by 6.2 MeV Quanta." *Physical Review* 59 (1941): 618–619.

Van Allen, James. "The Exploration of the Future." *Explorers Journal* 24 (Summer 1947): 1–5.

Van Allen, J., et al. "Methods in Cosmic Ray Measurements in Rockets." *Physical Review* 72 (1947): 173.

Van Allen, J., and H. Tatel. "The Cosmic Ray Counting Rate of a Single Geiger Counter from Ground Level to 161 Kilometers Altitude." *Physical Review* 73 (February 1948): 245–251.

Van Allen, J., et al. "The Aerobee Sounding Rocket: A New Vehicle for Research in the Upper Atmosphere." *Science* 108 (1948): 746–747.

Van Allen, James. "Exploratory Cosmic Ray Observations at High Altitudes by Means of Rockets." *Sky and Telescope* 7 (1948): 171–175.

Van Allen, James. "Introduction: Research by Rockets," *Physical Review* 73 (1948).

Van Allen, James. "Transition Effects of Primary Cosmic Radiation in Lead, Aluminum and the Atmosphere." *Proceedings of the Echo Lake Cosmic Ray Symposium, June 23–28, 1949*. Washington, D. C.: Office of Naval Research, Department of the Navy (November 1949): 95–102.

Van Allen, J., and A. Gangnes. "The Cosmic-Ray Intensity above the Atmosphere at the Geomagnetic Equator." *Physical Review* 78 (1950): 50–52.

Van Allen, J., and S. Singer, "On the Primary Cosmic-Ray Spectrum." *Physical Review* 78 (1950): 116.

Van Allen, James. "Rockets for Studying the Upper Atmosphere." *Aero Digest* 61 (September 1950): 20–23.

Van Allen, James. "The Rockets Report." *The Johns Hopkins Magazine* (October 1950): 1–8.

Van Allen, James. "Intensities of Heavy Cosmic-Ray Primaries by Pulse Ionization Chamber Measurements." *Physical Review* 84 (1951): 791–797.

Van Allen, J., and S. Singer, "Apparent Absence of Low-Energy Cosmic-Ray Primaries." *Nature* 170 (1952): 62–63.

Van Allen, James. "The Nature and Intensity of the Cosmic Radiation." In C. White and O. Benson, Jr., eds. *Physics and Medicine of the Upper Atmosphere*. Albuquerque: University of New Mexico Press, 1952. 239–266.

Van Allen, James. "The Angular Motion of High-Altitude Rockets." In C. White and O. Benson, Jr., eds. *Physics and Medicine of the Upper Atmosphere.* Albuquerque: University of New Mexico Press, 1952. 412–431.

Van Allen, James. "Pressures, Densities, and Temperatures in the Upper Atmosphere: The Rocket Panel." *Physical Review* 88 (1952): 1027–1032.

Van Allen, James. "Cosmic-Ray Intensity above the Atmosphere near the Geomagnetic Pole." *Physical Review* 89 (1953): 891.

Van Allen, James, and Melvin Gottlieb. "The Inexpensive Attainment of High Altitudes with Balloon-launched Rockets." In R. Boyd and M. Seaton, eds. *Rocket Exploration of the Upper Atmosphere.* London: Pergamon Press, 1954. 53–64.

Van Allen, James. "The Artificial Satellite as a Research Instrument." *Scientific American* 195 (November 1956): 41–47.

Van Allen, James, and Carl McIlwain. "Cosmic-Ray Intensity at High Altitudes on February 23, 1956." *Journal of Geophysical Research* 61 (1956): 569–571.

Van Allen, James. "Direct Detection of Auroral Radiation with Rocket Equipment." *Proceedings of the National Academy of Sciences* 43 (1957): 57–62.

Van Allen, James, et al. "Observation of High Intensity Radiation by Satellites 1958 Alpha and Gamma." *Jet Propulsion* (September 1958): 588–592.

Van Allen, James. "Radiation Belts around the Earth." *Scientific American* 200 (March 1959): 39–47.

Van Allen, James, et al. "Radiation Observations with Satellite 1958." *Journal of Geophysical Research* 64 (1959): 271–286.

Van Allen, James, and Louis Frank. "Radiation around the Earth to a Radial Distance of 107,400 km." *Nature* 183 (1959): 430–434.

Van Allen, James, et al. "Radiation Measurements from Explorer IV." *Proceedings, 9th International Astronautical Congress, Amsterdam 1958.* Wien: Springer-Verlag, 1959.

Van Allen, James, and Louis Frank. "Radiation Measurements to 658,300 km. with Pioneer 44." *Nature* 184 (1959): 219–224.

Van Allen, James, et al. "Satellite Observations of Electrons Artificially Injected into the Geomagnetic Field." *National Academy of Sciences Proceedings* 45 (1959): 1152–1170 and *Journal of Geophysical Research* 64 (1959): 877–891.

Van Allen, James. "The Geomagnetically-Trapped Corpuscular Radiation." *Journal of Geophysical Research* 64 (1959): 1683–1689.

Van Allen, James, et al. "The Aerobee Rocket." In Homer Newell, ed. *Sounding Rockets.* New York: McGraw-Hill, 1959. 54–70.

Van Allen, James. "Balloon-Launched Rockets for High-Altitude Research." In Homer Newell, ed. *Sounding Rockets.* New York: McGraw-Hill, 1959. 143–164.

Van Allen, James. "The Radiation Environment of the Earth." *General Semantics Bulletin* 24 and 25 (1959): 33–40.

Van Allen, James. "Origin and Nature of the Geomagnetically-Trapped Radiation." *Proceedings of the First International Space Science Symposium.* Nice, France: 1960. 749–750.

Van Allen, James. "On the Radiation Hazards of Space Flight." *The Physics and Medicine of the Atmosphere and Space.* New York: John Wiley and Sons, 1960. 1–13.

Van Allen, James. "The Earth and Near Space." *Bulletin of the Atomic Scientists* (May June 1961): 218–222.

Van Allen, James. "The Danger Zone." *Space World* 49 (December 1961): 22–23.

Van Allen, James, and W. Whelpley. "Radiation Observations with Satellite *Injun I*, September 28-October 4, 1961." *Journal of Geophysical Research* 67 (1962): 1660–1661.

Van Allen, James, and Louis Frank. "The Iowa Radiation Experiment (Mariner II Encounter with Venus)." *Science* 138 (1962): 1097–1098.

Van Allen, James. "Geomagnetically Trapped Radiation Produced by a High-Altitude Nuclear Explosion on July 9, 1962." *Nature* 195 (September 1962): 939–943.

Van Allen, James. "The Starfish Test." *Nuclear Information* 5 (January 1963): 1–12.

Van Allen, J., et al. "Absence of Martian Radiation Belts and Implications Thereof." *Science* 149 (1965): 1228–1233.

Van Allen, James. "Are We to Abandon the Planets to the Soviet Union?" *Science* 158 (1967): 1405.

Van Allen, J., et al. "Observed Absence of Energetic Electrons and Protons near Venus." *Journal of Geophysical Research* 73 (1968): 421–425.

Van Allen, J., et al. "The North Liberty Radio Observatory of the University of Iowa." *Solar Physics* 7 (1969): 159–163.

Van Allen, James. "The Trip to Jupiter." *Bulletin of the Atomic Scientists* (December 1973): 52–56.

Van Allen, J., et al. "Energetic Electrons in the Magnetosphere of Jupiter." *Science* 184 (1974): 309–311.

Van Allen, James, and Roger Randall. "Jupiter's Magnetosphere as Observed with Pioneer 10." *Journal of Geophysical Research* 79 (1974): 3559–3577.

Van Allen, James, "Investigation of Uranus, Its Satellites, and Distant Interplanetary Phenomena by Spacecraft Techniques." *Icarus* 24 (1975): 277–279.

Van Allen, et al. "Pioneer 11 Observations of Energetic Particles in the Jovian Magnetosphere." *Science* 188 (1975): 459–462.

Van Allen, James. "Interplanetary Particles and Fields." *Scientific American* 233 (1975): 160–173.

Van Allen, James. "The Magnetospheres of Jupiter, Saturn, and Uranus." In Edith Muller, ed. *Highlights of Astronomy.* Vol. 4. Dordrecht, Holland: D. Reidel Publishing Company, 1977. 195–224.

Van Allen, James. “Propagation of a Forbush Decrease in Cosmic Ray Intensity to 15.9 AU.” *Geophysical Research Letters* 6 (1979): 566–568.
Van Allen, J. et al. “Saturn’s Magnetosphere, Rings, and Inner Satellites.” *Science* 207 (1980): 415–421.
Van Allen, James. “An Analytical Solution of the Two Star Sight Problem of Celestial Navigation.” *Navigation* 28 (1981): 40–43.
Van Allen, James. “Radiation Belts of the Earth.” *Air & Space* 5 (1981): 10–11.
Van Allen, James. “Findings on Rings and Inner Satellites of Saturn by Pioneer 11.” *Icarus* 51 (1982): 509–527.
Van Allen, James. “Magnetospheres and the Interplanetary Medium.” In J. Kelly Beatty, et al, eds. *The New Solar System*. Cambridge, MA: Sky Publishing Corporation and Cambridge University Press, 1982. 23–32.
Van Allen, James. “The Beginnings of Magnetospheric Physics.” In L. Napolitano, ed. *Space: Mankind’s Fourth Environment*. New York: Pergamon Press, 1982, 419–433.
Van Allen, James. “On Being President of AGU.” *EOS* 64 (1983): 945.
Van Allen, James. “Genesis of the International Geophysical Year.” *EOS* 64 (1983): 977.
Van Allen, James. “Space Science, Space Technology and the Space Station.” *Scientific American* 254 (January 1986): 32–39.
Van Allen, James. “Myths and Realities of Space Flight.” *Science* 232 (May 1986): 1075–1076.
Van Allen, James. “Early Days of Space Science.” *Journal of the British Interplanetary Society* 41 (1988): 11–15.
Van Allen, James. “What Is a Space Scientist? An Autobiographical Example.” *Annual Review of Earth and Planetary Sciences* (June 1989): 1–26.
Van Allen, James, “The Magnetospheres of Eight Planets and the Moon.” Norwegian Academy of Science and Letters Nansen Memorial Lecture, Oslo, Norway, October 10, 1990.
Van Allen, James. “On the Future of Space Science and Applications.” Norwegian Academy of Science and Letters, Nansen/Birkeland Symposium, Oslo, Norway, October 11, 1990.
Van Allen, James. “Why Radiation Belts Exist.” *EOS: Transactions of the American Geophysical Union* 72 (1991): 361, 363.
Van Allen, James. “Space Station a.k.a. Project Vampire.” *Science* 265 (1994): 1017.
Van Allen: James. “Book Review of *Race to the Stratosphere: Manned Scientific Ballooning in America*, by David H. De Vorkin.” *Journal for the History of Astronomy* 26 (May 1995): 177–179.
Van Allen, James. “Book Review of *Science with a Vengeance: How the Military Created the U.S. Space Sciences after World War II*, by David H. De Vorkin.” *Journal for the History of Astronomy* 26 (May 1995): 179–181.

Van Allen, James. "Twenty-Five Milliamperes: A Tale of Two Spacecraft." *Journal of Geophysical Research* 101 (1996): 479–495.

Van Allen, James, and Bruce Randall. "A Durable Reduction of Cosmic Ray Intensity in the Outer Heliosphere." *Journal of Geophysical Research* 102 (1997): 4631–4641.

Van Allen, James. "My Life at APL." *Johns Hopkins APL Technical Digest* 18 (1997): 173–177.

Van Allen, James. "Update on Pioneer 10." *Eos: Transactions of the American Geophysical Union* 79 (1998): 123.

Van Allen, James. "Scott Ellsworth Forbush." *Biographical Memoirs*, Vol. 74. Washington, D.C.: The National Academy Press, 1998, 92–109.

Van Allen, James. "Magnetospheric Physics." In Johan Bleeker, et al, eds. *The Century of Space Science.* Dordrecht: Kluwer, 2001.

Van Allen, J., and W. R. Webber. "Observed Solar Modulation of Galactic Cosmic Ray Intensity in the Outer Heliosphere, 1997–2001." *Geophysical Research Letters* 29 (April 11, 2002): 1131.

Van Allen, James. "Gravitational Assist in Celestial Mechanics—A Tutorial." *American Journal of Physics* 71 (May 2003): 448–451.

Van Allen, James. "Is Human Spaceflight Now Obsolete?" *Science* 304 (May 2004): 822.

Van Allen, James. "Basic Principles of Celestial Navigation." *American Journal of Physics* 72 (November 2004): 1418–1424.

Van Allen, James, and Bruce Randall. "Projected Disappearance of the 11-Year Cyclic Minimum of Galactic Cosmic Ray Intensity in the Antapex Direction within the Outer Heliosphere." *Geophysical Research Letters* 32 (April 15, 2005).

Van Allen, James. "Encounter of an Asteroid with a Planet." *American Journal of Physics* 74 (August 2006): 717–719.

Van Allen, James. "Inference of Magnetospheric Currents from Multipoint Magnetic Field Measurements." *American Journal of Physics* 74 (September 2006): 809–814.

Vernov, S., et al. "Artificial Satellite Measurements of Cosmic Radiation." In L, Kurnosova, ed. *Artificial Earth Satellites.* New York: Plenum Press, 1960. 5–9.

Vernov, S. N., and A. E. Chudakov. "Terrestrial Corpuscular Radiation and Cosmic Rays." *Space Research I* (1960).

Von Braun, Wernher. "The Explorers." *Des Moines Sunday Register, This Week* magazine (April 13, 1958): 9.

Walsh, Tom. "Space Explorer James Van Allen: Pioneer of a New Frontier," Iowa City *Gazette* (February 4, 2002): front page.

Zavidonov, Ivan Vladimirovich. "Sputniks, Explorers and Propaganda: The Discovery of the Earth's Radiation Belts." *History and Technology: An International Journal* (December 2000): 99–124.

Index